Newnes Guide to
Television and Video Technology

K. F. Ibrahim

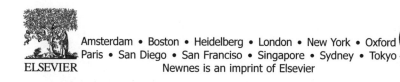

Amsterdam • Boston • Heidelberg • London • New York • Oxford
Paris • San Diego • San Franciso • Singapore • Sydney • Tokyo
Newnes is an imprint of Elsevier

Newnes

Newnes is an imprint of Elsevier
The Boulevard, Langford Lane, Kidlington, Oxford, OX5 1GB, UK
30 Corporate Drive, Suite 400, Burlington, MA 01803, USA

First edition 2007
Reprinted 2008

Notice
No responsibility is assumed by the publisher for any injury and/or damage to persons
or property as a matter of products liability, negligence or otherwise, or from any use
or operation of any methods, products, instructions or ideas contained in the material
herein. Because of rapid advances in the medical sciences, in particular, independent
verification of diagnoses and drug dosages should be made

British Library Cataloguing in Publication Data
A catalogue record for this book is available from the British Library

Library of Congress Cataloging-in-Publication Data
A catalog record for this book is available from the Library of Congress

ISBN: 978-0-7506-8165-0

For information on all Newnes publications
visit our website at www.elsevierdirect.com/newnes

Printed and bound in *Hungary*

08 09 10 10 9 8 7 6 5 4 3 2

Working together to grow
libraries in developing countries

www.elsevier.com | www.bookaid.org | www.sabre.org

ELSEVIER   BOOK AID International   Sabre Foundation

Newnes Guide to
Television and Video Technology

To my wife Valerie

Contents

Preface

This new edition of Newnes Guide to Television and Video technology continues the tradition of providing a comprehensive and up-to-date coverage of video and television technology first started in 1958 by J. F. Camm. In 1968 the book was completely re-written by Gordon J. King and updated by him in 1972. Eugene Trundle took over in 1988 with a further re-write and subsequent updates in 1996 and 2001. It is now my turn with a new rewrite to take this publication into the digital age of the 21st century, the age of flat-panel displays, high-definition television, Blu-ray technology and internet video streaming; the age of multimedia convergence in which the various strands of video, television and communication technologies overlap and complement each other.

The book starts with television fundamentals and builds up to multimedia convergence going through video displays, television receivers, recoding and playback systems. The first two chapters cover the principles of monochrome and colour television and the fundamentals of analogue TV broadcasting. This is followed by video compression for standard and high-definition and audio-compression techniques (Chapters 3–6) that are a prerequisite for digital video broadcasting and streaming. Chapter 7 deals with MPEG transport stream followed by Chapter 8 which explains the various types of modulation used in broadcasting digital television signals via satellite and terrestrial aerials.

The following four chapters deal with display devices. An overview of the requirements of a TV receiver is given in Chapter 9 together with a description of the traditional, and still widely used, cathode ray tube (CRT). Detailed explanation with extensive block diagrams and practical circuits of flat panel displays: Plasma, LCD, DLP and SED are given in Chapters 10–12. Chapters 13–18 use the knowledge acquired in the preceding chapters to give detailed description of television receivers using plasma panels, LCDs and DLPs as well as CRTs encompassing the latest commercially used techniques. Chapter 15 is exclusively dedicated to a generally neglected topic: power-supply generation. Power-supply requirements and circuitry for different systems are included in the relevant chapters. Projection systems including 3-dimentional techniques are covered in Chapter 19.

DVD playing systems (DVD, HD DVD and Blu ray) and magnetic tape recording are dealt with in Chapters 20 and 21 leading to digital recording and camcorder (Chapter 22) including recordable DVDs. The latest in cable and on-line television broadcasting are described in Chapter 23 and multimedia convergence in Chapter 24. Finally, the different types of ports and inter-connectivity are listed and explained in Chapter 25.

Throughout my years of teaching, I have found that looking at faulty systems and understanding their causes and symptoms is a powerful tool to understand their inner workings. It is only when you know the effect a faulty component or a malfunctioning chip can you truly say you understand how that system works. For this reason, I have included a section on common faults, their symptoms and their causes together with practical fault diagnosis procedures in relevant chapters. Eight appendices are also included to provide extensive background information from dBs to the seven-layer OSI communication model.

My appreciation and thanks go to my college, College of North West London and specially to the then head of my faculty, Frank Horan, for giving me the opportunity to develop and deliver courses on video and television technology that made writing this book possible at the time when colleges, encouraged by government funding mechanisms, were closing engineering courses in favour of hairdressing and beauty therapy. The neglect of engineering and technology and the obsession with the service sector is short sighted and a highly dangerous path to take for colleges as it is for governments. Just how many nail bars can a High Street sustain! It is my hope that this book will make a contribution towards the promotion and understanding of engineering and technology.

K. F. Ibrahim

1 Television fundamentals

For a reasonable understanding of colour television, it is essential that the fundamentals of television are known. As we shall see, all colour systems are firmly based on the original 'electronic-image dissection' idea which goes back to EMI in the 1930s, and is merely an extension (albeit an elaborate one) of that system.

Although there are few black and white TVs or systems now left in use, the compatible colour TV system used today by all terrestrial transmitters grew out of the earlier monochrome formats. In the early days, it was essential that existing receivers showed a good black and white picture from the new colour transmissions, and the scanning standards, luminance signal and modulation system are the same. What follows is a brief recap of basic television as a building block of the colour TV system to be described in later chapters.

At the television studio, the scene to be transmitted is projected on a photosensitive plate located inside the TV camera. The scene is repeatedly scanned by a very fast electron beam which ensures that consecutive images differ only very slightly. At the receiving end, a display device such as a plasma or LCD panel or a cathode-ray tube (c.r.t.) is used to recreate the picture by an identical process of scanning a screen by an electron beam. The phenomenon of persistence of vision then gives the impression of a moving picture in the same way as a cine film does. In the UK television broadcasting system, known as PAL, 25 complete pictures are scanned every second whereas in the US, NTSC is used with 30 pictures per second. Both systems will now be considered.

Scanning

The faceplate of the light-sensitive surface also known as pick-up device is made up of an array of hundreds of thousands of reversed-biased silicon photodiodes mounted on a chip, typically 7 mm diagonal, arranged in lines and columns. During the active field period, each photodiode acts as a capacitor, and acquires an electrical charge proportional to the amount of light falling on it. The image is sharply focused on the sensor faceplate by an optical lens system. Each diode is addressed in turn by an electron beam which is then addressed individually by the sensor's drive circuit so that (as viewed from the front) the charges on the top line of photodiodes are read out first, from left to right. Each line is read out in turn, progressing downwards, until the end of the bottom line is reached.

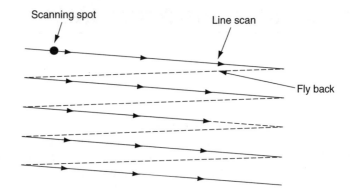

Figure 1.1 *Picture scanning*

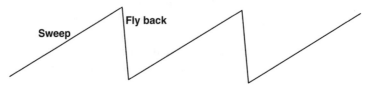

Figure 1.2 *Scanning sawtooth waveform*

The brightness of each element is thus examined line by line as shown in Figure 1.1 to form what is known as a video signal. A very large number of lines are employed to give adequate representation of the contents of the picture. In the UK's PAL, 625 lines are used while the USA's NTSC uses 525 lines. The waveform that provides the scanning movement of the electron beam is the sawtooth waveform shown in Figure 1.2 in which the sweep part provides the line scan and the flyback quickly takes the beam back to the starting position. At the end of each complete picture scan the electron beam moves back to the top of the scene and the sequence is repeated. In the UK, 25 complete pictures are scanned every second, chosen because it matched the 50 Hz frequency of the power supply. It is preferable to match the screen refresh rate to the power source to avoid wave interference that would produce rolling bars on the screen. With each picture containing 625 lines, the line frequency in the UK PAL system is therefore $25 \times 625 = 15{,}625$ Hz or 15.625 kHz. In the USA, the NTSC system was originally 30 pictures per second in the black and white system, chosen because it matched the nominal 60 Hz frequency of alternating current power used in the United States. However, with the introduction of colour in the USA, a problem arose as result of a beat frequency between the colour and the sound carriers which could produce a dot pattern on the screen. To avoid this, the original 30 Hz picture rate was adjusted down by the factor of 1000/1001 to 29.97 Hz. With each NTSC picture containing 525 lines, the line frequency is $29.97 \times 525 = 15{,}734$ Hz or 15.734 kHz. For high definition (HD), the number of lines is approximately doubled resulting in higher line frequencies as will be discussed in later chapters.

Interlacing

Normal sequential scanning, i.e. scanning complete pictures (625 or 525 lines) at one time followed by another complete picture scan, introduces unacceptable flicker when the picture is reproduced by a c.r.t. This is because once scanned, the brightness of a line fades away gradually as it awaits to be refreshed. Flicker is greatly reduced by a technique known as interlacing. Interlace scanning involves scanning the 'odd' lines 1, 3, 5, etc., first followed by the 'even' lines 2, 4, 6, etc. Only one-half of the picture, known as a field, is scanned each time. A complete picture therefore consists of two fields, odd and even, resulting in a field frequency of $2 \times 25 =$ 50 Hz for the UK system and or $2 \times 30 = 60$ Hz for the US system.

At the end of each field, the electron beam is deflected rapidly back to the beginning of the next scan. The odd flyback ends at the end of the last line of the odd field (point A in Figure 1.3) and the beam is then taken to the start of the first line of the even field (point B) travelling the distance of the screen height. To ensure the same flyback vertical distance and hence same field flyback time for both fields, the even flyback (dotted line) is ended halfway along the last line of the even field (point C) to take the beam to the start of the following odd field (point D) halfway along line 1. This is the reason the total number of lines is chosen to be an odd number.

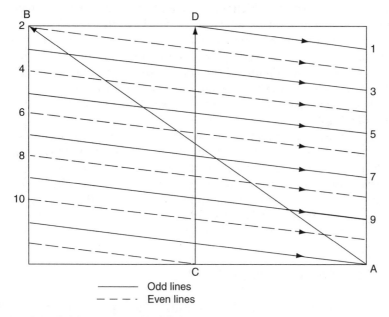

Figure 1.3 *Odd and even filed flyback*

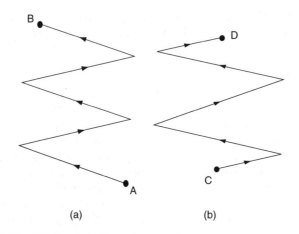

Figure 1.4 *Field flyback path*

Since the line scan continues to move the electron beam across the screen during the field flyback, the path traced by the beam during the flyback is as shown in Figure 1.4.

In the absence of picture information, scanning produces what is known as a raster.

Synchronisation pulses

The receiver or monitor which we shall use to display the picture has scanning waveform generators which must run in perfect synchronism with the readout of the image sensor at the transmitting end. This ensures that the video information picked up from the sensor is reproduced in the right place on the display. Plainly, if the camera sees a spot of light in the top right-hand corner of the picture and the monitor's scanning spot is in the middle of the screen when it reproduces the light, the picture is going to be jumbled up!

This is prevented by inserting synchronising pulses (sync pulses for short) into the video waveform at regular intervals, and with some distinguishing feature to enable the TV monitor to pick them out. Synchronising pulses are introduced at the end of each line to initiate the line flyback at the receiver; these pulses are called the line sync. Another set of synchronising pulse is introduced at the end of a field to initiate the start of the field flyback; this is called the field sync.

Composite video waveform

The picture information and the sync pulse of a single line constitute what is known as the composite video waveform. Each line is made up of the active video portion and the horizontal blanking portion. As shown in

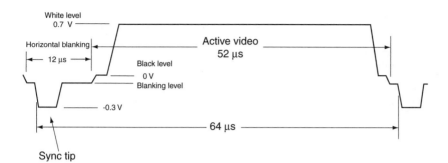

Figure 1.5 *Composite video (625-line)*

Figure 1.5, the picture information is represented by the waveform between the two-line sync pulses and thus may acquire any shape, depending on the varying picture brightness along the line. The waveform shown represents a line at peak white (maximum brightness).

Before and after every sync pulse, the voltage is held below the black level as shown for short periods of time, known as the front porch and the back porch, respectively. The front porch has a duration of 1.55 μs; it ensures the video brightness is completely blocked before the sync pulse is applied. The back porch has a longer duration of 5.8 μs; it provides time for the flyback to occur before the application of the video information. The back porch is also used for black level clamping. As can be seen, the front porch, the sync pulse and the back porch are at or below the black level. During this time, a total of 12.05 μs, the video information is completely suppressed; this is known as the line blanking period.

The total available peak-to-peak voltage of 1 V is divided into two regions:

- Below black level region 0 to −0.3 V reserved for the sync pulses (line and field)
- Above black level region 0–0.7 V (peak white), used for the video or picture information

The duration of one complete line of a composite video may be calculated from the line frequency:

$$\text{Line duration} = \frac{1}{\text{line frequency}} = \frac{1}{15.625\,\text{kHz}} = 64\,\mu s$$

Hence, the active video duration = 64 − 12 = 52 μs.

The composite video waveform consists of video, blanking and sync pulse, hence it is more commonly known as composite video, blanking and sync, CVBS.

The equivalent CVBS waveform and timings for the 525-line NTSC system is shown in Figure 1.6 with a line duration of 63.5 μs and active video duration of 52.6 μs.

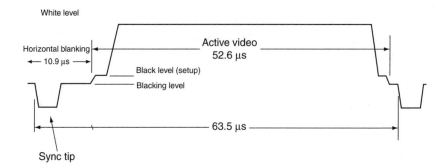

Figure 1.6 *CVBS (525-line)*

Common image format

Table 1.1 lists the properties of the 625- (PAL) and 525- (NTSC) line systems known as the common image format (CIF) that has been agreed by European and international standard organisations.

Active lines

While the specified number of lines is 625 and 525 for the respective PAL and NTSC, the actual active lines, i.e. the lines that are actually displayed at the receiving end are fewer. This is because a number of lines are 'lost' during the blanking period of the field flyback. These lines are called inactive, carrying no visible video information. For PAL, the number of inactive lines is 49 resulting in 567 active lines. The corresponding figures for NTSC are 45 and 480.

Table 1.1 *Common image format (CIF) for PAL and NTSC*

	PAL	NTSC
Lines/frame (total)	625	525
Lines/frame (active)	576	480
Pixels/line (active)	720	720
Line frequency (kHz)	15.625	15.75 (15.734)
Frame rate (Hz)	50	60 (59.94)
Line duration (total) (μs)	64	63.49 (63.55)
Line duration (active)	52 μs	52.6
Field duration (ms)	20	16.7
Aspect ratio	4:3	4:3
Interlace/progressive	Interlace (*i*)	Interlace (*i*)

Aspect ratio

For standard definition (SD) television, the image is normally broadcast with an aspect ratio, i.e. the ratio of width-to-height of 4:3 although the widescreen 16:9 format is also used by several digital television broadcasters. For high definition (HD) television, 16:9 is the specified format.

Pixels and bandwidth

A pixel, short for 'picture element', is the smallest active element or dot in an image. A well-defined picture will in general have more pixels than a less-defined one. The resolution of a picture is defined by the number of pixels in the horizontal and vertical directions. The number of pixels in the vertical direction is set by the number of lines. However, the same cannot be said of the number of pixels per horizontal line. This has been internationally set to 720 pixels/line for both PAL and NTSC. This gives a resolution for SD television of 720×576. For HD television, the figures are higher, namely 1920×1080 or 1280×720 (PAL) and 720×480 (NTSC).

Having determined the number of pixels per line, the total number of pixels in a single picture of a SD television may then be calculated as active pixels/line $\times$ active lines. This gives

$$720 \times 576 = 414,720 \, \text{pixels (PAL)}$$

and

$$720 \times 480 = 345,600 \, \text{pixels (NTSC)}$$

Video bandwidth

The frequency of the video waveform is determined by the change in the brightness of the electron beam as it scans the screen line by line. Maximum video frequency is obtained when adjacent pixels are alternately black and peak white; this represents the maximum definition of a TV image. For PAL, this means 414,720 alternating black and peak white pixels and for NTSC, 345,600 pixels. When an electron beam scans a line containing alternate black and white pixels, the video waveform is that shown in Figure 1.7, representing the variation of brightness along the line. As can be seen, for any adjacent pair of black and white pixels, one complete cycle is obtained. Hence, for the 10 pixels shown, five complete cycles are produced. It follows that, for a complete picture of alternate black and white pixels, the number of cycles produced is given by $1/2 \times$ number of pixels:

$$= \frac{1}{2} \times 414,720 = 207,360 \, \text{cycles/picture (PAL)}$$

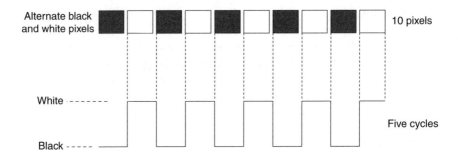

Figure 1.7 *Waveform for alternate peak white and black pixels*

and

$$= \frac{1}{2} \times 345,600 = 172,800 \, \text{cycles/picture (NTSC)}$$

Since there are 25 (PAL) complete pictures per second, then the number of cycles per second, i.e. the maximum video frequency is

$$\text{cycles/picture} \times 25 = 207,360 \times 25 = 5,184,000 \, \text{Hz} = 5.184 \, \text{MHz}$$

NTSC, with 30 pictures per second, has the same maximum video frequency as PAL namely,

$$\text{cycles/picture} \times 25 = 172,800 \times 30 = 5,184,000 \, \text{Hz} = 5.184 \, \text{MHz}$$

The minimum video frequency is obtained when the electron beam scans pixels of unchanging brightness. This corresponds to unchanging amplitude of the video waveform, a frequency of 0 Hz or d.c. The video bandwidth is, therefore, 0–5.184 MHz.

Notice here that the specifications outlined by the SD common interface format, results in a common video bandwidth for both systems. That is of course is the purpose of the CIF.

Television broadcasting

There are three methods of television broadcasting: terrestrial, satellite and cable. Each method may be used to broadcast analogue or digital TV programmes. Terrestrial is the traditional method of broadcasting television signals to the home employing UHF radio frequencies (400–800 MHz). Satellite broadcasting involves two stages. In the first place, the TV signals are sent to a satellite stationed at a distance of 35,765 km above the equator. The satellite sends these signals back to earth (on a different frequency) and they may be received using a simple satellite dish aerial. Cable broadcasting, as the name implies, uses a transmission cable to send TV signals to home subscribers. Let's first start with analogue terrestrial transmission.

Modulation

A variety of modulation techniques are used in TV broadcasting including amplitude modulation (AM) and frequency modulation (FM) are used in analogue terrestrial broadcasting, FM and quadrature phase shift keying (QPSK) in digital satellite broadcasting, quadrature amplitude modulation (QAM) and orthogonal frequency division multiplex (OFDM) in digital terrestrial television (DTTV) and digital audio broadcasting (DAB). Pulse code modulation (PCM) is also employed at both the transmitting and receiving ends. For analogue monochrome television, only two types are used: AM for video and FM for sound. For detailed explanation of amplitude and FM, refer to Appendix 8.

Terrestrial TV broadcasting uses AM for the video information. As described in Appendix 8, ordinary AM gives rise to two sets of sidebands on either side of the carrier, thus doubling the bandwidth requirement for the transmission. However, since each sideband contains the same video information, it is possible to suppress one sideband completely, employing what is known as single-sideband (SSB) transmission. However, pure SSB transmission demands a more complicated synchronous detector at the receiving end, making the receiver more expensive. The simple and cheap diode detector which is adequate for double-sideband (DSB) a.m. transmissions introduces a distortion known as quadrature distortion when used to demodulate a SSB AM transmission. This distortion is caused mainly by the lower end of the video frequency spectrum. To avoid this and still use the diode detector, a mixture of SSB and DSB modulation is employed, known as vestigial sideband transmission. In vestigial sideband modulation, DSB transmission is used for low video frequencies and SSB transmission for higher video frequencies.

Frequency spectrum, PAL channel

The frequency spectrum of vestigial sideband transmission used in the UK is shown in Figure 1.8, in which up to 1.25 MHz of the lower sideband is transmitted with the unsuppressed upper sideband.

As well as the composite video, it is also necessary to transmit a sound signal. Traditionally, only mono sound was broadcast using FM. A separate 6 MHz carrier with a bandwidth of 50 kHz (a deviation of ±25 kHz) is used. The 6 MHz carrier was chosen to fall just outside the highest transmitted video frequency (Figure 1.8) to avoid any inter-carrier interference. For instance, for a vision carrier of 510 MHz, the sound carrier is 510 + 6 = 516 MHz.

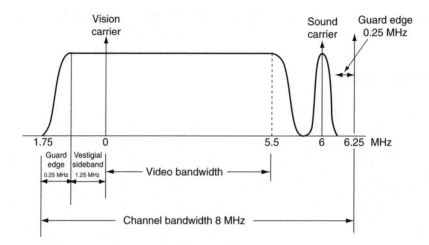

Figure 1.8 *Frequency spectrum PAL (UK)*

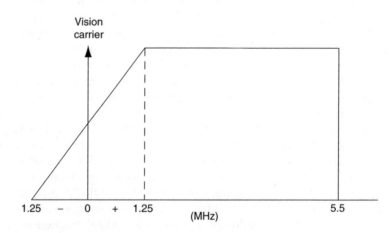

Figure 1.9 *Frequency response, PAL receiver*

It can be seen from Figure 1.8 that the frequency response remains constant over the range of video frequencies up to 5.5 MHz on the upper sideband and 1.25 MHz on the lower sideband. Above 5.5 MHz, a steep attenuation takes place to ensure that no video information remains beyond 6 MHz; this prevents any overlap with the sound information. An additional 0.25 MHz is introduced to accommodate the 50 kHz sound bandwidth and to provide a buffer space for the adjacent channel known as channel 'guard edge'. Similar attenuation is necessary at the lower sideband for frequencies extending beyond 1.25 MHz with a 0.25 MHz 'guard edge' to prevent any overlap with an adjacent channel on the other side. Gradual attenuation is necessary since it is not possible to have filters with instantaneous cut-off characteristics. Adding all of these frequencies together, we end up with an 8 MHz (1.75 + 6.25) bandwidth for a TV channel.

As a consequence of vestigial sideband transmission, video frequencies up to 1.25 MHz are present in both sidebands and frequencies above 1.25 MHz are present in one sideband only. When detected by a simple diode detector, the frequencies below 1.25 MHz will produce twice the output of the frequencies above 1.25 MHz. To compensate for this, it is necessary to shape the frequency response of the receiver so that frequencies that are present in both sidebands are afforded less amplification than those present in one sideband only. Such a response curve is shown in Figure 1.9.

Channel allocation

In order to cover a large area, a number of transmitting stations are used. Each transmitting station is assigned a number of 8-MHz channels. However, for analogue television broadcasting, to avoid inter-channel interference, adjacent channels on either side of the TV channel are not used resulting in a number of channels remaining empty, some of which have now been allocated for digital television. The UK is divided into several areas each served by a station transmitting UHF frequencies in bands IV and V. For example, Crystal Palace serves the Greater London area and uses the following analogue channels:

ITV channel 23	486–494 MHz
BBC1 channel 26	510–518 MHz
C4 channel 30	542–550 MHz
BBC2 channel 33	566–574 MHz
Channel 5 channel 37	600–608 MHz

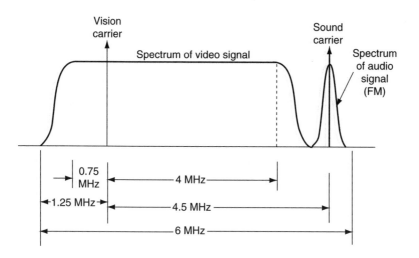

Figure 1.10 *Frequency spectrum NTSC (USA)*

Frequency spectrum, NTSC channel

The frequency spectrum for an NTSC channel as used in the USA is shown in Figure 1.10. The total channel bandwidth is 6 MHz with a sound carrier situated at 4.5 MHz away from the vision carrier. The vestigial sideband is ± 0.75 MHz and a maximum video frequency of 4.5 MHz which is adequate for SD television.

2 Colour television

The first problem facing the transmission of colour TV signals is compatibility with the existing monochrome transmission. When colour television was first introduced, it was agreed that colour TV signals must be capable of producing a normal black and white image on a monochrome receiver without any modification to the television set. Conversely, a colour receiver must be capable of producing a black and white image from a monochrome signal. A colour transmission system must therefore retain the monochrome information, sync pulses and the sound inter-carrier in the same form as those of normal monochrome transmission. The additional colour information has to be included within the composite video signal without interfering with it. Furthermore, the colour signal must fall within the same bandwidth as that allocated for monochrome transmission. To understand how this may be done, we must first look at the properties of light and colour.

Light and colour

We know radio and TV broadcasts are electromagnetic waves of various frequencies and various broadcast bands have already been mentioned in Chapter 1. As we go up in frequency we pass through the bands allotted to radio transmissions, followed by terrestrial TV broadcast and space communications. Way beyond these we come into an area where electromagnetic radiation is manifest as heat, and continuing upwards we find infra-red radiation, and then a narrow band (between approximately 4×10^8 and 8×10^8 MHz) of light energy. Beyond the 'light band' we pass into a region of ultra-violet radiation, X-rays, gamma-rays and then cosmic rays (Figure 2.1). Thus, light is just another electromagnetic wave.

The sensation of colour

Light waves falling on the eye pass through the pupil, which focuses the image on the retina at the back of the eye (Figure 2.2). The retina is sensitive to electromagnetic waves within the visible band; it can therefore translate the electromagnetic energy into suitable information, which is then passed to the brain via numerous optic nerve fibres.

The retina contains a large number of light-sensitive cells. Cells known as rods are sensitive to brightness (or luminance) only; cells known as

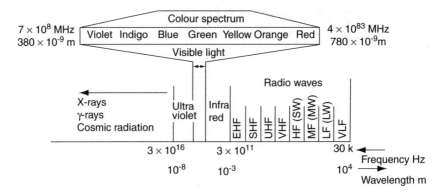

Figure 2.1 *Electromagnetic wave spectrum*

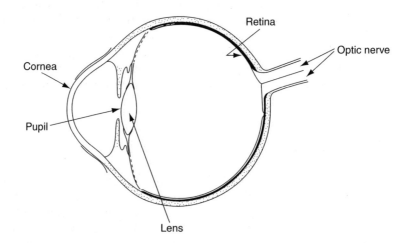

Figure 2.2 *The eye*

cones are sensitive to colour (or chrominance) only. Rods outnumber cones by a factor of 20 and they are 10,000 times more sensitive. The eye therefore reacts predominantly to the luminance content of an image, much more than to its chrominance. This is the reason for high video frequencies with fine picture details are perceived in black and white only.

Primary colours

Cones themselves are of three different types. One is energised by red, the second by green and the third by blue. These colours are known as primary colours. Colours other than the three primary colours are perceived through energising two or more types of cones simultaneously. For example, the sensation of yellow is produced by energising the red and green cones simultaneously. Other colours may be produced by mixing different proportions of primary colours. This is known as additive mixing. For example

yellow $= R + G$
magenta $= R + B$
cyan $= B + G$
white $= R + G + B$

Yellow, magenta and cyan are known as complementary colours, complementary to blue, green and red, respectively. A complementary colour produces white when added to the third primary colour. If, for instance, yellow is added to blue, then

yellow + blue $=$ (red + green) + blue $=$ white

Colours may also be produced by a process of subtractive mixing. Yellow, for example, may be produced by subtracting blue from white. Since

$W = R + G + B$, then
$W - B = (R + G + B) - B = R + G =$ yellow

Similarly,

$W - G = R + B =$ magenta
$W - R = G + B =$ cyan
$W - R - G - B =$ black (absence of colour)

The colour triangle

The chrominance content of a colour picture may be represented by the colour triangle shown in Figure 2.3. Pure white is represented by a point W at the centre of the triangle; other colours are represented by phasors (or vectors) extending from the centre W to a point on or within the triangle. Phasors going to the three corners of the triangle WR, WG and WB represent the primary colours, red, green and blue. Other colours such as yellow are represented by phasor WY with point Y falling between its two primary components, red and green. Phasors for cyan (WC) and magenta (WM) are constructed in a similar way.

Saturation and hue

Phasor WR in Figure 2.3 represents a pure red with no trace of any other colour present. It is said to be fully saturated. Desaturation is obtained when white is added to a pure colour. For instance, if white is added to

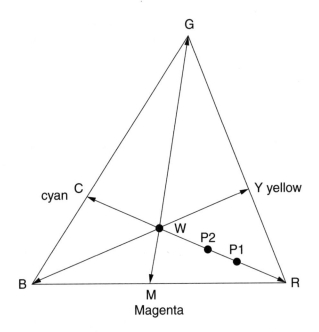

Figure 2.3 *Colour triangle*

red, desaturated red or pink is produced. On the colour triangle, this is represented by moving along phasor WR away from R (pure red) and towards W (pure white). Point PI thus represents desaturated red, or pink, represented by phasor WP1. A greater desaturation produces a shorter phasor; WP2 represents pale pink, and so on. The phasor length represents saturation; the phasor direction or angle represents hue. Hue denotes the principal primary component of a colour. For instance, the primary colour for pink is red; hence its phasor is in phase with the pure red phasor WR. A less saturated red would, say yellowish red; its phasor would be shifted towards WY or towards WM. Yellow itself has two primary colours and hence its phasor falls between red and green. Similarly for magenta. In this case, WM falls between red and blue.

Chromaticity diagram

The complete set of colours is represented by what is known as the chromaticity diagram illustrated in Figure 2.4. Real colours all fall along the curved part of horseshoe-shaped diagram which lies within the triangle. The curved part of the horseshoe is calibrated to show the wavelength in nanometres (nm) of the spectral colours, i.e. colours that are present on the visible electromagnetic spectrum. The colours between red and blue have no wavelength references, being 'non-spectral' colours resulting from a mix of components from opposite ends of the visible light spectrum.

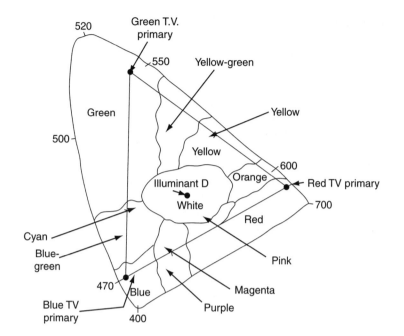

Figure 2.4 *Chromaticity diagram showing the TV colour triangle*

Colours produced by the television system described by the colour triangle falls within the horseshoe.

Colour temperature

It is difficult to define just what 'white' light is! The appearance of white light depends entirely on the strength of each of the primary components. What is construed as white light is invariably bluish. Many monochrome picture tubes glow with a bluish, rather cold white, the result of a predominance of high frequencies in the rendered light spectrum; this is because of the type of phosphor used.

Colour 'temperature' provides a means to define white and other colours. Colour temperature is defined as the temperature in Kelvin to which a 'black' radiating body has to be raised to emit a certain light such as white light. For instance, the reddish light of a candle has a colour temperature of 1200 K, white light has a temperature around 5500 K and the blue colour of a hazy sky has a temperature of 8000 K. It will be noticed that a red light has lower colour temperature than a blue light. For TV applications white light at 6500 K, which is 'standard daylight' referred to as illuminant D_{6500} has been adopted as a standard. Illuminant D_{6500} is shown in the centre of white area of the chromaticity diagram (Figure 2.4).

Principles of colour transmission

To meet the requirements of compatibility, the luminance signal Y is transmitted in the same way as in a monochrome system within which the chrominance is be contained. The next step is to produce the pure chrominance component from the RGB signal. To do this, the luminance had to be removed from the three primary colours, resulting in what is known as colour difference signals: $R - Y$, $G - Y$ and $B - Y$.

Since the luminance signal $Y = R + G + B$, and luminance Y is to be transmitted in full, only two colour difference signals need to be transmitted. Of the three colour difference signals, $Y - G$ normally has the smallest value. For this reason, difference signals $(R - Y)$ and $(B - Y)$ are selected. The missing $(G - Y)$ is recovered at the receiving end from the three transmitted components as follows:

$$R = (R - Y) + Y$$
$$B = (B - Y) + Y \quad \text{and since} \quad Y = R + G + B,$$
$$G = Y - R - B$$

The remaining issue that has to be resolved is the manner in which this additional information, $R - Y$ and $B - Y$, is added to the monochrome signal without causing it any interference. To do this, a separate colour subcarrier [4.43 MHz for Phase Alternate Line (PAL) and 3.58 MHz for National Television System Committee (NTSC)] is introduced which is then modulated by the two-chrominance components using quadrature amplitude modulation (QAM).

Frequency interleaving

Since the modulated colour subcarrier falls within the monochrome frequency spectrum, its sidebands naturally overlap with those produced by the original vision carrier. This overlap will result in pronounced patterning on a monochrome set receiving colour transmission. This is avoided by the choice of the subcarrier frequency resulting in what is known as frequency interleaving.

When the frequency spectrum of a TV signal is examined in detail, it is found that the distribution of frequencies is not uniform. Energy tends to gather in bunches centred on the harmonics, i.e. multiples of line frequency, known as monochrome clusters (Figure 2.5). This is not that surprising since the video information is sent out line by line. Furthermore, the amplitude of these side frequencies gets progressively smaller as we move away from the vision carrier. Similarly, colour information would also bunch into chrominance clusters. By choosing the subcarrier to fall between two monochrome clusters, the chrominance clusters may be arranged to fall

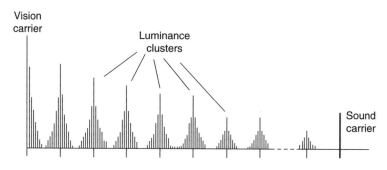

Figure 2.5 *Luminance clusters*

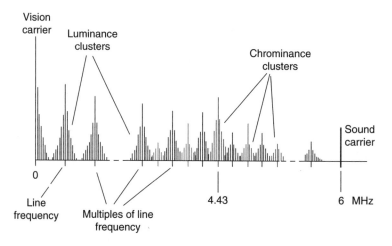

Figure 2.6 *Frequency interleaving used in colour television broadcasting*

neatly in the spaces between the clusters produced by the monochrome signal (Figure 2.6). The choice of the subcarrier frequency thus becomes very critical. Since the clusters are centred on multiples of line frequency, the colour subcarrier has to be an odd multiple of half the line frequency, known as half-line offset, for its clusters to interleave with the monochrome clusters. For NTSC, a multiple $n = 455$ is used giving a subcarrier frequency $f_{sc} = n \times f_h/2$ where f_h is the line or horizontal frequency

$$f_{sc} = 455 \times \frac{1}{2} 15.734 = 3.79485\,\text{MHz}$$

A high-value multiplier is used to position the colour clusters at the higher end of the video bandwidth where the monochrome clusters are small in amplitude. As for the reason for choosing an odd rather than an even multiplier, it is to remove the pattern of alternate black and white dots appearing along each line scan. The number of these dots corresponds to the

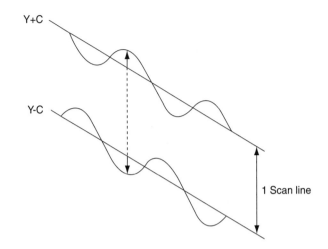

Figure 2.7 *Subcarrier phase inversion*

number of subcarrier cycles per line namely f_{sc}/f_h. Given that $f_{sc} = n \times f_h/2$, then

$$\frac{f_{sc}}{f_h} = \frac{(n \times f_h/2)}{f_h} = \frac{n}{2}$$

Since the multiple n is an odd number (455 for NTSC), then the number of subcarrier cycles per line will always be an odd half cycle at the end of the line. In other words, there will be subcarrier reversal line by line as illustrated in Figure 2.7 and the eye will see the average brightness of the line pairs cancelling out the dot pattern. Chrominance C will therefore change phase, line by line with $Y + C$ on one line and $Y - C$ on the following line and so on.

Furthermore, the choice half-line offset means that not only the colour clock cycles per line ends with a half cycle, but the number per frame also ends with a half cycle which for NTSC is

Subcarrier cycles/frame

= subcarrier cycles/line × no. of lines/frame

$$= \frac{n}{2} \times 525 = \frac{455}{2} \times 525 = 119{,}437.5$$

This ensures that the same point on the screen alternates in brightness from frame to frame about a value determined by the luminance signal.

While this will reduce the interference pattern to practically invisible level in the NTSC system, the nature of PAL is such that certain hues may experience an annoying vertical dot pattern. This is removed by using quarter-line offset making the subcarrier a multiple of one-quarter the line frequency which still reverses the chrominance on successive lines. However, a further modification is required, namely a phase reversal on successive fields if the dot pattern is to be removed completely. To do this, the colour subcarrier is further modified by adding the half-field frequency (25 Hz) to create a phase reversal on a successive field. The dot pattern will thus be reversed on successive fields to be cancelled out by the averaging process of the eye. PAL, in the UK, uses a multiple $n = 1135$ resulting in a subcarrier frequency of

$$f_{sc} = \left(\frac{1135 \times f_h}{4}\right) + \frac{f_v}{2}$$

$$= \frac{1135 \times 15.625}{4} + 0.025$$

$$= 4.43361875 \, \text{MHz (normally referred to as 4.43 MHz)}$$

As far as line and frame reversal of the subcarrier, PAL is exactly the same as NTSC.

Quadrature amplitude modulation

Now that the subcarrier frequency has been set, we can have a look at how it is modulated by the chrominance components. For this, quadrature amplitude modulation, QAM is used for both PAL and NTSC. In QAM, two carriers at right angles (quadrature) to each other are modulated by two separate signals (Figure 2.8): $B - Y$ modulating the in-phase carrier

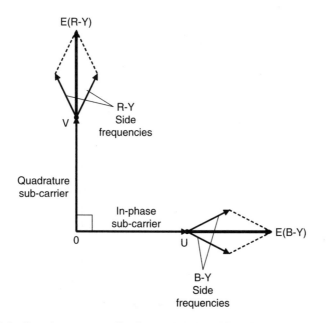

Figure 2.8 *Quadrature amplitude modulation (QAM)*

OU and R − Y modulating the quadrature carrier OV. As with ordinary amplitude modulation, each modulated carrier produces two bands of side frequencies, one on each side of the carrier represented by the pair of phasors shown. Each pair of phasors produces a resultant colour difference phasor: E(B − Y), known as I for in-phase and E(R − Y), known as Q for quadrature, respectively.

As was mentioned in Chapter 1, the information in a modulated carrier is contained in the side frequencies only. The quadrature carriers may thus be suppressed to obtain the two colour difference signals, I and Q illustrated in Figure 2.9 and by so doing significantly reduce the severity of the dot pattern introduced at the receiver by the presence of the colour signal. These in turn are added to produce a resultant chrominance phasor. The chrominance phasor which has the same frequency as the suppressed carriers corresponds to the phasor associated with the colour triangle with its length (or amplitude) representing saturation and its angle (or phase), θ representing hue. I and Q may have a positive or a negative value which enables all colours and hues to be represented.

The bandwidth of each chrominance component is limited to approximately 1 MHz on each side of the colour carrier (Figure 2.10). The relatively

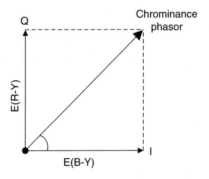

Figure 2.9 *The I and Q colour phasor diagram*

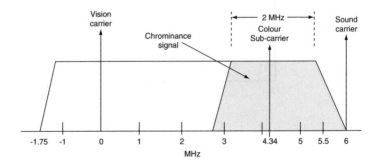

Figure 2.10 *Colour television frequency components, PAL (UK)*

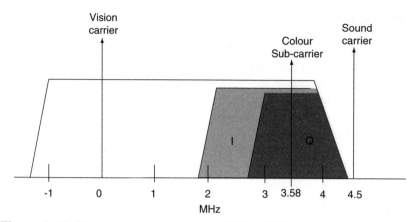

Figure 2.11 *Frequency components, NTSC (USA)*

narrow bandwidth is quite sufficient for an adequate reproduction of a colour image at the receiving end. This is because as was mentioned earlier in this chapter, the eye is not very sensitive to colour and perceives high video frequencies in black and white only.

NTSC colour components

In the NTSC 525-line system, the two colour difference signals, I and Q, are given different bandwidths. For Q, double sideband is used with a bandwidth of ±400 kHz. For the I component, vestigial sideband is used with a bandwidth of 400 kHz for the upper sideband and 1.3 MHz for the lower sideband. This reduced bandwidth is a direct result of the reduced NTSC channel bandwidths. Figure 2.11 illustrates the frequency components of the 525-line NTSC system as used in the USA showing the 3.58 MHz colour subcarrier and the I and Q chrominance components.

NTSC refresh rate

It was stated in Chapter 1 that the 60 Hz nominal refresh rate for NTSC has been slightly modified to 59.94 Hz. This came about with the introduction of colour television. In an NTSC broadcast, the vision carrier is amplitude modulated by the video signal as described above, while sound is transmitted by frequency modulating a carrier 4.5 MHz higher. If the video signal is affected by non-linear distortion, the 3.58 MHz colour carrier may beat with the sound carrier to produce a dot pattern on the screen. To minimise this and avoid any interference between the chrominance signal and the audio carrier, the original 60 Hz field rate was adjusted down by the factor of 1000/1001, to 59.94059 fields per second.

Composite colour signal

When the modulated subcarrier is added to the luminance (monochrome) signal to form the composite colour signal, the modulated subcarrier appears as a sine wave superimposed on the monochrome signal; it changes in amplitude and phase. The amplitude of the subcarrier represents saturation. Thus a fully saturated colour is represented by maximum subcarrier amplitude; black and white is represented by zero subcarrier amplitude. Hue, by contrast, is represented by the phase angle of the subcarrier. A typical waveform is shown in Figure 2.12.

Colour burst

To ascertain the phase angle of the chrominance signal and therefore hue, a 'burst' of about 10 cycles of the original subcarrier is transmitted for use as a reference at the receiver. This colour burst is mounted on the back porch of the line sync as shown in Figure 2.13. At the receiving end, the phase of

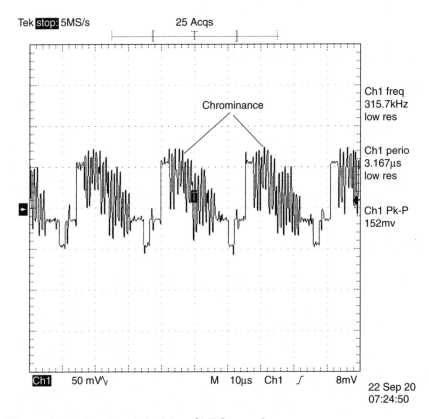

Figure 2.12 *A colour television CVBS waveform*

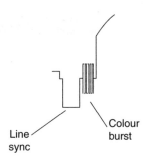

Figure 2.13 *Colour burst mounted on back porch*

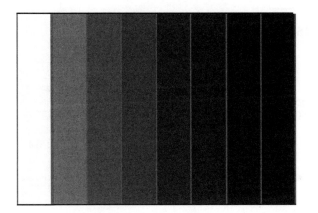

Figure 2.14 *Grey scale display*

the modulated colour signal is compared with the phase of the regenerated subcarrier to provide a measure of the phase angle and therefore hue. The absence of a colour burst indicates a black and white transmission.

The standard colour bar display

There have been several test displays since the TV broadcasting began; among them is one of the more important namely the colour bar. A standard colour bar waveform produces eight vertical bars of uniform width which include three primary colours, three complementary colours, black and white in the following order:

White Yellow Cyan Green Magenta Red Blue Black

On a monochrome receiver this is known as the grey scale display shown in Figure 2.14 with peak white on the left of the screen followed by grey stripes changing in luminance growing progressively darker from left to right. The luminance steps are not uniform as illustrated by the video waveform in Figure 2.15.

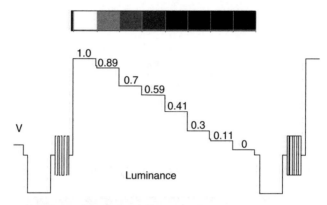

Figure 2.15 *Grey scale video waveform*

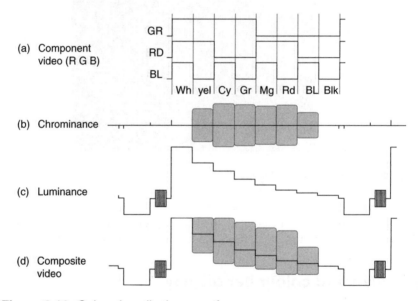

Figure 2.16 *Colour bar display waveforms*

The waveforms of the primary components red, green and blue of the colour bar display are shown in Figure 2.16a together with the chrominance signal in (b). When the chrominance signal (b) is added to the grey scale waveform (c), the standard colour bar waveform in (d) is obtained.

Gamma-correction

In a cathode ray tube, the beam current represents the brightness of the display. Changes in brightness are produced by changes in the voltage between the cathode and the grid of the tube and with it the beam current.

This relationship is non-linear and if not corrected, severe deterioration in the quality of the picture will result. This non-linearity at the receiving end is compensated by the introduction of an equal and opposite non-linearity at the transmitting end; it is known as gamma-correction (γ-correction). The voltage E from the camera is raised to a power of $1/\gamma$ (i.e. $E^{1/\gamma}$), where for UK TV transmission $\gamma = 2.2$ and $1/\gamma = 0.45$. Gamma-corrected signals are indicated using a prime, e.g. Y', R' and gamma-corrected colour difference signals are indicated as Y' − R' and Y' − B'.

Weighting factors

The subcarrier has maximum amplitude when 100% or fully saturated colour is transmitted. Since the subcarrier is added to the luminance signal, the amplitude of the composite colour signal may exceed the maximum possible voltage. To avoid this, the peak amplitude of the chrominance signal, i.e. the peak amplitude of the colour difference signals, is reduced by a factor known as the weighting factor. The new weighted components of the chrominance signal are called U and V where

$$U = 0.493(B' - Y')$$
$$V = 0.877(R' - Y')$$

The resultant chrominance phasor is now produced by the phasor sum of U and V. The values of U and V and their respective phasors representing a colour bar display is shown in Table 2.1 and Figure 2.17, respectively, and Figure 2.18 shows the expected waveforms for the colour difference signals.

PAL colour system

There are three main systems of colour transmission: NTSC, PAL and SECAM (sequential colour with memory). All three systems split the

Table 2.1 *U and V and other values for colours in a colour bar display*

Colour	Y	B − Y	R − Y	U	V	Phasor amplitude	Angle (°)
Yellow	0.89	−0.89	+0.11	−0.4388	+0.0965	0.44	167
Cyan	0.7	+0.3	−0.7	+0.1479	−0.6139	0.63	283
Green	0.59	−0.59	−0.59	−0.2909	−0.5174	0.59	241
Magenta	0.41	+0.59	+0.59	+0.2909	+0.5174	0.59	61
Red	0.3	−0.3	+0.7	−0.1479	−0.6139	0.63	103
Blue	0.11	+0.89	−0.11	+0.4388	−0.0965	0.44	347
White	1.0	0	0	0	0	0	−
Black	0	0	0	0	0	0	−

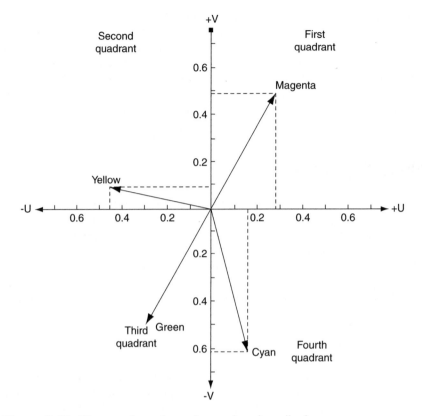

Figure 2.17 *Phasors for colours in a colour bar display*

colour picture into luminance and chrominance; all three use colour difference signals to transmit the chrominance information. The difference between them lies in the way in which the subcarrier is modulated by the colour difference signals. SECAM (used in France, Russia, Eastern Europe and parts of Africa) transmits the colour difference signals U and V on alternate lines: U on one line, V on the next, and so on. The other two systems, NTSC (used mainly in North and Central America, parts of South America, Japan and south east Asia) and PAL (used in the UK, western Europe with the exception of France, Asia, most of Africa, Australia and New Zealand), transmit both chrominance components simultaneously using QAM.

However, it is found that, due to a phase shift either in the transmission path or in the receiver itself, the relative phases of the chrominance signal and the colour subcarrier burst suffer a change in the form of phase delay or advance (Figure 2.19). This effect, known as differential phase distortion, can be quite severe, and to counter it, NTSC receivers are fitted with a hue control with which the phase of the subcarrier generator can be adjusted to somewhere near the correct axis, as subjectively judged on

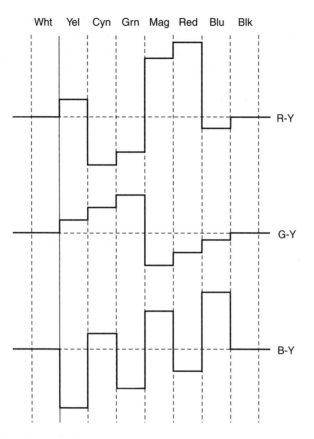

Figure 2.18 *Colour difference waveforms*

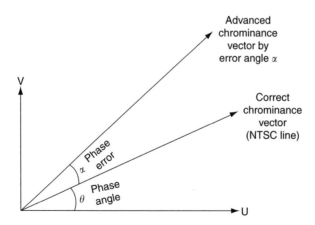

Figure 2.19 *Hue phase error*

flesh-tones. This is somewhat haphazard, and certainly inconvenient, as hue adjustment is often necessary on channel-changing.

Such errors are almost completely corrected by the PAL system. With PAL, the V signal is reversed line by line, V on one line followed by −V on the next, and so on, hence the name 'Phase Alternate Line'. The first line is called an NTSC line and the second line is called a PAL line. Phase errors are thus reversed from one line to the next. At the receiving end, a process of averaging consecutive lines by the eye cancels out the errors to reproduce the correct hue. This simple method is known as PAL-S. A more accurate method is the use of a delay line to allow for consecutive lines (NTSC and PAL) to be added to each other, thus cancelling phase errors physically. This is known as PAL-D.

In order to distinguish between the two types of lines, the colour burst is made to change its phase by approximately 180, as an NTSC line is followed by a PAL line and back again. For this reason, the PAL colour burst is known as the swinging burst in the PAL system.

3 Digital television

The transmission of television signals involves an analogue carrier waveform that is modulated by the video (and audio) information. While the carrier is analogue, the video information may be analogue (analogue television) or digital known as digital television (DTV). In analogue television, the totality of the composite video, blanking and sync is transmitted in its original analogue format. In DTV, the video and audio information are first converted into a digital format composed of a series of zeros and ones (bits). The series of bits is then used to modulate an analogue carrier before broadcasting via an aerial. At the receiving end, the digitised video and audio information is converted back to their original analogue formats for viewing and listening by the user.

Broadcasting a high volume of information requires a very wide bandwidth which for analogue television is between 5 and 6 MHz. For digital video broadcasting (DVB), a bandwidth of 10 or more times wider is necessary. For this reason, data compression techniques are used to reduce the bandwidth to manageable proportions. In fact, data compression is so effective that more than one programme is made to fit within the bandwidth allocated for a single analogue channel. This is just one advantage of DTV broadcasting. Here are some more:

- Very good picture quality
- Increased number of programmes mentioned above
- Lower transmission power – reduces adjacent channel interference
- Lower signal-to-noise ratio
- No ghosting

Principles of digital video broadcasting

Broadcasting of DTV signals involves three steps as illustrated in Figure 3.1:

- Digitisation
- Compression
- Channel encoding

Digitisation is the process of converting the analogue video and audio signals into a series of bits using an analogue to digital converter (ADC). To reduce the bandwidth requirements, data compression is used for both the video and audio information. This is carried out by the video and audio MPEG encoder which produces a series of video and audio packets known as the packetised elementary stream (PES). These are further broken up

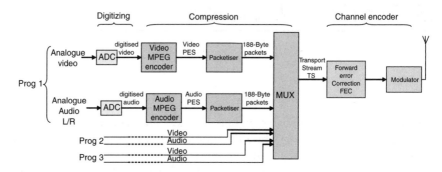

Figure 3.1 *Digital video broadcasting (DVB)*

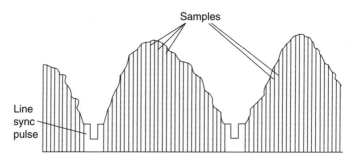

Figure 3.2 *Sampling*

into smaller 188-byte packets. Packets belonging to a number of different programmes are fed into a multiplexer to produce what is known as a transport stream (TS). Following the addition of error correction data by the FEC (forwards error correction) processor, the TS is used to modulate a carrier ready for broadcasting. Satellite DVB makes use of frequencies between 10,700 and 12,750 MHz using quadrature frequency shift keying (QFSK) while terrestrial DVB is restricted to the UHF band of frequecies currently available for analogue TV broadcasting with a channel band-width of 8 MHz in the UK and 6 MHz in the USA. A DVB multiplex occupying one analogue channel spectrum can accommodate from 3 to 10 different TV 'channels' or programmes.

Digitising the TV picture

Digitising a TV picture means sampling the contents of a picture frame by frame and, scan-line by scan-line (Figure 3.2). In order to maintain the quality of the picture, there must be at least as many samples per line as there are pixels, with each sample representing one pixel. For DTV, the picture frame is a matrix of pixels: horizontal and vertical. The total number of pixels is the product of horizontal pixels (pixels/line) × vertical pixels (number of lines). The number of pixels will depend on the format used, e.g. standard definition television (SDTV; PAL or NTSC) or high definition television (HDTV). Let's start with SDTV.

As we know from Chapter 2, PAL standard television uses 625 lines of which 576 are 'active' in that they may be used to carry video information and NTSC uses 525 lines system with 480 active lines. As for the number of pixels per line, SDTV specifies 720 pixels per line for both systems, giving a total number of pixels per picture of

$$576 \times 720 = 414,720 \, \text{pixels (PAL) and}$$
$$480 \times 720 = 345,600 \, \text{pixels (NTSC)}$$

Each scan line will therefore be represented by 720 samples, and each sample will represent one pixel. Sample 1 represents pixel 1, sample 2 represents pixel 2, etc. The process is repeated for the second line, and so on until the end of the frame and then repeated all over again for the next frame. To ensure that the samples are taken at exactly the same point of the frame, the sampling frequency must be locked to the line frequency 15.625 kHz for PAL and 15.734 kHz for NTSC. For this reason, the sampling rate must be wholly divisible by either line frequency.

SDTV sampling rate

The analogue video waveform contains the video information together with line sync pulses. Only the video information needs to be digitised and converted into a video bitstream. As was described earlier in Chapter 1, of the total 64 μs period of one line of a PAL composite video, 12 μs are used for the sync pulse, the front porch and the back porch leaving 52 μs to carry the video information. With 720 pixels per line,

$$\text{The sampling rate} = \frac{\text{number of pixels per line}}{52 \, \mu s}$$
$$= \frac{720}{52} = 13.8 \, \text{MHz}$$

This would be the sampling frequency if sampling was to start at the end of one sync pulse and finish at the start of the next. However, to allow for analogue drift, sampling is extended to a period slightly longer than the active line duration. In this way, sampling begins just before the end of a sync pulse and ends just after the start of the next sync pulse. For this reason, a frequency lower than 13.8 MHz is necessary. A sampling rate of *13.5 MHz* was thus selected to satisfy this as well as other criteria. One of these is that the sampling rate must be divisible by line frequency: $13.5 = 864 \times 15.625$ for PAL and 858×15.734 for NTSC. The number of active pixels per line is therefore $= 13.5 \times 52 = 702$ for PAL and $13.5 \times 52.6 = 710$ pixels for NTSC.

The other criterion for adequate sampling is the Nyquist rate. This states that to ensure that the reconstructed waveform contains all of the information of the original analogue waveform, the minimum sampling rate must not be less than twice the highest frequency of the analogue input. This is known as the Nyquist rate. In practice, the sampling rate is chosen to be 10–12% higher than the Nyquist rate to cater for the inevitable slope of the filter's cut-off curve. With a nominal maximum video frequency of 6 MHz for SDTV, the Nyquist rate is $2 \times 6 = 12$ MHz. The chosen sampling rate of 13.5 MHz is 12.5% higher than the Nyquist rate. For HDTV, a sampling frequency of 74.25 MHz which is 5.5×13.5 MHz is chosen to cater for the increased number of pixels per line. More on this in later chapters.

Video sampling

As outlined in Chapter 1, colour TV broadcasting involves the transmission of three components: luminance Y and colour differences $C_R = R - Y$ and $C_B = B - Y$. In the analogue TV system, the luminance is transmitted directly using amplitude modulation (terrestrial broadcasting) or frequency modulation (satellite broadcasting). For the chrominance components, quadrature amplitude modulation is used with a colour subcarrier of 4.43 MHz. In DVB, the three components are independently sampled, converted into three digital data streams before compression, modulation and subsequent transmission. For the luminance signal, which contains the highest video frequencies, the full sampling rate of 13.5 MHz is used. As for the chrominance components C_R and C_B, which contain lower video frequencies, a lower sampling rate or subsampling is acceptable. CCIR (Comite Consultatif International Radiocommunication) recommends a subsampling rate of half that used for the luminance rate, i.e. $0.5 \times 13.5 = 6.75$ MHz. Following the multiplexer where all three streams are combined into a single stream, a total sampling rate of $13.5 + 6.75 + 6.75 = 27$ MHz is obtained which is known as the *system clock* (Figure 3.3).

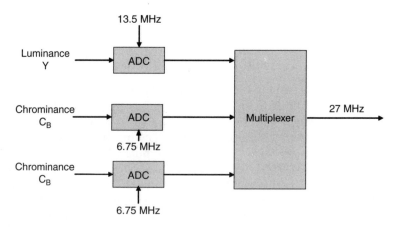

Figure 3.3 *Luminance and chrominance signals are sampled at 13.5 and 6.75 MHz respectively*

4:2:2 sampling structure

There are several structures for subsampling the chrominance components. One way is to sample the chrominance components every other pixel, known as 4:2:2 sampling structure (Figure 3.4). The 4:2:2 reduces the chrominance resolution in the horizontal dimension only, leaving the vertical resolution unaffected. The ratio 4:2:2 indicates that both C_R and C_B are sampled at half the rate of the luminance signal. Notice that the chrominance samples are evenly distributed across the picture, producing alternate Y-only columns and alternate co-sited (Y, C_R and C_B) columns. Where interlacing is used, the distribution of the chroma samples remains evenly distributed in each field (Figure 3.5).

4:1:1 sampling structure

To reduce the bandwidth, and hence lower the bit rate, a 4:1:1 sampling structure may be used. Here, the chrominance components are sampled at a quarter of the luminance rate (every forth pixel), hence the ratio 4:1:1 (Figure 3.6). The 4:1:1 sampling structure was used in early digital applications with good results. However, it can be seen from Figure 3.6 that there is a large imbalance between the vertical and horizontal chrominance resolution. To overcome this while maintaining the same bit rate, the 4:2:0 sampling was introduced.

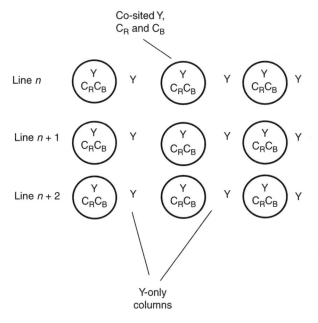

Figure 3.4 *4:2:2 sampling structure*

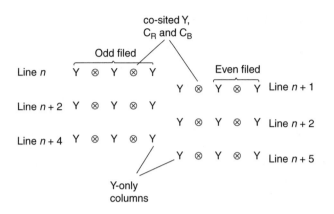
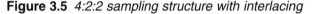

Figure 3.5 *4:2:2 sampling structure with interlacing*

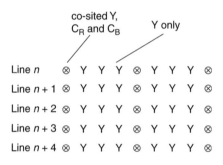

Figure 3.6 *4:1:1 sampling structure*

4:2:0 sampling structure

In this technique, subsampling at half full rate is used in both the horizontal and vertical direction, i.e. sampling every other pixel as well as every other line. The result is alternate Y-only columns as well as alternate Y-only rows as illustrated in Figure 3.7.

The above distribution of Y and co-sited samples are adequate for a sequentially scanned TV picture whereby chrominance samples are distributed evenly across the complete TV picture. However, this technique fails with interlaced pictures. The reason is as follows. Since the chrominance samples are taken every other line, lines 1, 3, 5, and so on, they belong to one field only (the top field) with none taken from the next field (the bottom field). To overcome this, sample interpolation is used to obtain an even spread of the chroma information. Interpolation involves deducing a single chrominance sample by averaging the chrominance values of two adjacent lines in one field, and inserting the sample halfway between two lines from successive fields. Even distribution of the chrominance samples may thus be realised when the two fields are combined.

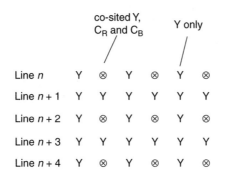

Figure 3.7 *4:2:0 sampling*

The bit rate

Sampling is followed by quantisation where sample values are rounded up or down to quantum values before they are converted into a multi-bit code. The precise number of quantums or levels is determined by the bit depth, the number of bits in the code. For studio applications, 10-bit coding is used. For domestic applications such as TV broadcasting and DVD, 8-bit coding is regarded as adequate. Given a bit depth of 8, the number of discrete signal levels or quantums available is $2^8 = 256$.

The bit rate may then be calculated as follows:

- Bit rate = number of samples per second $\times$ number of bits per sample
- But, the number of samples per second = number of samples per picture $\times$ number of pictures per second
- And the number of samples per PAL picture = $720 \times 576 = 414{,}720$
- Then, given a picture rate of 25,

$$\text{Number of samples per second} = 720 \times 576 \times 25 = 1{,}036{,}800$$

For NTSC, the number of samples is the same, namely

$$720 \times 480 \times 30 = 1{,}036{,}800$$

The bit rate generated by the luminance component using an 8-bit code is therefore

$$= 720 \times 576 \times 25 \times 8$$
$$= 82{,}944{,}000$$
$$= 82.944\,\text{Mbps}$$

The bit rate for the chrominance components depends on the sampling structure used. For a 4:2:2 sampling structure with horizontal subsampling only, the number of samples $= 360 \times 576 = 207{,}360$ per picture.

This gives a chrominance bit rate of

$$360 \times 576 \times (\text{picture rate})$$
$$\times (\text{number of bits}) = 360 \times 576 \times 25 \times 8$$
$$= 41.472 \, \text{Mbps for each chrominance}$$
component which is half the luminance bit rate.

Total chrominance bit rate is therefore
$$41.472 \times 2 = 82.944 \, \text{Mbps}$$

Giving a total bit rate of
$$82.944 + 82.944 = 166 \, \text{Mbps}$$

For a 4:2:0 structure where both horizontal and vertical subsampling is used, the chrominance bit rate for each signal is
$$360 \times 288 \times 25 \times 8 = 20.736 \, \text{Mbps}$$

Giving a total bit rate of
$$82.944 + 20.736 + 20.736 = 124.416 \, \text{Mbps}$$

It will noted that using a 4:1:1 structure in which quarter subsampling in the vertical direction is used, would result in the same chrominance bit rate, namely

$$180 \times 576 \times 25 \times 8 = 20.736 \, \text{Mbps for each chrominance component}$$
with a total bit rate of
$$20.736 \times 2 = 41.472 \, \text{Mbps and a total bit rate of}$$
$$82.944 + 41.472 = 124.416 \, \text{Mbps}$$

The actual bandwidth requirements depend on the type of modulation used. However, for pure pulse code modulation (PCM), the bandwidth is half the bit rate, which in this case is in the region of 60 MHz. This is impractically high and clearly illustrates the need for data compression which is the subject of the next chapter.

4 MPEG encoding

As we have seen from the previous chapter, when an ordinary analogue video is digitised, it requires as much as 124 Mbps for standard definition (SD) television, an inhibitive bandwidth of up to 62 MHz. To circumvent this problem, a series of data compression techniques have been developed to reduce the bit rate. The ability to perform this task is quantified by the compression ratio. The higher the compression ratio is, the lower the bit rate and with it the lower the bandwidth requirements. However, there is a price to pay for this compression as compression inevitably leads to increasing degradation of the image. This is called artefacts. However, advanced compression techniques are so sophisticated that they almost completely avoid the perception of artefacts. Inevitably, such techniques are complex and expensive.

The two basic compression standards are JPEG and MPEG. In broad terms, JPEG is associated with still digital images whilst MPEG is dedicated to digital video. The most popular MPEG standards are MPEG-2 and MPEG-4 with the former associated with SD and the latter with high definition (HD) television.

A digital television programme consists of three components: video, audio and service data (Figure 4.1). The original video and audio information is analogue in form and has to be sampled and quantised before being fed into the appropriate coders. The service data, which contains additional information such as teletext and network-specific information including electronic programme guide (EPG), is generated in digital form and requires no encoding.

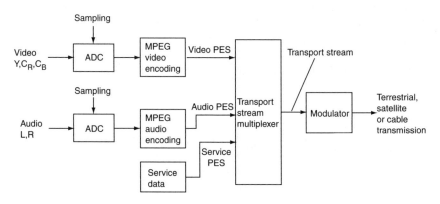

Figure 4.1 *Components of DTV*

The encoders compress the data by removing non-essential or redundant parts of the picture and sound signals and perform bit reduction operations to produce individual video and audio packetised elementary streams (PESs). Similarly, service data is also organised into similar packets to form part of the programme PESs. For television broadcasting purposes, MPEG-2 is used for SD and MPEG-4 for HD television. Let's start with MPEG-2.

Video MPEG-2 coding

There are two distinguishing features of a video clip, both of which are utilised by MPEG in its data compression technique. The first is that a video piece is a sequence of still images and as such can be compressed using the same technique as that used by JPEG. This is known as *spatial inter-frame* compression. The second feature is, in general, successive images of a video piece differ very little, making it possible to dispense with the unchanging or redundant part and send only the difference. This type of compression which is time-related, inter-frame compression known as *temporal* DCT compression; DCT (discrete cosine transform) being the name of the mathematical process used.

Video MPEG coding consists of three major parts: data preparation, compression (temporal and spatial) and quantisation (Figure 4.2).

Video data preparation

The purpose of video data preparation is to ensure a raw-coded sample of the picture frame organised in a way that is suitable for data compression. The video information enters the video encoder in the form of line-scanned coded samples of luminance Y, and chrominance C_R and C_B. Video preparation involves regrouping these samples into 8×8 blocks to be used in spatial redundancy removal. These blocks are then rearranged into 16×16 macroblocks to be used in temporal redundancy removal. The macroblocks are then grouped into slices which are the basic units for data compression. The make up of a macroblock is determined by the chosen MPEG-2 profile. Using 4:2:0 sampling, a macroblock will consist of four

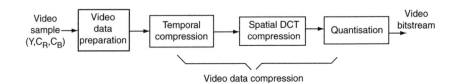

Figure 4.2 *MPEG video compression*

blocks of luminance and one block of each of the chrominance components C_R and C_B. Theoretically, a slice may range from one macroblock up to the whole picture. But in practice a slice will cover a complete picture row or part of a picture row.

Temporal compression

Temporal compression, or inter-frame compression, is carried out on successive frames. It exploits the fact that the difference between two successive frames is very slight. Thus, it is not necessary to transmit the full contents of every picture frame since most of it is merely a repetition of the previous frame. Only the difference needs to be sent out. Two components are used to describe the difference between one frame and the preceding frame: *motion vector* and *difference frame*. To illustrate the principle behind this technique, consider a sequence of two frames shown in Figure 4.3. The contents of the cells in the first frame are scanned and the contents are described as follows: *lion, horse, frog, globe, chair, bulb, leaves, tree and traffic lights*. The second frame is slightly different from the first and if described in full in the same way as the first frame: *plane, horse, frog, globe, lion, bulb, leaves, tree and traffic lights*.

However, such an exercise involves a repetition of most of the elements of the first frame, namely horse, frog, globe, bulb, leaves, tree and traffic lights. The repeated elements are known as redundant because they do not add anything new to the original composition of the frame. To avoid redundancy, only the changes of the contents of the picture are described instead. These changes may be defined by two aspects: the movement of the tiger from cell A1 to cell B2 and the introduction of a plane in cell A1. The first is the motion vector. The newly introduced *plane* is the difference frame and is derived by a slightly more complex method. First the motion vector is added to the first frame to produce a predicted frame (Figure 4.4).

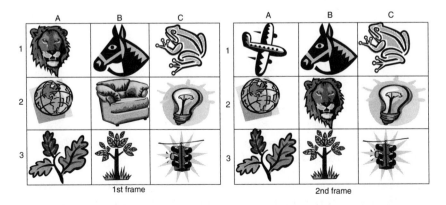

Figure 4.3 *Two successive frames: Current frame 2 and previous frame 1*

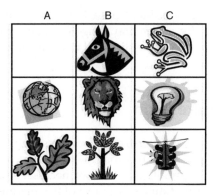

Figure 4.4 *Predicted frame obtained by adding the motion vector to the first frame*

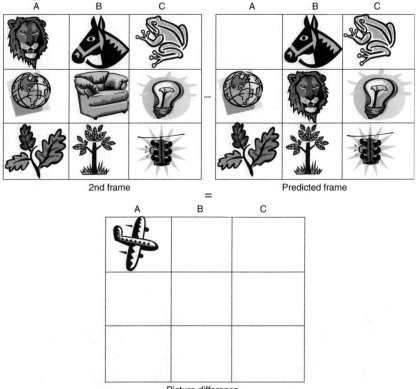

Figure 4.5 *Difference frame is obtained by subtracting the predicted frame from the second or current frame*

The predicted frame is then subtracted from the second frame to produce the difference frame (Figure 4.5). Both components (motion vector and frame difference) are combined to form what is referred to as a P-frame (P for predicted).

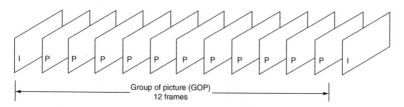

Figure 4.6 *Group of pictures, GOP*

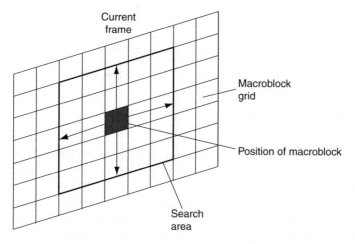

Figure 4.7 *Block matching*

Group of pictures

Temporal compression is carried out on a group of pictures (GOP) normally composed of 12 non-interlaced frames. The first frame of the group (Figure 4.6) acts as the anchor or reference frame known as the *I-frame* (I for inter). This is followed by a *P-frame* obtained by comparing the second frame with the I-frame. This is then repeated and the third frame is compared with the previous P-frame to produce a second P-frame and so on until the end of the group of 12 frames when a new reference I-frame is then inserted for the next group of 12 frames and so on. This type of prediction is known as *forward prediction*.

Block matching

The motion vector is obtained from the luminance component only by a process known as block matching. Block matching involves dividing the Y component of the reference frame into 16 × 16 pixel macroblocks, taking each macroblock, in turn, moving it within a specified area within the next frame and searching for matching block pixel values (Figure 4.7). Although the sample values in the macroblock may have changed slightly

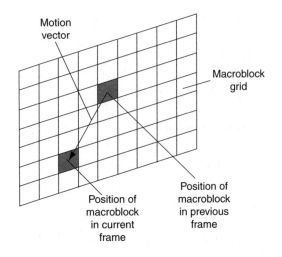

Figure 4.8 *Motion compensation vector*

from one frame to the next, correlation techniques are used to determine the best location match which is down to a distance of one half-pixel in the case of MPEG-2 and quarter-pixel in the case MPEG-4. When a match is found, the displacement is then used to obtain a motion compensation vector that describes the movement of the macroblock in terms of speed and direction (Figure 4.8). Only a relatively small amount of data is necessary to describe a motion compensation vector. The actual pixel values of the macroblock themselves do not have to be retransmitted. Once the motion compensation vector has been worked out, it is then used for the other two components, C_R and C_B. Further reductions in bit count are achieved using differential encoding for each motion compensation vector with reference to the previous vector.

Predicted and difference frames

The motion compensation vector alone is not sufficient to define the video contents of a picture frame. It may define a moving block but it fails to define any new elements such as the background that may have been revealed by the movement of the block. Further information is therefore necessary. This is obtained by first predicting what a frame known as the P-frame would look like if it were reconstructed using only the motion compensation vector and then comparing this with the actual frame. The difference between the two contains the necessary additional information which, together with the motion compensation vector, fully defines the contents of the picture frame. The P-frame is constructed by adding the motion vector to the same frame that was used to obtain the very same motion vector. The P-frame is then subtracted from the current frame to

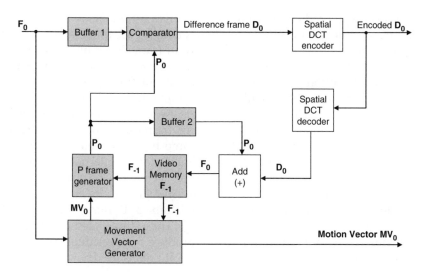

Figure 4.9 *Temporal prediction*

generate a difference frame, which is also known as the residual or pre-diction error. The difference frame now consists of a series of pixel values, a format suitable for subsequent spatial data compression.

Referring to Figure 4.9, current frame F_0 is fed into Buffer 1 and held there for a while. It is also fed into the movement vector generator which uses the contents of the previous frame F_{-1} stored in video memory to obtain the motion vector MV_0. The motion vector is then added to F_{-1} to produce predicted frame P_0 which is compared with the contents of the current frame F_0 in Buffer 2 to produce residual error or difference frame D_0. Residual error D_0 is fed into the spatial DCT encoder and sent out for transmission. Encoded D_0 is decoded back to reproduce D_0 as it would be reproduced at the receiving end. D_0 is then added to P_0 which has been waiting in Buffer-2 to reconstruct the current frame F_0 for storage in the video memory for the next frame and so on.

Bidirectional prediction

The bit rate of the output data stream is highly dependent on the accuracy of the motion vector. A P-frame that is predicted from a highly accurate motion vector will be so similar to the actual frame that the residual error will be very small, resulting in fewer data bits and therefore a low bit rate. By contrast, a highly speculative motion vector will produce a highly inaccurate prediction frame, hence a large residual error and a high bit rate. Bidirectional prediction attempts to improve the accuracy of the motion vector. This technique relies on the future position of a moving matching block as well as its previous position.

Bidirectional prediction employs two motion estimators to measure the forward and backward motion vectors, using a past frame and a future frame as the respective anchors. The current frame is simultaneously fed into two motion vector estimators. To produce a forward motion vector, the forward motion estimator takes the current frame and compares it macroblock by macroblock with the past frame that has been saved in the past-frame memory store. To produce a backward motion vector, the backward motion estimator takes the current frame and compares it macroblock by macroblock with a future frame that has been saved in the future frame memory store. A third motion vector, an interpolated motion vector, also known as a bidirectional motion vector, may be obtained using the average of the forward and backward motion vectors. Each vector is used to produce three possible predicted frames: P-frame, B-frame and average or bidirectional frame (Bi-frame). These three predicted frames are compared with the current frame to produce three residual errors. The one with the smallest error, i.e. the lowest bit rate, is used.

Spatial compression

The heart of spatial redundancy removal is the DCT processor. The DCT processor receives video slices in the form of a stream of 8×8 blocks. The blocks may be part of a luminance frame (Y) or a chrominance frame (C_R or C_B). Sample values representing the pixel of each block are then fed into the DCT processor (Figure 4.10), which translates them into an 8×8 matrix of DCT coefficients representing the spatial frequency content of the block. The coefficients are then scanned and quantised before transmission.

The discrete cosine transform

The DCT is a kind of Fourier transform. A transform is a process which takes information in the time domain and expresses it in the frequency domain. Fourier analysis holds that any time domain waveform can be represented by a series of harmonics (i.e. frequency multiples) of the original fundamental frequency. For instance, the Fourier transform of a 10-kHz square wave is the series of sine waves with frequencies 10, 30 and 50 kHz, and so on (Figure 4.11). An inverse Fourier transform is the process of

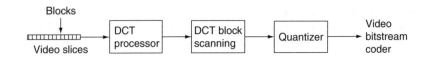

Figure 4.10 *Spatial DCT compression*

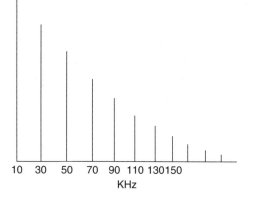

Figure 4.11 *Frequency components of 10l kHz square wave*

adding these frequency components to convert the information back to the time domain.

In common usage, frequency, measured in hertz, refers to temporal (i.e. time-related) frequency, such as the frequency of audio or video signals. However, frequency need not be restricted to changes over time. Spatial frequency is defined as changes in brightness over the space of a picture frame and can be measured in cycles per frame. In the picture frame (Figure 4.12), the changes in brightness along the horizontal direction can be analysed into two separate spatial frequency components:

* Zero or d.c.: grey throughout the frame representing the average brightness of the frame.

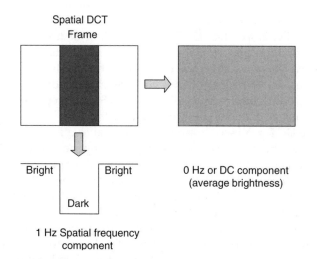

Figure 4.12 *Spatial frequency analysis*

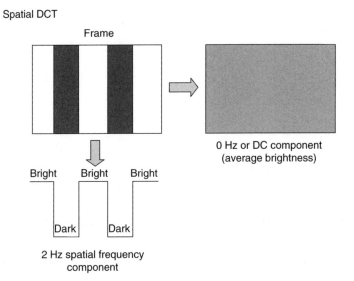

Figure 4.13 *2 Hz spatial frequency*

- 1 Hz: brightness changing horizontally from bright to dark then back to bright, a horizontal spatial frequency of 1 cycle per frame, equivalent to 1 Hz.

If the picture content is changed to that shown in Figure 4.13, a 2-Hz frequency component is produced together with a d.c. component. More complex video elements would produce more frequency components as illustrated in Figure 4.14. A similar analysis may be made in the vertical direction resulting in vertical spatial frequencies.

Normal pictures are two-dimensional, and following transformation, they will contain diagonal as well as horizontal and vertical spatial frequencies. MPEG-2 specifies DCT as the method of transforming spatial picture information into spatial frequency components. Each spatial frequency is given a value, known as the DCT coefficient. For an 8×8 block of pixel samples, an 8×8 block of DCT coefficients is produced (Figure 4.15). Before DCT, the figure in each cell of the 8×8 block represents the value of the relevant sample, i.e. the brightness of the pixel represented by the sample. The DCT processor examines the spatial frequency components of the block as a whole and produces an equal number of DCT coefficients to define the contents of the block in terms of spatial frequencies.

The top left-hand cell of the DCT block represents the zero spatial frequency, equivalent to 0 Hz or d.c. component. The coefficient in this cell thus represents the average brightness of the block. The coefficients in the other cells represent an increasing spatial frequency component of the block, horizontally, vertically and diagonally. The values of these coefficients are determined by the amount of picture detail within the block. The spatial frequency represented by each cell is illustrated in Figure 4.16.

Spatial DCT

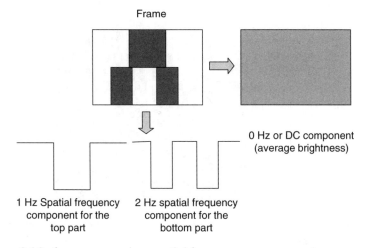

Figure 4.14 *A more complex spatial frequency components*

8x8 pixel values

146	144	149	153	155	155	155	155
150	151	153	156	159	156	156	156
155	155	160	163	158	156	156	156
163	161	162	160	160	159	159	159
159	160	161	162	162	155	155	155
161	161	161	161	160	157	157	157
161	162	161	163	162	157	157	157
160	162	161	161	163	158	158	158

DCT processor

8x8 DCT coefficients

314.91	-0.26	-3.02	-1.30	0.53	-0.42	-0.68	-0.33
-5.65	-4.37	-1.56	-0.79	-0.71	-0.02	0.11	-0.30
-2.74	-2.32	-0.39	0.38	0.05	-0.24	-0.04	-0.02
-1.77	-0.48	0.06	0.36	0.22	-0.02	-0.01	0.08
-0.16	-0.21	0.37	0.39	-0.03	-0.17	0.15	0.32
0.44	-0.05	0.41	-0.09	-0.19	0.37	0.26	-0.25
-0.32	-0.09	-0.08	-0.37	-0.12	0.43	0.27	-0.19
-0.46	0.39	-0.35	-0.46	0.47	0.30	-0.14	-0.11

Figure 4.15 *8 × 8 pixel block transformed into 8 × 8 DCT coefficient block*

A block containing identical luminance (or chrominance) throughout, e.g. part of a clear sky, will be represented by the d.c. component only, with all other coefficients set to zero. A block that contains different picture detail will be represented by various coefficient values in the appropriate cells. Coarse picture detail will utilise a number of cells towards the left top corner and the cells and fine picture detail will utilise a number of cells towards the bottom right-hand corner. The grey scale pattern in Figure 4.17a has horizontal spatial frequencies only and hence the DCT coefficients shown in (c).

DCT does not directly reduce the number of bits required to represent the 8 × 8 pixel block. Sixty-four pixel sample values are replaced by 64 DCT

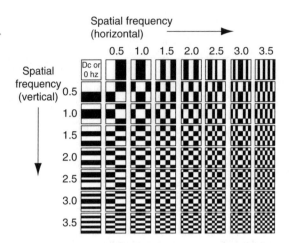

Figure 4.16 *DCT wavetable illustrating the spatial frequencies represented by each DCT cell*

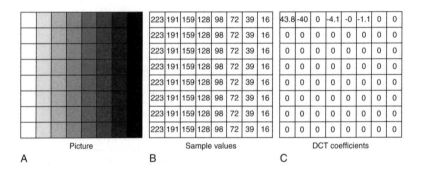

Figure 4.17 *DCT coefficients of a grey scale 8 × 8 block*

coefficient values. The reduction in the number of bits follows from the fact that, for a typical block of a natural image, the distribution of the DCT coefficients is not uniform. An average DCT matrix has most of its coefficients, and therefore energy, concentrated at and around the top left-hand corner; the bottom right-hand quadrant has very few coefficients of any substantial value. Bit rate reduction may thus be achieved by not transmitting the zero and near-zero coefficients. Further bit reduction may be introduced by weighted quantising and special coding techniques of the remaining coefficients.

Quantising the DCT block

After a block has been transformed, the DCT coefficients are quantised (rounded up or down) to a smaller set of possible values to produce a simplified set of coefficients.

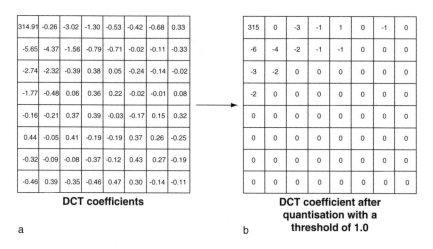

314.91	-0.26	-3.02	-1.30	-0.53	-0.42	-0.68	0.33
-5.65	-4.37	-1.56	-0.79	-0.71	-0.02	-0.11	-0.33
-2.74	-2.32	-0.39	0.38	0.05	-0.24	-0.14	-0.02
-1.77	-0.48	0.06	0.36	0.22	-0.02	-0.01	0.08
-0.16	-0.21	0.37	0.39	-0.03	-0.17	0.15	0.32
0.44	-0.05	0.41	-0.19	-0.19	0.37	0.26	-0.25
-0.32	-0.09	-0.08	-0.37	-0.12	0.43	0.27	-0.19
-0.46	0.39	-0.35	-0.46	0.47	0.30	-0.14	-0.11

DCT coefficients

a

315	0	-3	-1	1	0	-1	0
-6	-4	-2	-1	-1	0	0	0
-3	-2	0	0	0	0	0	0
-2	0	0	0	0	0	0	0
0	0	0	0	0	0	0	0
0	0	0	0	0	0	0	0
0	0	0	0	0	0	0	0
0	0	0	0	0	0	0	0

**DCT coefficient after
quantisation with a
threshold of 1.0**

b

Figure 4.18 *Quantisation*

For instance, the DCT block in Figure 4.18a may be reduced to a very
few coefficients shown in Figure 4.18b if a threshold of 1.0 is applied.
Further compression is introduced by a non-linear or weighted quantisation. The video samples are given a linear quantisation but the DCT coefficients receive a non-linear quantisation; a different quantisation level
is applied to each coefficient depending on the spatial frequency it represents within the block. High quantisation levels are allocated to coefficients representing low spatial frequencies; this is because the eye is
most sensitive to low spatial frequencies. Lower quantisation is applied to
coefficients representing high spatial frequencies. This will increase the
quantisation error at these high frequencies, introducing error noise that is
irreversible at the receiver. However, these errors are tolerable since high
frequency noise is less visible than low frequency noise. The d.c. coefficient
at the top left hand is treated as a special case and is given the highest
priority. A more effective weighted quantisation may be applied to the
chrominance frames since quantisation error is less visible in the chrominance component than in the luminance component.

Quantisation error is more visible in some blocks than in others; one
place where it shows up is in blocks that contain a high contrast edge
between two plain areas. Then the quantisation parameters can be modified to limit the quantisation error, particularly in the high frequency cells.

Zigzag scanning of the DCT matrix

Before coding the quantised coefficients, the DCT matrix is reassembled into
a serial stream by scanning the DCT cells in the zigzag pattern shown in
Figure 4.19, starting at the top left-hand cell. The zigzag scan pattern makes
it more likely that the coefficients having significant values are scanned first

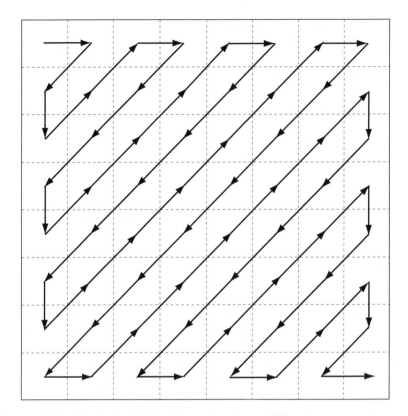

Figure 4.19 *Zigzag scanning of non-interlaced video*

followed by low value coefficients. For the example in Figure 4.18, the scanned order is 315, 0, −6, −3, −4, −3, −1, −2, −2, 0, 0, 0, −1, 1 and −1. No further transmissions are necessary since the remaining coefficients are zero and thus contain no information. This is indicated by a special *end of block* (EOB) code, appended to the scan. Sometimes a significant coefficient may be trapped within a block of zeros, then other special codes are used to indicate a long string of zeros.

The zigzag pattern illustrated above will optimise the number of successive zero coefficients for a progressively scanned picture frame. A different pattern (Figure 4.20) has to be used when optimising the DCT for an interlaced picture scan. This is because, in a block with an interlaced field, the DCT block contains lines from one field only, and these lines must have come from a screen area of 16 lines high. In a progressive scan, the DCT block is obtained from a screen area of eight lines high. Thus, in the case of an interlaced picture, a DCT coefficient representing a vertical spatial frequency is taken over a vertical dimension that is twice as large as the horizontal dimension. The probability of non-zero or significant vertical frequency coefficients is therefore twice as high as the corresponding

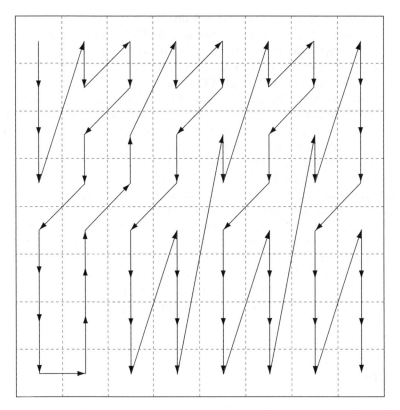

Figure 4.20 *Scanning pattern for interlaced video*

probability for horizontal frequencies. Hence, the distribution of the inter-laced coefficients is different from the distribution of the progressive coef-ficients. This requires a DCT scan pattern that will favour vertical frequency coefficients twice as much as horizontal frequency coefficients and hence the pattern in Figure 4.20.

Coding of DCT coefficients

The coding of the quantised DCT coefficients employs two compression techniques: *run-length coding* (RLC) and *variable-length coding* (VLC). RLC exploits the fact that among the non-zero DCT coefficients, there are likely to be several successive occurrences of zero coefficients. Instead of transmitting these coefficients as zeros, the number of zero coefficients is encoded as part of the next non-zero coefficient. Consider the following set of DCT values:

14, 6, 0, 4, 3, 0, 0, 5, 7, 0, 0, 0, 0, 0

RLC will form the series of DCT values into the following groups: (14) (6) (0, 4) (3) (0, 0, 5) (7) (0, 0, 0)

The number of codes required to transmit these values has thus been reduced from 14 to 7 by grouping any zero coefficient or a run of zero coefficients together with the following non-zero coefficient, (0, 4) and (0, 0, 5). Each group is then given a unique code. The final run of zeros is grouped together and replaced by a single EOB code.

The actual code allocated for each group is determined by the probability of its occurrence. Those occurring most frequently are given a shorter code word than those that occur infrequently. This is the principle of VLC, also known as *entropy* coding. The most well-known method for VLC is the Huffman code, which assumes previous knowledge of the probability of each DCT value. For instance, a DCT value 3 which occurs frequently may be allocated a 6-bit code word; the infrequent DCT value 12 is allocated a 14-bit code word and a zero followed by a 4 is allocated a 9-bit code word. EOB is the most frequently occurring string and it may be allocated a mere 2-bit code word. The code words are held in a lookup table in read-only memory (ROM). At the receiving end, the bitstream has to be resolved into its original code words. Both RLC and VLC are known as lossless coding techniques. Lossless codes, as the name suggests, do not introduce any losses and they are fully reversible at the receiving end.

Buffering

Quantisation, RLC and VLC produce a bit rate that depends upon the complexity of the picture content as well as the amount and type of movement involved. A variable bit rate would occupy a varying amount of bandwidth and may exceed the total available bandwidth with detrimental effect on picture quality. To avoid this, a constant bit rate is necessary. This is obtained by dynamically changing the quantisation of the DCT matrix block (Figure 4.21). The bit-stream is first fed into a memory store before being fed out at a constant rate for transmission. If the bit rate increases, and the buffer begins to overflow, the bit rate control unit is activated; this causes the quantisation level to be reduced, thus decreasing the data bit rate.

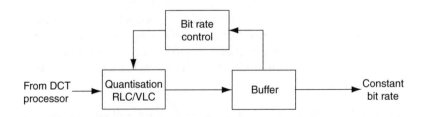

Figure 4.21 *Buffering used to control the bit rate*

The complete DCT coder

Figure 4.22 shows a block diagram of a basic DCT coder. Luminance or chrominance frame pixel values are first converted from line scan to 8×8 block scan before each block is transformed into a DCT matrix of 8×8 coefficients. The coefficients are quantised into a number of quantum levels, determined by the bit rate control, and then scanned to produce a stream of $8 \times 8 = 64$ DCT coefficients for each block. This is followed by RLC and VLC. The compressed bitstream is then fed into a RAM buffer. The resulting bit rate can be varied by the broadcaster at will. This is carried out by simply adjusting the bit rate control. If the output bit rate is reduced to say VHS quality from standard broadcasting PAL quality, there will be fewer quantisation levels with the subsequent deterioration in picture quality. However, more programmes may then be squeezed into a single RF channel.

Forward prediction coder-decoder, codec

An MPEG-2 coder (which embodies a decoder, hence the name codec) using forward prediction is shown in Figure 4.23. P-coded frames in a group of pictures may be obtained by using an I-frame or a reconstructed

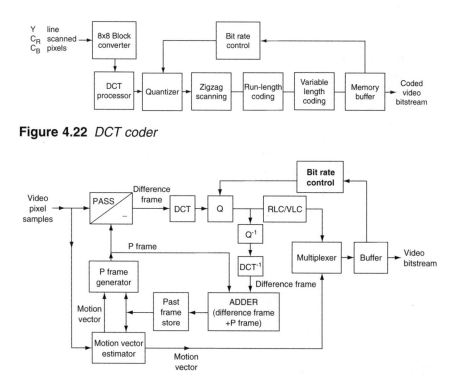

Figure 4.22 *DCT coder*

Figure 4.23 *Forward prediction codec*

P-frame as the reference frame. The past-frame memory store must therefore be continually updated as the pictures in the GOP arrive, frame by frame. The video input, consisting of Y, C_R and C_B pixel sample values, first enters a memory buffer to be fed into the codec, frame by frame as required. Each frame is identified as a Y, a C_R or a C_B as well as an I- or P-frame in a group of pictures.

For the I-frame, which cannot be compared with previous frames and hence can suffer only DCT compression, the pixel samples bypass the comparator and go directly to the DCT encoding section to produce a spatially compressed DCT frame. The quantised DCT coefficients form the only input to the multiplexer since there is no motion vector. The DCT coefficients are also fed into an inverse quantiser (Q^{-1}) and an inverse DCT (DCT^{-1}) decoder to reconstruct the picture back to its original pixel format in exactly the same way as at the receiver. The reconstructed frame is the only input to the adder since there is no motion vector generated by the I-frame. The reconstructed frame is then saved into the past-frame memory store. When the next picture frame arrives, it is first fed into the motion vector generator. At the same time, the I-frame stored in the past-frame memory is made available to the motion vector generator to produce a motion vector for the frame.

The motion vector is then fed into the P-frame generator. At the same time, the saved I-frame is made available to the P-frame generator. A predicted P-frame is therefore produced which takes two separate paths. One path leads to the comparator unit (with subtract selected) to produce a difference frame or residual error, which after DCT, quantisation and RLC/VLC is fed into the multiplexer. The second path takes the P-frame to the adder to reconstruct the current (second) picture frame and save it in the past-frame memory store to be available for the processing of the next (third) frame. The third frame will follow the same process as the second except that the saved frame, which is used to generate the motion vector and the P-frame, is the frame that has been stored in memory during the processing of the previous frame, namely a P-coded picture frame, and so on to the end of the group of pictures.

GOP construction

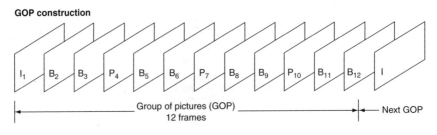

Figure 4.24 *Typical GOP construction*

GOP construction

Incoming frames within a group of pictures may be coded in one of three ways: I, P or B. For a given picture quality, a coded I-frame needs three times more bits than a coded P-frame, which itself requires 50% more bits than a coded B-frame. To reduce the bit rate, a larger number of B-frames is employed. In a typical group of pictures of 12 frames, there are one I-frame, three P-frames and eight B-frames (Figure 4.24). The composition of the GOP is described with two parameters: the number of pictures in the group and the spacing between anchor frames (I- or P-frames). For example, Figure 4.24 shows a GOP with $M = 12$ (12 pictures in the group) and $N = 3$. It is also described as IBBPBBPBBPBB construction.

5 High definition television

High definition television (HDTV) is arguably the most important innovation in television since the introduction of colour. Commercial HDTV services have become a reality and HD broadcasts are now commonplace.

HDTV is not new and HDTV broadcasts have been going on in the USA and Japan for a number of years. However, what was lacking was an internationally agreed standard. Europe did not enter the HDTV market till recently and that delay was utilised to reach an agreement on a unified standard. The first step was to agree to a common interface format (CIF) for HDTV. HDTV is invariably associated with digital television, and although this is not necessarily the only way it can be delivered, without the processing capabilities of digital technology, HDTV would be impractical.

Why HDTV?

The two most prominent television systems, PAL and NTSC produce a moving picture with a quality referred to as standard definition (SD). While picture quality using SDTV is very good and has served television users very well over decades, it does suffer from noticeable degradation in quality in large size displays. To keep up with customer expectations and maintain picture quality, higher picture definition, i.e. high resolution must be used. The following summarises the advantages of HDTV:

- HDTV offers a step change in picture quality, delivering a home viewing experience similar to that offered in cinemas.
- Viewers will see greater depth and tone of colours and textures. The improved clarity of HDTV broadcasts will bring an even greater sense of excitement and drama to a wide range of programme genres.
- HDTV can provide more than four times more data on a display screen than a SD broadcast. This is because of the increased picture resolution that HD broadcasts deliver and HD display devices will be able to receive.
- Availability of multi-channel audio and surround sound.
- Widescreen as standard.

HDTV common interface format

An ITU-R BT.709 standard provides three different specifications for HDTV: 720/50p, 1080/50i and 1080/50p. For North America 1080/60i and 720/60p are available. In each case the number of lines (720 or 1080) is the

Table 5.1 *HDTV common interface format*

Property	1080/50i	1080/60i	720/50p	1080/50p
Lines/frame (total)	1250 (2 × 625)	1125	750	1125
Lines/frame (active)	1080	1080	720	1080
Pixels/line (active)	1920	1920	1280	1920
Line frequency (kHz)	31.25 (2 × 15.625)	33.75	37.5	62.5
Frame rate	25 Hz	30 Hz	50 Hz	50 Hz
Field rate	50	60	—	—
Line duration (total) (μs)	32	29.63	26.67	16
Sampling rate (MHz)	74.5	74.5	74.5	148.5
Aspect ratio	16:9	16:9	16:9	16:9
Interlace/progressive	Interlace (*i*)	Interlace (*i*)	Progressive (*p*)	Progressive (*p*)

active lines for the display. These were chosen in order to have a CIF with the SMPTE standards used in North America and Japan. For Japan and Korea it is 1080/60*i* and for China and Australia 1080/50*i* (Table 5.1).

HDTV is widescreen television with an aspect ratio of 16:9 (1.78:1). This is the same ratio as the pixel ratios: 1920 × 1080 for the 1080*i* and 1280 × 720 for the 720*p* formats. This means that for HDTV, the pixels are 'square' in the same way as computer-based graphics making integration between the two easier.

The road to MPEG-4/H.264/AVC

In January 1988, MPEG gathered together some experts from the ISO/IEC in order to define a standard for the encoding of motion pictures. The standard was MPEG-1 for applications in the multimedia field and its successor MPEG-2 for broadcasting applications. MPEG-2 was successfully implemented in SD digital video broadcasting (DVB), and DVD. The first successor to MPEG-2 was MPEG-4 published in 1999 taking an object-based approach to video compression. Conventional images became object planes and where an object intersects an object plane, it is described using intra-coding, forward projection and bi-directional predictions. MPEG-4 also introduced wavelet coding to still objects. It was defined for a host of new applications in the low-bit multimedia field including inter-active mobile multimedia communications, videophone, mobile audio–visual communication, multimedia electronic mail, remote sensing, video conferencing, games and many others. In 2001, with the aim of developing a more efficient compression system, the standardisation bodies ISO/IEC and ITU combined their efforts in a working group charged with developing a coding system which was to be known as *advanced video coding* (AVC) suitable for broadcasting high definition video. In 2003, the AVC system was established and integrated as part 10 of the MPEG-4 and assumed the name H.264. In 2004, AVC/H.264 was included as part of DVB transport system.

The AVC system is not compatible with MPEG-2 in that it does not produce a compliant bitstream and thus its introduction will require the use of new encoders and decoders.

MPEG-4 profiles

The H.264/AVC scheme includes a different set of profiles than MPEG-2 to support the extensive list of multimedia applications. The following is a summary of MPEG-4 profiles:

- baseline profile for low-delay end-to-end applications;
- extended profile for mobile application;
- main profile for broadcasting applications at SD at a bit rate of 1.5–2 Mbps;
- high profile for broadcasting at high definition as well as studio applications.

The high profiles has several sub-sections including 8-bit and 10-bit samples, 4:2:0, 4:2:2 and 4:4:4 sampling structures.

H.264/AVC features

The MPEG-4 H.264/AVC standard represents the advanced evolution of familiar MPEG-2 technology. It was designed specifically for carriage over existing MPEG-2 transport and modulation infrastructures. The result is a standard that includes powerful data compression. The aim is to make it possible to produce SDTV quality pictures at a bit rate of 1.5–2 Mbps and HDTV at a bit rate of 6–10 Mbps. Its audio counter part is *advanced audio coding (AAC)* system which supports multi-channel and surround sound.

The higher compression efficiency is paid for in terms of increased complexity in both the encoder at the transmitting end and the decoder at the receiving end. The AVC encoder is about eight times more complex than that used in MPEG-2 requiring extremely fast and intensive computing processing power. Such processing powers are now available which made the use of AVC in HDTV broadcasting a commercial viability.

AVC does not support object-based coding. It is intended for use with entire pictures and as such it builds upon MPEG-2 adding some refinements to existing coding tools and introducing new compression techniques (Table 5.2).

Table 5.2 *MPEG-4 H.264/AVC new and improved features*

MPEG-4 H.264/AVC *new techniques*
- Intra-frame (within a frame) prediction
- New 4×4 DCT like block transformation
- In-loop de-blocking filtering

MPEG-4 H.264/AVC *improved techniques*
- Quarter-pixel motion prediction compared with the 1/2 pixel in MPEG-2
- The use of smaller and more flexible block sizes when computing the motion compensation vector compared with the fixed 16×16 block of MPEG-2
- The availability of more than one reference frame in place of the single reference I frame in MPEG-2
- The use of more powerful CAVLC/CABAC entropy encoding in place of VLC

Intra-frame (spatial) prediction

One of the main innovations of AVC is the introduction of intra-frame (within a frame) prediction. *Intra-prediction* as it is more commonly known, is used where inter-prediction cannot be used, namely on the I frame. It is applied to all three components: Y, C_R and C_B. Intra-prediction makes use of the fact that adjacent blocks within a single frame display a degree of similarities. A block is predicted using data of previously scanned block or blocks of the same frame.

Intra-blocks and modes

For the luminance samples, intra-prediction may be carried on each 4×4 sub-block or for a 16×16 macroblock. The process involves copying pixel values of previously coded blocks (sub-blocks or macroblocks) into the current block. Since block coding is performed in a raster scan order (left to right, top to bottom); the previously encoded blocks used for intra-prediction are those above and to the left of the block being predicted as illustrated in Figure 5.1, namely:

- the lower pixel row of the block immediately above: A, B, C, D;
- the left pixel column of the block immediately to the left: I, J, K, L;
- the lower pixel row of the block above and to the right: E, F, G, H;
- lower right pixel of the block above and to the left (M).

There are a total of nine prediction modes for each 4×4 luma block; four modes for a 16×16 luma block; and one mode that is always applied to each 4×4 chroma block. For example 4×4 luma Mode 0 (vertical) copies

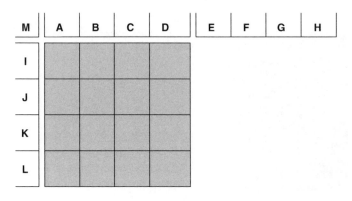

Figure 5.1 *Data blocks for intra prediction*

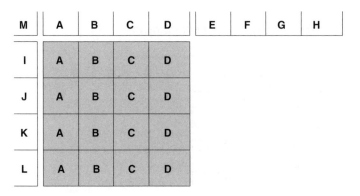

Figure 5.2 *Mode 0 (vertical)*

lower pixel line of block above (A B C D) to all pixel lines in current block as shown in Figure 5.2. This mode may be used in a picture with a left to right gradient (Figure 5.3).

Mode 1 (horizontal) copies the rightmost column of the block to the left (I J K L) suitable for a picture with a left to right gradient (Figure 5.4).

In Mode 2 (DC or average) Pixels values of previous blocks A B C D E F G H I J K L are averaged and the average (DC) value is copied in all 16 locations of the predicted block as illustrated in Figure 5.5 where P is the calculated average.

The other modes are: Mode 3 (diagonal down-left); Mode 4 (diagonal down-right); Mode 5 (vertical-right); Mode 6 (horizontal-down); Mode 7 (vertical-left) and Mode 8 (horizontal-up).

For a 16 × 16 block, only four modes are available: vertical (mode 0), horizontal (mode 1), DC (mode 2) and plane. Plane mode is a refinement to the DC mode which looks for change in the horizontal brightness in the top row and left column to work out the average value.

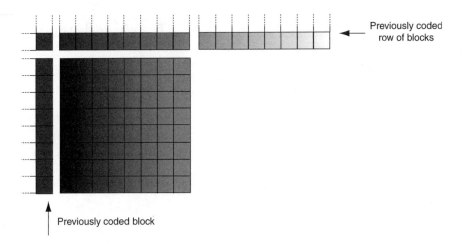

Previously coded
row of blocks

Previously coded block

Figure 5.3 *Mode 0 for a left to right gradient*

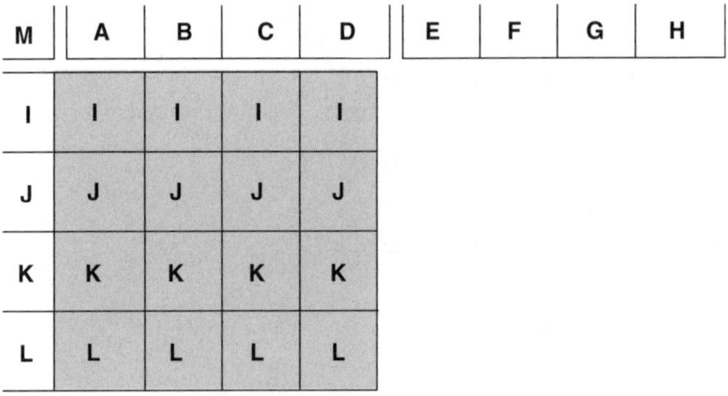

M	A	B	C	D	E	F	G	H
I	I	I	I	I				
J	J	J	J	J				
K	K	K	K	K				
L	L	L	L	L				

Figure 5.4 *Mode 1 (horizontal)*

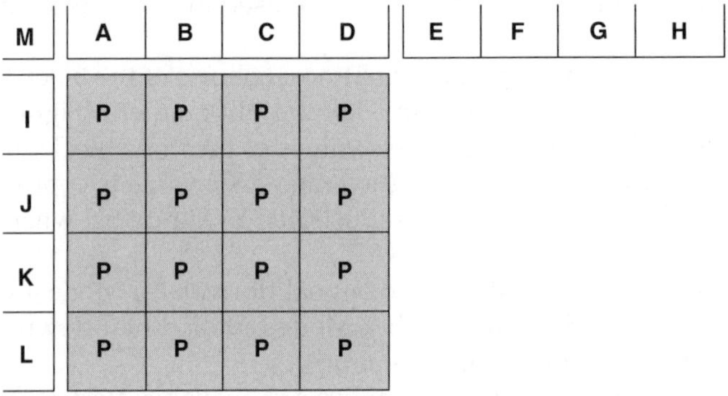

M	A	B	C	D	E	F	G	H
I	P	P	P	P				
J	P	P	P	P				
K	P	P	P	P				
L	P	P	P	P				

Figure 5.5 *Mode 2 (DC or average)*

Size and mode selection

The size of the block to be used for prediction is chosen by the coder which identifies edges and their direction (horizontal, vertical, diagonal, etc.) in order to select the most appropriate block size. The mode used to encode the prediction of a block is chosen based on the textures and gradients in the video source data. Different modes are tried and the one producing the least amount of residual error is selected.

The choice of intra-prediction mode for each 4×4 block must be signalled to the decoder and this could potentially require a large number of bits. However, intra-modes for neighbouring 4×4 blocks are highly correlated. For example, if previously encoded 4×4 blocks X and Y in Figure 5.6 were predicted using mode 2, it is likely that the best mode for block Z (current block) is also mode 2.

Intra-prediction operation

The principle of intra-prediction is the same as that of inter-prediction (Figure 5.7). A predicted P block is produced based on previously scanned and encoded block or blocks of the same frame. The predicted block is

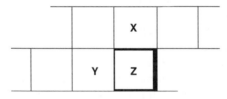

Figure 5.6 *Adjacent 4 × 4 intra-coded blocks*

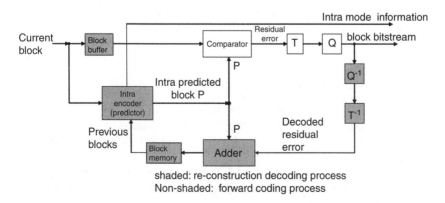

Figure 5.7 *AVC intra (spatial) prediction process*

then compared with the current block or more precisely with the current block as it would be when decoded at the receiving end to produce an intra-residual error which is transformed into coefficients and quantised to form the bitstream. In Figure 5.7, the block memory holds the pixel data of the previous block or blocks which are fed into the intra-encoder to produce a predicted intrablock P. The predicted block is then fed into a comparator to produce the residual error. Following transformation and quantisation, the bit stream is sent to the channel encoder for transmission. Data containing information of the intra-prediction mode parameters is also sent out to form part of the bit stream. The block memory is refreshed with the currently encoded block. This is carried out by decoding the bitstream (inverse transformation, T^{-1} and inverse-quantisation, Q^{-1}) to re-produce the residual error as it would be produced at the receiving end. This residual error is then added to the predicted block P to reconstruct the current block and refresh the block memory store.

AVC motion compensation

AVC achieves its most significant gains over MPEG-2 through substantial improvements to the motion compensated prediction process. Table 5.3 gives a summary of these improvements.

AVC uses smaller picture areas with vectors that have accuracy down to 1/4 pixel. In interpolating the motion vector, more than one previous picture frame may be used. This provides an advantage in a situation where a non-typical picture such as a flash from a firing gun is inserted in a normal sequence. MPEG-2 which relies on a singles previous reference picture, cannot handle such situations adequately. AVC on the other hand deals with this simply by referring to the picture before the gun flash when calculating the motion vector. Bidirectional inter-frame coding is also enhanced with this capability since it improves the accuracy of the motion vector which can now take account of few earlier pictures as well as one later picture. While in MPEG-2, a B frame could not be used as a reference picture frame, with AVC, this is now possible.

Table 5.3 *AVC improved properties*

- Supports a number of block sizes from 16 × 16 to 4 × 4 luminance pixel samples compared with the fixed 16 × 16 used in MPEG-2
- Uses more than one reference frame to search for a good motion predictive match
- Efficient bi-directional predicted B frames, are used more extensively
- Double the accuracy of the motion prediction (AVC offers interpolated quarter pixel prediction)

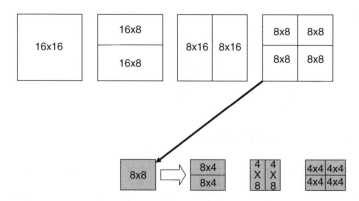

Figure 5.8 *Macroblocks partitioning options for motion vector processing*

Motion compensation block sizes

AVC 16×16 macroblocks may be coded with one and up to sixteen motion vectors. With MPEG-2 prediction is carried out using a 16×16 macroblock which is highly inefficient if the edge of a moving object such as a football goes across the macroblock resulting in a larger residual error. In such cases, better results are obtained if the macroblock is divided into different sizes and shapes according to the angle and position of the edge of the object. The 16×16 macroblock may be partitioned into four high level partitions: 16×16, 16×8, 8×16 and 8×8. When the 8×8 option is selected, it may be further subdivided into three low-level finer partitions: 8×4, 4×8 and 4×4 as illustrated in Figure 5.8.

A separate motion vector is required for each partition or sub-partition. Each motion vector must be coded and transmitted; in addition, the choice of partition(s) must be encoded in the compressed bitstream. Choosing a large partition size (e.g. 16×16, 16×8, 8×16) means that a small number of bits are required to signal the choice of motion vector(s) and the type of partition; however, the motion compensated residual may contain a significant amount of bits in frame areas with high detail. Choosing a small partition size (e.g. 8×4, 4×4) may give a lower-energy residual after motion compensation but requires a larger number of bits to signal the motion vectors and choice of partition(s). The choice of partition size therefore has a significant impact on compression efficiency. In general, a large partition size is appropriate for homogeneous areas of the frame and a small partition size may be beneficial for detailed areas. This method of partitioning macroblocks into subpartitions of varying sizes in order to cater to the shape of an edge is known as *tree structured motion compensation*.

Once the motion vector is obtained from the luminance component, it is then used for the chrominance components. Each chroma block is partitioned in the same way as the Y component, except that since the resolution of each chroma component in a macroblock (C_R and C_B) is half that of the luminance Y component, the chrominance partition sizes are halved. An 8×16 partition in Y corresponds to a 4×8 partition in chroma and an 8×4 partition corresponds to 4×2 in chroma; and so on. The horizontal and vertical components of each motion vector (one per partition) are therefore halved when applied to the chroma blocks.

Motion vector prediction

Encoding a motion vector for each partition can take a significant number of bits, especially if small partition sizes are chosen. Motion vectors for neighbouring partitions are often highly correlated and so each motion vector may be predicted from vectors of nearby, previously coded partitions. *A predicted vector, MVp*, is formed based on previously calculated motion vectors. The difference between the current vector and the predicted vector, *MVD* is then encoded and transmitted. The method of forming the prediction MVp depends on the motion compensation partition size and on the availability of nearby vectors. At the decoder, the predicted vector MVp is formed in the same way and added to the decoded vector difference MVD.

New transforms and quantisation

Once the motion vectors have been identified, the next stage is to produce the frame difference or the residual error. In AVC this is done using enhancements of proven MPEG-2 mechanisms. MPEG-2 uses a discrete cosine transform (DCT) based on an 8×8 pixel block. This is effective for some applications, but imperfect, since errors in the math would result in loss of data. To get around this weakness, the new AVC technology uses a new *pseudo DCT* 4×4 integer transform that is designed for accuracy, ease of processing and can be implemented using 16-bit integer values with addition and bit-shifting operations. Coding using this transform is fully reversible at the decoding stage of the receiver.

   Another area that will yield significant gains relates to the bit-allocation process. Quantisation or bit rate control is the key part of the process, enabling the system to determine how to use bits wisely to attain the desired bit rate. MPEG-2's DCT is fully defined, with no room for improvement. In contrast, the quantisation rate control process offered by AVC has the potential for continued advancement over time.

Adaptive de-blocking filter

One of the main deficiencies of the MPEG-2 coding process is *'blocking artefacts'*. These are caused by the division of the image into blocks for the purposes of transform coding. The artefacts occur at the boundaries of

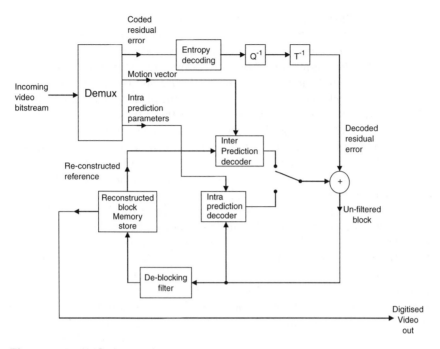

Figure 5.9 *AVC decoder used at the receiving end*

these blocks and are most visually prominent with high degree of quantisation at periods of extreme compression stress. To overcome this, AVC
incorporates a *de-blocking filter* as part of the decoding process as illustrated in Figure 5.9. The incoming bitstream contains video information on
a block by block basis. Each block is defined by three coded components:
residual error, motion vector and intra-prediction mode parameters. The
purpose of the demultiplxer is to separate these three components. The
coded residual error is decoded in three stages in reverse order of the coding process at the transmitter stage: entropy decoding, inverse quantisation
(Q^{-1}) and inverse transformation (T^{-1}). For the reference intra-predicted
blocks, the intra-parameters are fed into the intra-decoder to obtain an
intra-predicted block which is added to the residual error to produce the
original block. These blocks are fed back into the intra-prediction decoder
to be used as a basis for decoding subsequent blocks. They are also fed
into the de-blocking filter to smooth the boundaries and reconstruct the
blocks into a digitised I frame. The reconstructed I frame is stored in memory to be used as a basis for decoding the P frames. For decoding the
P frames, the decoded residual error is now added to the inter-predicted
blocks that were obtained using the motion compensation vector. These
blocks are de-blocked and used to reconstruct a digitised video which is
used for display purposes. It is also fed into a memory store for decoding
subsequent P frames.

De-block filtering is an adaptive process which must function in the same way in all decoders and that includes the decoding part of the AVC codec at the transmitting end where it is known as an in-loop de-blocking as illustrated in Figure 5.10. For the I frame, the intra-prediction coder produces predicted video, block by block and sends out the intra-mode parameters (block size and mode) to the multiplexer for inclusion in the video bitstream. The predicted blocks from the intra-coder are compared with the current blocks and a residual error is produced which is then transformed and quantised and following entropy coding is fed to the multiplexer. Following inverse quantisation (Q^{-1}) and inverse transformation (T^{-1}), the residual error is added to the intra-predicted blocks from the intra-coder to produce the original block. These are fed back to the intra-coder for the purposes of encoding the remaining I-frame blocks. They are also fed into the de-blocking filter to smooth the edges and re-construct the complete I frame in the same way as it is produced by the decoder at the receiving end. For the P frames, the inter-prediction coder uses the reference I frame in the re-constructed de-blocked frame buffer to produce motion vectors and inter-predicted blocks. The motion vector parameters are fed directly to the multiplexer while the inter-predicted blocks are compared with the actual block of the current frame and a residual error is produced which is transformed, quantised and entropy coded before going into the multiplexer. In the same way as the intra-predicted frame, the P frame is reconstructed for use in subsequent inter-coding.

The de-blocking filter has two benefits: block edges are smoothed, improving the appearance of the image and reduced residual error as a result of the use of filtered macroblocks for motion compensation prediction.

Filtering is applied to vertical or horizontal edges of 4×4 blocks affecting up to three pixels on either side of the boundary. The choice whether to filter or not and the strength of the filtering depends on the type of blocks under consideration and the gradient across them. Filtering is stronger at places where there is likely to be significant blocking distortion such as the boundary of intra-coded macroblocks. To assess the gradient, the de-blocking filter examines a set of vertical and horizontal pixels across the block boundary and assesses changes in their values. A step change could indicate a blocking artefact or may be a genuine transition in the image. The threshold is dynamically set taking into account the degree of quantisation. A low level of quantisation indicates a very small gradient across the boundary which most likely is due to image features and should be left alone. On the other hand, when the degree of quantisation is high, blocking distortions are likely and the filter is turned on. Thus, if the step change is smaller than the specified threshold, it is assumed to be genuine and the filter is switched off. Otherwise, the filter is brought in and the filtering strength is set according to the boundary type and the calculated gradient.

De-blocking filters have to be modified when an interlaced video is used because the vertical separation of the pixels in a field is twice that of a frame.

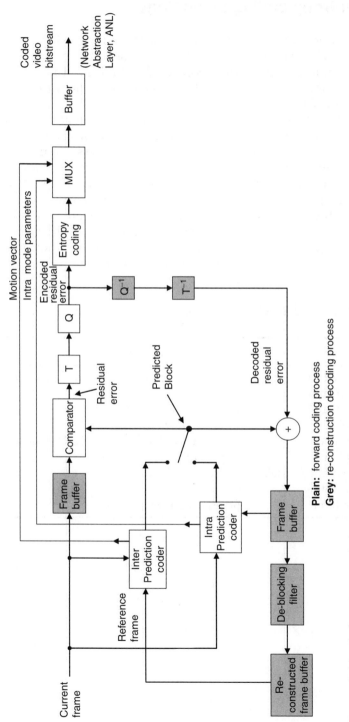

Figure 5.10 *AVC codec used at the transmitting side*

New entropy coding technology

The final stage in the compression process is entropy coding. Here, AVC adds two alternative methods for a more powerful entropy coding compared with the variable length coding, VLC used in MPEG-2. These are *context adaptive variable length coding* (CAVLC) and *context adaptive binary arithmetic coding* (CABAC).

Recall the VLC employed in MPEG-2 which provides an extremely efficient coding scheme based on a table indicating which symbols are much more likely to occur than others. The CAVLC process is a more refined version of this coding method employing a number of codeword tables to encode transform coefficients. The difference is that the best table is selected adaptively based upon statistics of already processed data.

The CABAC process offers substantial gains of about 20% compared with VLC at the expense of intensive computing power for both the encoder and decoder. It continually updates the statistics of the incoming data and adaptively adjusts the algorithm in real time.

Summarised features of AVC

Technical item	MPEG4/AVC features
Block division	16×16, 16×8, 8×16, 8×8, 8×4, 4×8, 4×4
Inter-prediction	P and B
Intra-prediction	Intra 16×16: 4 types
	Intra 4×4 : 9 types
Quadrature conversion	4×4 pseudo-DCT (without calculation errors)
Quantisation	Exponential transformation
Entropy coding/variable length coding	CAVLC, CABAC
De-block filter	Standardised loop filter

6 Audio encoding

Audio compression was developed some time before the formation of MPEG and its Eureka 147 project. Eureka 147 developed the *MUSICAM* (Masking pattern adapted Universal Sub-band Integrated Coding and Multiplexing) jointly with some European manufacturers. MUSICAM was designed for *digital audio broadcasting (DAB)*. In parallel with this, another system was being developed called ASPEC (Adaptive Spectral Perceptual Entropy Coding) jointly by AT&T Bell Labs, Thomson and others. ASPEC included high compression for audio transmission on ISDN lines. Following a comprehensive subjective testing on both systems by the Swedish Broadcasting Corporation, MPEG audio group combined both systems into a common standard with three levels known as layers: I, II and II. The three layers differ in coding complexity and performance in terms of bit rate reduction. *Layer I* is a simplified version of MUSICAM which provides low compression rates at low cost. *Layer II* employs MUSICAM technology in full; it provides high compression rates and is generally employed in DAB and digital television (DTV). *Layer III* (commonly known as *MP3*) combines the best features of both techniques, providing extremely high rates of compression. It is mainly employed in music download and telecommunication applications where high compression ratios are necessary. Further development has produced *advanced audio coding (AAC)* with increased rates of compression making multi-channel surround sound a practical possibility for home users.

Principles of MPEG-1 audio

The process of audio compression starts with digitising the L and R audio signals before going into the MPEG audio encoder as illustrated in Figure 6.1. Digitising involves sampling the L and R channels separately and then converts the samples into multi-bit pulse-coded modulation (PCM) codes by the quantiser. The output is a series of PCM pulses representing the stereo audio channels. This is followed by MPEG encoding.

Before MPEG encoding, the audio is uncompressed PCM with a bit rate that is dependent on the chosen sampling rate. MPEG audio supports three sampling rates, 32, 44.1 and 48 kHz. At a sampling rate of 48 kHz, the bit rate of an uncompressed PCM audio may be as high as 480 kbps per channel, i.e. almost 1 Mbps for stereo. Audio compression will reduce the bit rate by up to a factor of 7 or 8, depending on the coding layer used. At a sampling frequency of 48 kHz, these are the typical bit rates for hi-fi quality sound:

- Layer I: 192 kbps per channel (384 kbps for stereo)
- Layer II: 128 kbps per channel (256 kbps for stereo)
- Layer III: 64 kbps per channel (128 kbps for stereo)

MPEG audio basic elements

MPEG coding consists of five basic steps as illustrated in Figure 6.2:

- *Framing*: The process of grouping the PCM audio samples into PCM frames for the purposes of encoding.
- *Sub-band filtering*: The use of poly-phase filter bank to divide the audio signal into 32 frequency sub-bands.
- *Masking*: To determine the amount of redundancy removal for each band using a psychoacoustic model.
- *Scaling and quantisation*: To determine number of bits needed to represent the samples such that noise is below the masking effect.
- *Formatting*: The organisation of the coded audio bitstream into data packets.

Framing

Before encoding can take place, the PCM audio samples are grouped together into frames, each one containing a set number of samples. MPEG encoding, is applied to a fixed-size frame of audio samples: 384 samples

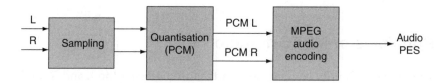

Figure 6.1 *Basic elements of audio encoding*

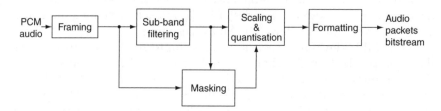

Figure 6.2 *Elements of MPEG audio encoding*

for Layer I audio encoding and 1152 samples for Layer II. At a sampling rate of 48 kHz, using Layer I, 384 samples correspond to an audio time span known as *window* of $1/48 \times 384 = 8$ ms. In Layer II coding, frames of 1152 samples are used which correspond to a window size of $1/48 \times 1152 = 24$ ms. This size window is adequate for most sounds which are normally periodic. However, it fails where short duration transients are present. To overcome this, Layer III provides for a smaller window size to be employed whenever necessary. Layer III specifies two different window lengths: long and short. The long window is the same as that used for Layer II, 24 ms (1152 samples at a sampling frequency of 48 kHz). The short window is 8 ms duration containing 384 samples. The long window provides greater frequency resolution. It is used for audio signals with stationary or periodic characteristics. The short window provides better time resolution at the expense of frequency resolution used in blocks with changing sound levels containing transients.

Sub-band filtering

The human auditory system, what we commonly call hearing, has a limited, frequency-dependent resolution (Figure 6.3). This frequency dependency can be expressed in terms of critical frequency bands which are less than 100 Hz for the lowest audible frequencies and more than 4 kHz at the highest. The human auditory system blurs the various signal components within these critical bands. MPEG audio encoding attempts to simulate this by dividing the audio range into 32 equal-width frequency sub-bands to represent these critical bands using a poly-phase filter bank.

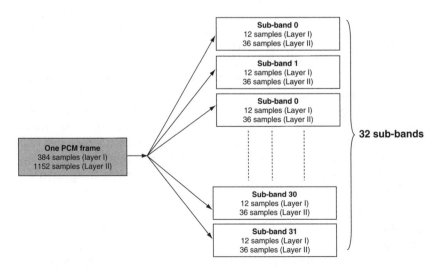

Figure 6.3 *Sub-band filtering*

Each sub-band block, known as a *bin*, consists of $384/32 = 12$ PCM samples for Layer I and $1152/32 = 36$ samples for Layers II and III.

This technique suffers from a number of shortcomings:

- The equal widths of the sub-bands do not accurately reflect the human auditory system's frequency-dependent behaviour.
- The filter bank and its inverse are not lossless transformations. Even without quantisation, the inverse transformation cannot perfectly recover the original signal.
- Adjacent filter bands have a major frequency overlap. A signal at a single frequency can affect two adjacent filter bank outputs.

These shortcomings are mostly overcome with the introduction of AAC, MPEG AAC and MPEG surround.

Masking

The human auditory system uses two transducers on either side of the head, namely the ears. Disturbances in the equilibrium of the air, what we call sound waves are picked up by the ears and using the mechanics of the ear and numerous nerve ends. They are passed to the brain to produce the sensation of hearing. Our hearing sense is a very complex system which informs us of the source of the sound, its volume and its distance as well as its character such as its pitch. To do this, the hearing system works in three domains: time, frequency and space. The interaction within and between these domains, creates a possibility for redundancy, i.e. sounds that are present but not perceived. This is the basis of audio masking.

It is well known that the human ear can perceive sound frequencies in the range 20 Hz–20 kHz. However, the sensitivity of the human ear is not linear over the audio frequency range. Experiments show that the human ear has a maximum sensitivity over the range 2–5 kHz and that outside this range its sensitivity decreases. The uppermost audio frequency is 20 kHz and lowest that can be heard is about 40 Hz. However, reproduction of frequencies down to 20 Hz is shown to improve the audio ambience making the experience more natural and real. The manner in which the human ear responds to sound is represented by the threshold curve in Figure 6.4. Only sounds above the threshold are perceived by the human ear; sounds below the curve are not and therefore they do need not to be transmitted. However, the hearing threshold curve will be distorted in the presence of multiple audio signals which could 'mask' the presence of one or more sound signals.

There are two types of audio masking:

- *spectral* (or frequency) masking where two or more signals occur simultaneously and
- *temporal* (i.e. time-related) masking where two or more signals occur in close time proximity to each other.

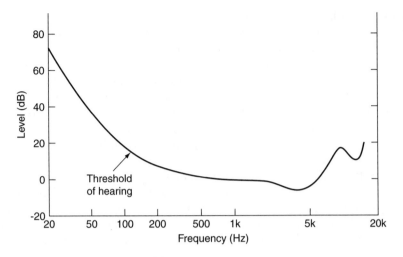

Figure 6.4 *Hearing threshold curve*

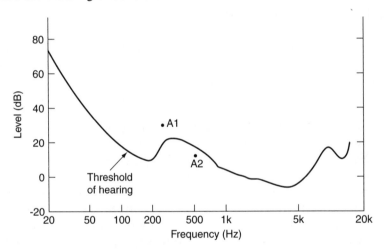

Figure 6.5 *Spectral masking*

It is common experience that a very loud sound, say a car backfiring will render a low-level sound, such as that from a television or radio receiver inaudible at the time of the backfiring. A closer examination would show that the softer sound is also made inaudible immediately before and immediately after the louder sound. The first is spectral masking and the second is temporal masking. Consider two signals, A1 at 500 Hz and A2 at 300 Hz, having the relative loudness shown in Figure 6.2. Each individual signal is well above the hearing threshold shown in Figure 6.4 and will be easily perceived by the human ear. But, if they occur simultaneously, the louder sound A1 will tend to mask the softer sound A2, making it less audible or completely inaudible. This is represented by the loud sound A1 shifting the threshold as shown in Figure 6.5. Two other signals with different frequencies and different relative loudness will result in a different distortion of the masking shape.

In temporal masking, a sound of high volume will tend to mask sounds immediately preceding it (pre-masking) and immediately following it (post-masking). Temporal masking represents the fact that the ear has a finite time resolution. Sounds arriving over a period of about 30 ms are averaged whereas sounds arriving outside that time period are perceived separately. This is why echoes are only heard when the delay is substantially more than 30 ms.

The effects of spectral and temporal masking can be quantified through subjective experimentation to produce a model of human hearing known as a psychoacoustical model. This model is then applied to each sub-band to determine which sounds are perceived and therefore need coding and transmitting and which fall below the threshold and are masked.

Scaling and quantisation

The purpose of scaling, also known as companding is to improve the signal-to-noise ratio (SNR) caused by the quantising errors which are most pronounced with low-level sounds. Quantising errors are inherent in the digitising process itself in which a number of discrete quantum steps are used as determined by the number of bits. For instance, for three bits, there would be $2^3 = 8$ quantum steps; 8-bit codes have $2^8 = 256$ steps and so on. Within a given amplitude range the fewer the number of quantum steps that are available, the larger the height of the step, known as the quantum. Sample levels which may have any value within the given amplitude range will invariably fall between these quantum levels resulting in an element of uncertainty or ambiguity in terms of the logic state of the least significant bit (LSB). This ambiguity is known as the quantising error which is equal to 1/2 the quantum step. With large signals, there are so many steps involved, the quantum is therefore small and with it the quantising error. For low-value samples representing low-level sound signals where fewer steps are needed, the quantising error is large and may fall above the noise level. The quantising error may be reduced by scaling.

Scaling is achieved by amplifying the samples representing low-level sound by a factor known as the *scale factor*. This will increase the number of bits used for the samples resulting in a smaller quantum and reduced quantising error. The scale factor is specified for each sub-band block, which in the case of Layer I coding contains 12 samples. The amplitudes of the 12 samples are examined and the scale factor is then based on the highest amplitude present in the block. Scaling of a sub-band is determined by the power of each sub-band and its relative strength with respect to its neighbouring sub-bands. If the power in a band is below the masking threshold, then it won't be decoded.

The small number of samples is such that the sample amplitudes are unlikely to vary very much within the sub-band block. This technique cannot be used for Layers II and III with 36 samples per block as large variations in sample levels are more likely to occur. First, the sample with the highest amplitude in the block is used to set the scale factor for the whole block. However, if there are vast differences between the amplitudes within each group of 12 samples within the block, a different scale factor may be set for each 12-sample group. The scale factor is identified by a 6-bit code, giving 64 different levels (0–63). This information is included with the audio packet before transmission.

Once the samples in each sub-band block are companded, they are then 're-quantised' in accordance with the predefined masking curve. The principle of MPEG audio masking is based on comparing a spectral analysis of the input signals with a predefined psychoacoustical masking model to determine the relative importance of each audio component of the input. For any combination of audio frequencies, some components will fall below the masking curve, making them redundant as far as human hearing. They are discarded and only those components falling above the masking curve are re-quantised. The spectral analysis of the input fed into the masking processor may be derived directly from the 32 sub-bands generated by the filter. This is a crude method which is used at Layer I audio encoding. A more accurate analysis of the audio spectrum is to use a fast Fourier transform (FFT) processor. Layer I provides an option to use a 512-point FFT. Layer II uses a 1024-point FFT.

Once the components that fall below the hearing threshold are removed by the masking processor, the remaining components are allocated appropriate quantising bits. The masking processor determines the bit allocation for each sub-band block, which is then applied to all samples in the block. Bit allocation is dynamically set with the aim of generating a constant audio bit rate stream over the whole 384-sample (Layer I) or 1152-sample (Layers II and III) blocks. This means that some sub-bands can have long code words provided others in the same block have shorter code words.

Formatting

Coded audio information together with the necessary parameters for the decoding process are grouped into blocks known as frames. A frame (Figure 6.6) is a block of data with its own header and a payload which contains the coded audio data as well as other information such as bit allocation and scaling factor that are necessary for the decoder to re-produce the original sound. Figure 6.7 shows the construction of MPEG-1 Layers I, II and III and MPEG-2 audio frames. The cyclic redundancy count (CRC)

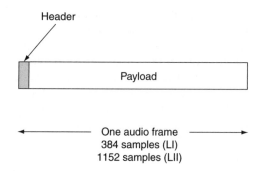

Figure 6.6 *Audio frames*

decoder calculates the CRC checksum of the payload and compares it with the checksum stored in the CRC field. A difference indicates that the data contained in the frame has altered during transmission, in which case, the frame is dropped and replaced with a 'silent' frame.

The auxiliary data at the end of the frame is another optional field which is used for information such as multi-channel sound. In Layer II, an additional field SCFSI (scale factor select information) is included to determine which of the three scale factors has been sent for each sub-band. In Layer III, all the information regarding scale factor, bit allocation and SCFSI are included in single field called the side information field which also include information about surround sound.

The header is 32 bits (4 bytes) in length with 13 different fields as illustrated in Figure 6.8. The first field is the frame sync comprising of 11 bits (31-21) which are always set to 1 indicating the start of a new frame. The next field (bits 20-19) indicates the MPEG version (MPEG-1 or MPEG-2) followed by the layer type (bits 18-17). The presence of CRC field in the payload is indicated by bit 16 set to logic 0. Four bits (15-12) are allocated to indicate the bit rate of the transmission with the sampling frequency indicated by bits 11 and 10. Bit 9 of the header tells the decoder if padding has been used in the frame. Padding is used if the frame is not completely filled with payload data. Bit 8 is for private use followed by 2 bits (7,6) to indicate the channel mode, namely stereo, joint stereo, dual channel or single channel. The next field (5,4) is only used when the channel mode is joint stereo. The remaining three fields contain information on copyright (bit 3), original media (bit 2) and emphasis to tell the decoder that the data must be de-emphasised (1,0).

In the case of Layers I and II, the frames are totally independent of each other. In the case of Layer III, frames are not always independent. This is due to the use of the buffer reservoir, which ensures a constant bit rate by changing frame sizes. Layer III frames are thus often dependent on each other. In the worst case, nine frames may be needed before it is possible to decode one frame.

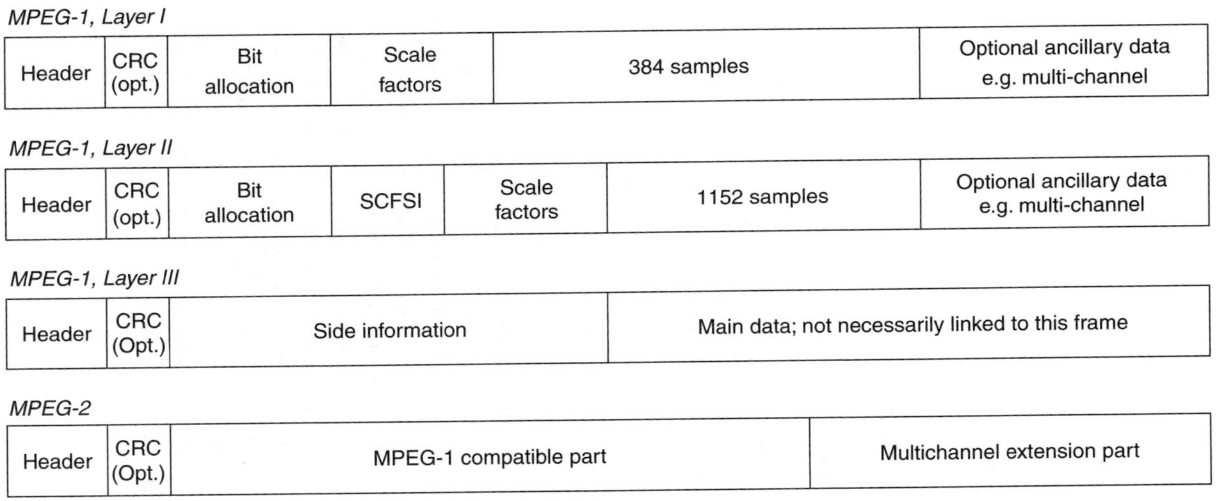

MPEG-1, Layer I

Header	CRC (opt.)	Bit allocation	Scale factors	384 samples	Optional ancillary data e.g. multi-channel

MPEG-1, Layer II

Header	CRC (opt.)	Bit allocation	SCFSI	Scale factors	1152 samples	Optional ancillary data e.g. multi-channel

MPEG-1, Layer III

Header	CRC (Opt.)	Side information	Main data; not necessarily linked to this frame

MPEG-2

Header	CRC (Opt.)	MPEG-1 compatible part	Multichannel extension part

Figure 6.7 *Audio frame construction*

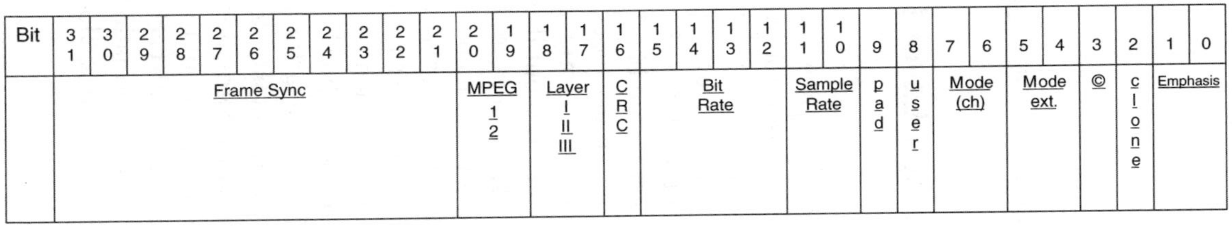

Bit	31	30	29	28	27	26	25	24	23	22	21	20	19	18	17	16	15	14	13	12	11	10	9	8	7	6	5	4	3	2	1	0
	Frame Sync										MPEG 1 2		Layer I II III		CRC	Bit Rate				Sample Rate		padding	user	Mode (ch)		Mode ext.		© Copyright	Original	Emphasis	

Figure 6.8 *The audio frame header*

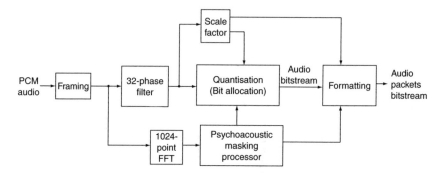

Figure 6.9 *MPEG Layer II coder*

The full Layer II audio coder

Analogue audio L and R channels are sampled separately at a predetermined sampling rate. The samples are then encoded as a PCM stream, which is used for MPEG audio encoding (Figure 6.9). Before MPEG coding, the stream is organised into the basic PCM frames of 1152 samples. These frames are then filtered into 32 frequency sub-band blocks, each containing $1152/32 = 36$ samples. This is the basic encoding audio blocks. The audio sub-band blocks take two different paths: one path examines the individual blocks and allocates a scale factor for companding purposes. The second path takes the PCM audio stream to the quantiser which carries out its bit allocation function in accordance with the masking algorithm from the psychoacoustic processor. The FFT processor prepares a spectral analysis of the PCM input for the masking processor. The masking processor removes redundant audio components and sets the quantising levels for the remaining audio samples. The companded audio samples are then re-quantised in accordance with the bit allocation set by the masking threshold processor to generate a fixed bit rate bitstream. Both the scale factor and the bit allocation are varied as necessary to maintain a constant bit rate at the output. The bitstream by itself does not contain sufficient information for the receiving end to decode the audio signals. Information about the sampling rate, scale factor and bit allocation has to be included with each coded audio bitstream, along with a variety of other data. This information is incorporated within the audio packet produced by the formatting block.

Layer III coding

The MPEG-2 Layer III (MP3) coding structure retains the 1152-sample frame size and the 32-phase sub-band filter. However, the output from the filter is further processed by *modified discrete cosine transform (MDCT)*

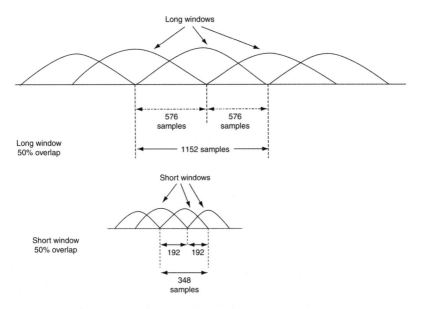

Figure 6.10 *MP3 long and short windows*

processor. The main purpose of MDCT is to compensate for some of the filter bank deficiencies mentioned earlier.

MP3 specifies two different size overlapping MDCT windows: a long window with 1152 Samples and a short window of 384 samples as shown in Figure 6.10. The overlap is 50% which means that the number of samples per window is actually $^1/_2 \times 1152 = 576$ and $^1/_2 \times 384 = 192$ for the long and short windows. Given the 32 sub-bands from the filter bank, the number of samples per sub-band is $576/32 = 18$ samples or $1152/32 = 36$ with 50% overlap and $192/32 = 6$ sample or $384/32 = 12$ with 50% overlap for long and short windows, respectively. Note the short block length is one third that of a long block. In the short block mode, three short blocks replace a long block so that the number of samples for a frame of audio samples is unchanged regardless of the block size selection. For a sampling frequency of 48 kHz, the lengths of the respective windows are $1152/48 = 24$ ms and $384/48 = 8$ ms. Switching between the long and short windows decreased the frequency resolution by a factor of 3 but increases the temporal or time resolution by the same factor. The long block is used for audio signals with stationary or periodic characteristics while the short block option is used in blocks containing transients which require better time resolution.

Pre-echo

The combination of band coding and window size causes a strange phenomenon known as *pre-echo* which is not removed by switching window sizes. Consider a transient occurring towards the end of a block.

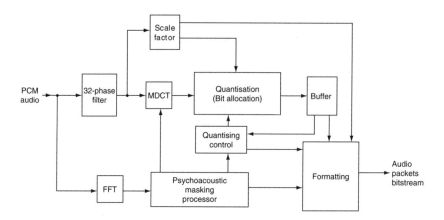

Figure 6.11 *Layer III coder*

Quantising will take account of this and set the quantisation level for the whole block to a higher level than without the transient. Quantising noise will occur at the beginning of the block which is audible before the transient itself, hence the name pre-echo. Pre-echo is eliminated by the use of the buffer shown in Figure 6.11.

Full Layer III audio coder

In the Layer III audio coder (Figure 6.11), the output from the 32-phase filter is fed into a MDCT processor before going into the bit allocation block. Apart from compensating for the deficiencies of the poly-phase filter, the MDCT ensures critical sampling in which the number of coefficients is the same as the number of samples. A simple discrete frequency transform (DFT) will produce twice as many coefficients than samples because the 50% overlap of the windows results in the same samples falling into two adjacent windows. For this reason, sub-sampling is used to ensure that the number of samples reverts back to its non-overlapping level.

The FFT processor drives the psychoacoustic masking processor which drives the MDCT to determine the window size for each individual sub-band. The masking processor also produces a masking threshold for the quantising control unit. Following bit allocation, the bitstream is fed into a buffer to ensure a constant bit rate. When the buffer overflows, it sends a control signal to the quantising control unit to change the quantisation level to reduce the bit rate and vice versa. This process will also serve to remove pre-echo. During stationary sound material, the buffer contents are deliberately reduced by the quantiser. If a transient arrives, the increased number of coefficients may be handled by filling the buffer without increasing the quantisation level thus avoiding pre-echo. The audio bitstream as well as information on the quantisation level, scale factor and masking are formatted into an audio packet.

Advanced audio coding

AAC supports up to 48 audio channels incorporating mono, stereo and 5.1 audio. It was developed by MPEG to deliver the highest possible quality using newly developed compression tools. The driving force to develop AAC was the quest for an efficient coding method for surround sound like those being used in cinemas today. There have been algorithms for these signals in MPEG-2 for some time but further considerable reduction in bit rates was necessary.

MPEG-2 AAC was developed first and declared as an international standard in April 1997. It introduced *temporal noise shaping (TNS)* and *inter-block prediction*. MPEG-2 AAC was followed by MPEG-4 audio. MPEG-4 standardises natural audio coding at bit rates ranging from 2 kbps up to and above 64 kbps. When variable rate coding is allowed, coding at less than 2 kbps, such as an average bit rate of 1.2 kbps, is also supported. The presence of the MPEG-2 AAC standard within the MPEG-4 tool set provides for general compression of audio in the upper bit rate range. MPEG-4 AAC extends these tools by adding new techniques such as *perceptual noise substitution (PNS), twin vector quantisation (TVQ)* and *long-term prediction (LTP)*.

MPEG-2 AAC

Like all perceptual coding schemes, a psychoacoustic model is used to simulate the ability of the human auditory system to perceive different frequencies. Tones at different frequencies with equal power are not perceived with equal power. The perceptual model is also used to model the masking effect of loud tones that mask quieter tones and quantisation noise around its frequency. The perceivable frequencies are divided into several frequency bands; this part of the signal spectrum is then analysed and a masking threshold is calculated.

Although AAC has a similar structure to MP3, compatibility with other MPEG audio layers has been removed and AAC has no granule structure within its frames whereas MP3 might contain one or two granules per frame. *Granules* are used where, in order to reduce the number of bits used to describe a sample, a number of samples are quantised as a group. Furthermore, direct MDCT processing is performed over the PCM sample frames before the audio signal is divided into 32 sub-bands. The same tools (psychoacoustic filters, scale factors and Huffman coding) are applied to reduce the number of bits used for encoding. Another important difference is that AAC has a better frequency resolution up to 1024 frequency lines compared with 576 for MP3. Similar to MP3 coding scheme, the two-window options are available before MDCT is performed in order to achieve a better time/frequency resolution. In the long window mode, MDCT is directly applied over 1024 PCM samples. In short windowing mode, an AAC frame is first divided into eight short windows each of which contains 128 PCM samples and MDCT is applied to each short window individually. Thus, in the short window mode, there are 128 frequency lines decreasing the spectral resolution by eight times whilst increasing the temporal resolution by the same factor. With a 48-kHz sampling rate, the length of the two windows are $1024/58 = 21.3$ ms and $128/48 = 2.7$ ms. With a 50% overlap, the windows sizes are $2 \times 1024 = 2048$ and $2 \times 128 = 256$ samples. Table 6.1 compares the 2-window options of MP3 and AAC.

The basic structure of MPEG-2 AAC is illustrated in Figure 6.12. It introduces improvements to existing tools and few new tools. The crucial differences between MPEG-2 AAC and its predecessor MPEG audio Layer III are as follows:

Filter bank

A direct MDCT transformation is performed over the samples before dividing the audio signal in 32 sub-bands as in MP3 encoding. Similar to MP3 coding scheme, two 50% overlapping windows are used before MDCT is performed. At a sampling rate of 48 kHz, the window sizes correspond to a window of 21 and 2.6 ms.

Window shape

In AAC, the encoder can select the optimal shape for the windows between a *Kaiser–Bessel-derived* (*KBD*) window with improved far-off rejection of its filter response and a sine window with a wider main lobe.

Table 6.1 *MP3 and AAC windows properties*

Encoder	Window type	Number	Samples per frame	Samples per frame with 50% overlap	Window length (ms)
MP3	Long	1	576	1152	24
	Short	3	192	384	8
AAC	Long	1	1024	2048	21
	Short	8	128	256	2.6

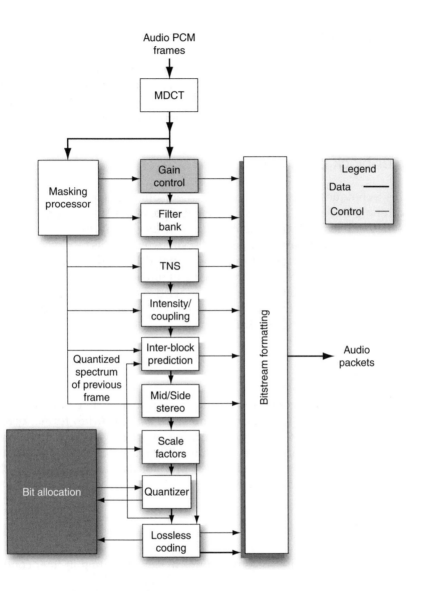

Figure 6.12 *MPEG-2 AAC encoder*

Temporal noise shaping

This tool is an intra-block (within a block) compression technique which uses the values of previously filtered 20 or more coefficients to predict the current coefficient. The prediction is subtracted from the actual value and the prediction error or residual thus obtained is transmitted. At the decoder, an identical predictor is used to reverse the process.

Intensity/coupling

This is used where stereo or surround sound is transmitted at very low bit rate. It discards the spatial information related to the stereo and surround sound and transmits mono with amplitude codes to allow the signal to be panned out in the spatial domain at the receiving end.

Inter-block prediction

This technique exploits the fact that when sound is stationary or periodic with no transient, adjacent blocks exhibit great similarities in their quantised coefficients. A coefficient in a given block may then be predicted from the confidents at the same location in two previous blocks. As before, the prediction is subtracted from the actual value and a residual error is obtained and transmitted. The predictor only operates on coefficients below 16 kHz. Prediction can only be used over a specified number of frames after which they have to be reset. Protracted use of prediction would result in errors and drift.

Mid-side stereo

This is a facility for converting multi-channel sound (stereo or surround) to the sum and difference format known as *mid-side* (*M/S*) format before quantising in cases where quality can be improved. In M/S stereo, the middle (sum of left and right) and side (difference of left and right) channels are encoded. In surround sound, M/S format can be applied to the front and rear L/R pairs separately.

Quantisation

MPEG-2 AAC quantiser uses non-uniform steps resulting in finer control of quantisation resolution and improved coding gain.

Referring to Figure 6.12, the MPEG-2 AAC coder provides a good and consistent quality by dynamically switching between window sizes, intra-prediction (TNS) and inter-block prediction and the control of buffer occupancy to deal with peaks and transients.

MPEG-4 audio

The MPEG-4 audio standard provides a universal toolbox for transparent and efficient audio coding for many different application areas. Its universality makes it possible to use the same standard (MPEG-4 audio) for different applications, so it is no longer dedicated to specific applications. It can be adapted to different applications by selecting only the required tools out of its toolbox. Another advantage in comparison to older standards is the expandability of the standard. There are still new developments made for MPEG-4 to provide new tools for even more applications. The predefined profiles optimised for certain important applications define the tools used for these applications. Possible applications for MPEG-4 audio are internet streaming or downloads, digital radio broadcast, digital satellite and cable broadcast, portable players, data storage (audio), third generation mobile phone and wireless networks multimedia services and bidirectional communications.

MPEG-4 AAC

MPEG-4 AAC combines the tools for general audio coding such as that used in digital audio and high definition television (HDTV) broadcasting. It builds and extends the compression tools used in MPEG-2 AAC. The following are the additional tools provided by MPEG-4 AAC:

Perceptual noise substitution

Noise-like sound which may be part of a normal sound material is very difficult to encode as it has very little if any redundancy. Experiments have shown that under certain circumstances, a listener would not be able to distinguish between the original noise-like signal and general noise that has the same amplitude that has been generated by the decoder. PNS sends the parameters of the noise-like signal (which requires fewer bits to describe) in place of the signal itself. In a band where there is no dominant tone and no transients, the coefficients representing the band are replaced by a noise substitution flag and the total power of the coefficients. The decoder recreates the noise-like signal by generating random coefficients (general noise) with the same power as that of the original signal.

Long-term prediction

This is a developed version of the prediction used in MPEG-2 AAC described earlier. This tool is especially effective for the parts of a signal which have clear pitch property (i.e. with many tonal components like a solo violin as well as speech). It exploits time redundancy between the current and the preceding frame (backward prediction).

Referring to Figure 6.13, the spectral coefficients of the preceding frame are fed through a decoder to be matched with the current frame to get the best prediction parameters. Then, the spectral representations of

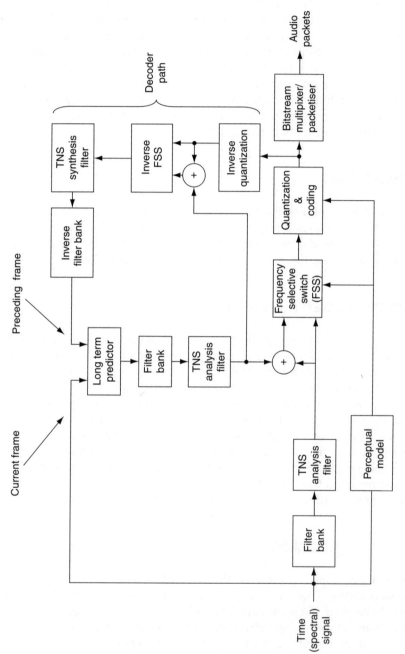

Figure 6.13 *Long-term prediction*

the predicted frame are fed into a filter bank (TNS), filtered and sub-tracted from the current frame which has also gone through identical fil-ter bank and TNS filter to get a residual error signal. A frequency selective switch is used to choose either the residual or the original signal for each band, for further coding with the signal needing the smaller bit rate is chosen.

Twin vector quantisation

In MPEG-2 AAC, the coefficients are quantised using Huffman coding techniques. In cases where the channel bit rate is low, coarse quantisation takes place resulting in errors. For this reason, MPEG-4 AAC provides an alternative coding system in cases where the bit rate is below 16 kbps, namely *twin vector quantisation* (TVQ). TVQ works on blocks of coefficients rather than individual coefficients, with one symbol representing a num-ber of coefficients. Error is minimised by the use of interleaving.

Low-delay AAC

MPEG-4 audio toolbox contains a number of other techniques that may be used for other applications. One of these tools is low-delay AAC (AAC-LD).

While the MPEG-4 AAC provides very efficient coding of general audio signals at low bit rates, it has an algorithmic encoding/decoding delay of up to several 100 ms which is tolerable for broadcasting applications. As an example, for the general audio coder operating at 24 kHz sampling rate and 24 kbps, this results in an algorithmic coding delay of about 110 ms plus up to additional 210 ms for the use of the bit reservoir. Such long delays are unacceptable for such applications as real-time bidirectional communication. To enable coding of general audio signals with a delay not exceeding 20 ms, MPEG-4 specifies a low-delay audio coder. Compared with speech coding schemes, this coder allows compression of general audio signal types, including music, at a low delay. It operates at up to 48 kHz sampling rate and uses a frame length of 512 or 480 samples, com-pared with 1024 or 960 samples used in standard MPEG-2/4 AAC. Also, the size of the window used in the analysis and synthesis filter bank is reduced by a factor of 2. No block switching is used to avoid the 'look-ahead' delay due to the block switching decision. To reduce pre-echo arte-facts in case of transient signals, window shape switching is provided instead. For non-transient parts of the signal a sine window is used, while the low overlap KBD window is used in case of transient signals. Use of the bit reservoir is minimised in the encoder in order to reach the desired tar-get delay. As one extreme case, no bit reservoir is used at all. Verification tests have shown that the reduction in coding delay comes at a very mod-erate cost in compression performance.

Surround sound

Stereo has been a mainstream consumer format for more than 40 years, and so it is not surprising that there has been search for new technologies that further enhance the listener experience. Along with other types of refinements, such as longer audio sample word lengths and higher sampling rates, what is known as high resolution audio, the move towards multi-channel audio or surround sound has become a practical possibility. Consumers can buy inexpensive 5.1-channel playback systems and even 7.1 channel systems are becoming common. However, a non-disruptive transition from stereo to multi-channel audio requires media formats that can serve both those using conventional stereo equipment and those using next-generation multi-channel equipment. While some recent consumer media, such as DVD-video, DVD-audio and super audio CD, resolve the problem by storing both stereo and multi-channel versions of the sound material, this is not a viable option for applications that have to work under severe channel bandwidth limitations, such as digital audio and TV broadcasting or Internet streaming.

The new MPEG surround provides an efficient bridge between stereo and multi-channel presentations in low-bit rate applications with complete backward compatibility with non-multi-channel audio systems. Legacy receivers decode an MPEG surround bitstream as stereo, enhanced receivers provide multi-channel output. It employs the tools available in AAC but introduces a technique for incorporating multi-channel surround sound.

Delivering more than two channels with analogue systems is difficult, however, the Dolby surround system managed to do this by encoding the two rear channels into the standard stereo signals. In the digital field, any number of channels may be encoded and multiplexed. The problem is the bit rate requirements.

MPEG surround coding techniques overcome these shortcomings by including spatial parameters as part of the sound bitstream.

Table 6.2 *Multi-channel sound formats*

1/0	Mono
2/0	Stereo (right, left)
3/0	Right, left, centre
2/1	Right, left, surround
3/1	Right, left, centre, surround
2/2	Right, left, right surround, left surround
3/2 (also known as 5)	Right, left, centre, right surround, left surround
5/2 (also known as 7)	Right, left, centre right, centre, centre left, right surround, left surround

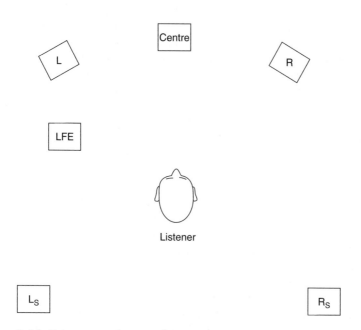

Figure 6.14 *5.1 surrounds sound arrangement*

Multi-channel formats

Multi-channel sound can take many formats as listed in Table 6.2.

In addition, a low frequency effect (LFE), also known as a sub-woofer channel dedicated 120 Hz and lower frequencies is available with all combinations, such as Dolby Digital 5.1 which adds a sub-woofer to the 5-channel 3/2 mode and the 7.1 adding a sub-woofer the 7-channel 5/2 mode. The LFE channel provides the theatre effect. A typical 5.1 surround sound arrangement is shown in Figure 6.14.

Perception of sounds in space

We live in a reverberant world with sounds coming from all direction, some direct, some reflected. If the auditory system is able to distinguish between every different sound, we would hear a confusing cacophony of sounds. As was stated earlier, the ear has finite temporal discrimination as well finite frequency discrimination. When two or more versions of the same sound (direct and reflected) arrive at the ear, they would not be treated as separate sounds unless they are separated by more than 50–60 ms in which case they may be heard as echoes. This is the reason why 'echoes' are only heard in a valley or a dome where the reflecting wall is some distance away. Versions of a sound that are separated by about 30 ms or less are assumed to be a single sound. This, however, does not impair the ability of the auditory system to locate the source of the sound based on the version that arrives to the ear first. This is the sound that travelled the shortest distance and that must be the source of the sound. This phenomenon is known as the precedent effect. The reflected sounds are the reverberations which give the effect of the ambience of real-life surround sound. They are at a relatively low level and as such are assumed to be inaudible by most coding systems and are therefore not transmitted. This is what has to be re-created at the decoding end.

Spatial perception is primarily attributed to three parameters, or *cues*, describing how humans localise sound in the horizontal plane: *inter-aural level differences* (ILD), *inter-aural time differences* (ITD) and *inter-aural coherence* (IC). These three concepts are illustrated in Figure 6.15, which schematically shows a human head and a distant sound source. Direct or first-arrival sound from the source impinges on the left ear while direct sound received by the right ear is diffracted around the head, with associated time delay and level attenuation. These two effects result in the ITD and ILD cues associated with a given source. If the sound is from a point source in a reverberant environment, reflected sound may impinge on both ears, or if the sound is from a diffuse source, non-correlated sound may impinge on both ears, either of which gives rise to the IC cue.

MPEG surround exploits inter-channel differences in level, phase and coherence equivalent to the ILD, ITD and IC cues to capture the spatial image of a multi-channel audio signal relative to a stereo (or mono) signal constructed from the original multi-channel signals. The cues are encoded in a very compact form and included into the side data portion of a MPEG

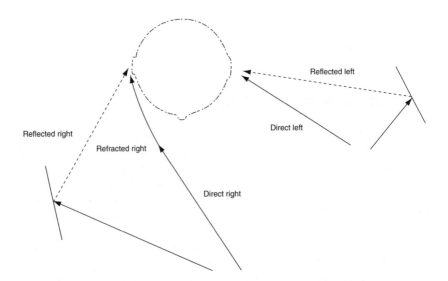

Reflected left

Direct left

Reflected right

Refracted right

Direct right

Figure 6.15 *Perception of spatial sound by human auditory system*

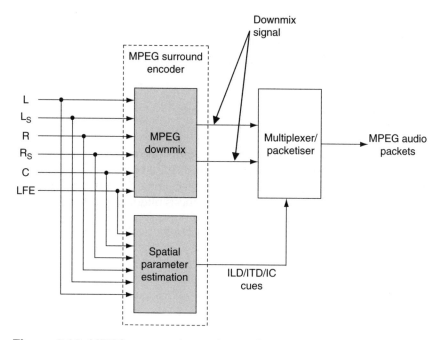

Figure 6.16 *MPEG surround sound encoder*

audio packet or in a separate auxiliary packet. Figure 6.16 illustrates the principle of MPEG surround sound encoding. The MPEG surround encoder receives a multi-channel audio signal, e.g. 5.1, a total of six channels. They are fed into a downmixer to produce a 2-channel downmix signal for stereo (or one-channel downmix signal for mono). The downmix signal is a faithful representation of the original multi-channel signal in stereophonic (or in the monophonic) spheres. It is this downmix signal that is compressed for transmission rather than the original multi-channel signal. The multi-channel signal is also fed into a spatial parameter estimation block to extract the ILD, ITD and IC cues of the input surround sound for inclusion in the audio packet bitstream.

A key aspect of the MPEG surround technique is that the transmitted downmix (e.g. stereo) is an excellent stereo version of the multi-channel signal. This is vital, since stereo presentation remains one of the main listening modes primarily via headphones, such as portable music players. Additionally, MPEG surround supports a mode in which the downmix is compatible with popular matrix surround decoders, e.g. Dolby surround.

At the decoding stage, the cue parameters are used to expand the downmix signal into a high-quality multi-channel output (Figure 6.17). The operation involves a filterbank analyser for high-resolution time/frequency

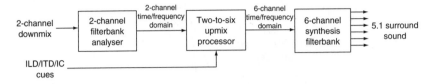

Figure 6.17 *MPEG surround sound decoder*

transformation in preparation for the 2–6 upmix process. The upmix processor using the transmitted spatial cues converts the 2-channel time/frequency representation of the input downmix into a 6-channel time/frequency representation which, following a 6-channel synthesis filterbank, is converted into the original 6-channel surround sound.

7 MPEG-2 transport stream

Broadcasting digital television signals involves multiplexing and formatting the coded audio and video elementary bitstreams, into packetised elementary streams (PESs) and multiplexing them together with relevant service PESs as well as PESs of other programmes. The multiplexed PESs are then formatted into 188-byte packets to form a transport stream as illustrated in Figure 7.1. This is followed by the channel encoder in which the transport stream is used to modulate a suitable carrier. The transport stream is fully defined by MPEG-2. MPEG-4 on the other hand is designed to be transport agnostic. This means that MPEG-4 coded data can be carried over different transport layers and could move from one transport layer to another. In the case of HDTV, an amendment to MPEG-2's transport stream has been defined to carry MPEG-4 data. This involves introducing additional identification and other flags specifically for MPEG-4 packets. Transport of MPEG-4 over IP has also been defined.

Transport stream multiplexing

Video and audio coders deliver their outputs in the form of an *elementary stream (ES)*. Raw uncompressed pieces of video or audio, known as *presentation units* are fed into their respective coders to produce video and audio access units. A video access unit could be an I, P or B coded picture. The audio access units contain coded information for a few milliseconds of sound window: 24 ms (layer II) and 24 or 8 ms in the case of window switching (layer III). The sequence of the video and audio access units forms the video and audio elementary streams respectively. Each ES is then broken up into packets to form the video or audio PES. Service data and other data are similarly grouped into their own PESs. The PES packets are then sliced into smaller 188-byte transport packets.

The transport stream is intended for broadcasting applications in which the communication medium is error prone, be it satellite or terrestrial. An error-prone transmission medium is one that has a potential bit error rate (BER) greater than 10^{-14}. To ensure that errors caused by the transmission medium can be corrected at the receiving end, and to facilitate the multiplexing of several programmes, the transport stream packet is set to a standard, relatively short length of 188 bytes. For applications on DVD, where the communication medium has a lower potential BER, a packet length in kilobytes can be used to form what is known as a programme stream.

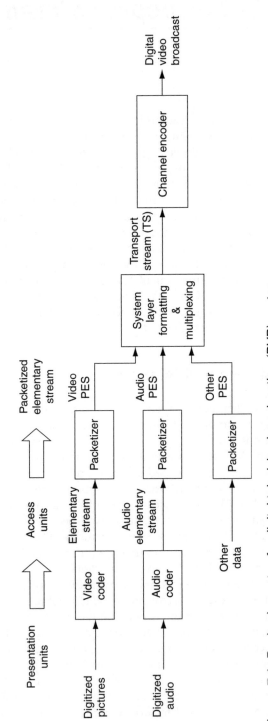

Figure 7.1 *Basic elements of a digital television broadcasting (DVB) system*

MPEG-2 PES packet

Like any data packet, the PES packet starts with a header followed by the payload (Figure 7.2). The PES packet may vary in length up to a maximum of 64 KB. A typical length however is 2 KB. The payload consists of access units taken sequentially from the original ES. There is no requirement to align the start of access units and the start of the payload (Figure 7.3). Thus a new access unit may start at any point on the payload of a PES packet, and it is also possible for several small access units to be contained in a single PES packet.

The essential components of the header are the start code prefix (3 bytes), the start code stream_id (1 byte), programme time stamp, PTS (33 bits) and the decoding time stamp, DTS (33 bits). PTS and DTS need not be included in every PES packet as long as they are included at least once every 100 ms in transport stream applications (DTV) or 700 ms in programme stream applications (DVD). The header includes other fields containing important parameters such as the length of the PES packet, the length of the header itself and whether PTS and DTS fields are present in the packet. Besides these, there are several other optional fields, 25 in total, which may be used to convey additional information about the

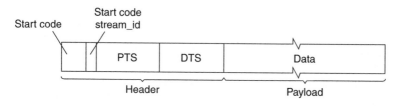

Figure 7.2 *PES packet construction*

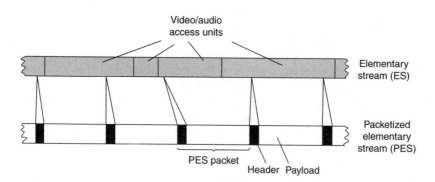

Figure 7.3 *The video and audio elementary access units used to con-struct a series of PESs*

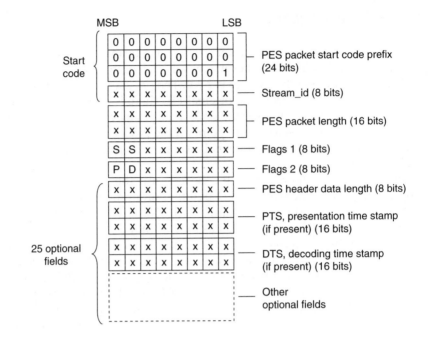

Figure 7.4 *PES header*

PES such as whether scrambled or not, relative priority and copyright information.

The contents of the PES header are outlined in Figure 7.4. The first four fields comprise the PES packet start code. It consists of a 24-bit prefix and 8-bit stream identification (stream_id). This combination of 32 bits is guaranteed not to arise in the PES other than at the start of a PES packet. The prefix is set to 00 00 01 in hex. The stream_id byte distinguishes packets belonging to one ES from those of another within the same programme. MPEG specifies the permitted values for this field, which include 32 different values for audio and 16 different values for video elementary streams. The next field, packet length field, is another mandatory field indicating the length of the packet in bytes after the end of this field. This is followed by two flag fields, flags 1 and flags 2, containing bits which show the presence or absence of various optional fields. The two most significant bits of flags 1, marked SS, indicate the type of scrambling if any, and the two most significant bits of flags 2, marked P and D, indicate the presence or absence of PTS and DTS fields respectively. The last mandatory field, PES_header_length, gives the number of bytes of optional data present in the header before the first byte of the payload is reached. The remaining fields are optional including the PTS and the DTS fields.

Time stamps

At the receiving end, video and audio access units belonging to different programmes arrive at different times. A demultiplexer picks out the video and
audio access units belonging to the selected programme and stores them into
a buffer. When instructed to do so, the video decoder takes a complete access
unit from the buffer, decodes it and displays the picture on the screen.
Similarly, the audio decoder decodes audio access units to provide a few milliseconds of sound in sync with the displayed picture. This process can only
be carried out with a common time base if the video access units are displayed in the correct order and in sync with the audio pieces, what is known
as *lip sync*. This is provided by *time stamps*. It is generated at the multiplex-
ing/formatting stage at the transmitting end. There are two types of time
stamp: *presentation time stamp (PTS)* and *decoding time stamp (DTS)*. The PTS
specifies the time when an access unit should be removed from the buffer,
decoded and displayed at the receiving end. A PTS is adequate for decoding
audio and other data. However, a second time stamp is required to decode
the video elementary stream; this is the DTS. A DTS specifies the time when
a video access unit should be removed from the buffer and decoded but not
displayed to the viewer. Instead, the decoded picture is held temporarily in
memory for later presentation. This is necessary for I- and P-coded pictures
where they are separated by a B-coded picture. In these cases, both DTS and
PTS are necessary. The PTS determines the time when the video access unit
is decoded and the DTS determines the time when the decoded picture is
released from the temporary store for presentation to the viewer. This means
the PTS will always be longer than the DTS. The transfer of data from the
buffer and the decoding process itself take a certain amount of time, and this
has to be compensated in the design of the decoder. Failure to do this would
result in lip-sync fault as experienced by some cheap decoder boxes.

It is not necessary for every access unit to be allocated a time stamp. A
decoder will normally know the rate at which access units are to be decoded
and it is therefore sufficient to provide time stamps on an occasional basis
to ensure the decoding process maintains long-term synchronisation. For
DTV broadcasting, MPEG specifies that a time stamp must occur at least
every 100 ms in an audio or video PES.

Program clock reference

For the time stamps to have meaning at the receiving end, some common
measure of time must be available. This is provided by the *programme clock
generated* at the multiplier stage of the transmitter. The programme clock is
based on the 27 MHz video sampling clock. There is no requirement that
the system clock should be related to any real-time standard. It is purely a
notional time. In the transport stream multiplex, which carries a number of

programmes, each programme has its own independent programme clock but it need not be synchronised with the clocks of other programmes, although several programmes may share a single programme clock. Access units are assigned time stamps based on the programme clock. The 27 MHz clock is divided by a factor of 300 to generate a standard time unit of 90 kHz expressed as a 33-bit binary number. A similar process is carried out for the 47.5 MHz system clock for an HD transport stream. Samples of this clock, known as the *programme clock reference* (PCR), are included in the transport stream. A PCR for each programme clock in the multiplex must appear in the transport stream at least every 100 ms.

At the receiving end, the PCR arriving on the transport stream is used to speed up or slow down the local 27 MHz (47.5 MHz for HDTV) voltage-controlled oscillator. This ensures the two transmitter and receiver clocks are fully synchronised.

Transport stream packet

The MPEG-2 transport stream is a series of 188-byte packets with each packet consisting of a header and a payload. The payload carries data belonging to different PES packets from the various components of one of a number of programmes. This process is subject to two constraints. The first is that only data from one PES packet may be carried in any one transport packet. The second is that a PES packet should always start at the beginning of the payload part of a transport packet and end at the end of a transport packet. The PES packet itself, typically 2048 KB in length, is larger than the transport packet and will thus be spread across a large number of transport packets. Furthermore, since it is unlikely that a single PES packet will fill the payloads of an integer number of transport packets, the last transport packet that carries the residue of the PES will only be partially occupied. To avoid breaking the two constraints mentioned earlier, the excess space is deliberately 'wasted' by filling it with an *adaptation field*; the field length is the difference between 184 bytes and the PES residue (Figure 7.5). The adaptation field is never completely wasted. It is structured so it can carry some useful data such as the PCR.

Null transport packets

When all the PESs are converted into transport packets, these packets are then sent out sequentially at a constant bit rate to form the MPEG-2 transport stream. The bit rate of the transport stream is determined by the number of transport packets produced by the multiplexer. If the number of programmes is few with fewer packets, the bit rate may fall below the value set by the broadcaster. In such a situation, service information and

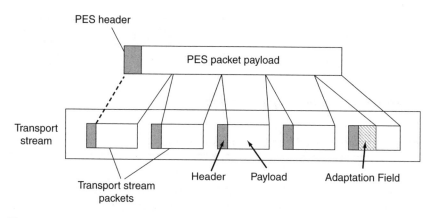

Figure 7.5 *Adaptation field*

Table 7.1 *The use of the adaptation field*

Field	Description	Number of bits
Adaptation_field_length	Total length of adaption field	8
Flags	Various information on subsequent fields	8
Optional fields	Various fields, including a 48-bit PCR field	unspecified
Stuffing	Bytes fo fill in reaming space	unspecified

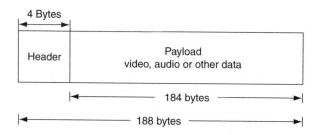

Figure 7.6 *The 188-Byte transport packet*

'null' transport packets are used to soak up any spare multiplex capacity and ensure a constant bit rate.

Transport packet header

Each transport packet comprises a 4-byte header followed by a 184-byte payload of actual video, audio or other coded data, and a total of 188 bytes (Figure 7.6). The header provides the information needed by the receiver

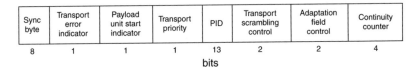

Sync byte	Transport error indicator	Payload unit start indicator	Transport priority	PID	Transport scrambling control	Adaptation field control	Continuity counter
8	1	1	1	13	2	2	4

bits

Figure 7.7 *The transport packet header*

Table 7.2 *Contents of transport packet header*

Field	Function	Bits
Syn_byte	Header start sequence (hex code 47)	8
Error_indicator	Indicates error in previous stage	1
PUSI (payload_unit_ start_indicator)	Indicates start of payload	1
Priority	Indicates transport priority	1
PID (packet identifier)	Indicates content of packet	13
Scrambling_control	Indicates type of scrambling used	2
Adaptation_field_ control	Indicates the presence of an adaptation field	1
Payload flag	Indicates the presence of payload data in the packet	1
Continuty_counter	Keeps count of truncate PES portions	4

to unpack the various programmes to generate the ES of the selected programme. The structure of the transport packet header is illustrated in
Figure 7.7 and listed in Table 7.2.

The header commences with a 1-byte sync word, 01000111 binary
(47 hex), which provides a run-in clock sequence for the packet. This
sequence is not unique and it can quite naturally occur in other fields of the
transport packet. However, the fact that a sync word will occur every 188
bytes within the transport stream enables the decoder to lock onto it. This
is another reason for making sure that all payloads are fully filled with
actual data or stuffed with useless data. The payload_unit_start_indicator
is a 1-bit flag which when set to 1 indicates that the first byte of the
payload part of the transport packet is also the first byte of the PES packet.
The 13-bit *packet identifier* (*PID*) is used to indicate the ES its payload
belongs to. There may be many ESs comprising many different programmes. With 13 bits there are $2^{13} = 8192$ possible values or codes (0–8191)
available to be allocated. Of these, 17 are reserved for special purposes. The
remaining 8175 codes may be allocated to ESs as necessary. MPEG does not
specify any constraints on which PID code of the available 8175 is assigned
to which ES, except that ESs must be allocated unique PID values. This allocation of PID codes is carried out by the broadcaster at the system layer
multiplexer stage. The continuity counter keeps track of how a single PES

packet is spread across transport packets. It is incremented between successive transport packets belonging to the same ES.

Programme-specific information

As stated earlier, an MPEG-2 transport stream can be used to carry information for more than one programme, where each programme is composed of several ESs: audio, video and other data packets. These ESs are identified by a unique PID. At the receiving end, the decoder must be able to identify the ESs that comprise the selected programme. This is the purpose of the programme-specific information (PSI). PSI consists of four tables:

- *The programme association table (PAT)* contains a list of all the programmes together with the PID of the transport packets that carry their *programme map table (PMT)*.
- The PMT contains a list of all ESs belonging to the selected programme with the PIDs of the relevant transport packets.
- The *network information table (NIT)* provides information about the physical network carrying the transport stream, such as channel frequencies and service name; this table is optional.
- The *conditional access table (CAT)* provides information on the scrambling system, if any, and the PID of the transport packets carrying the conditional access management and entitlement information; this table is present only where conditional access is applied.

When a receiver is turned on and a programme selected, the decoder is first directed to the transport packets with a PID of zero to retrieve the PAT. It examines the PAT to identify the PID of the transport packet that contains

PAT (PID 0)

Programme number	PID
0 (NIT)	16
1 (BBC)	306
2 (ITV)	3F5
5	17
18	244
10	87

PMT for programme 1

PID for programme clock reference	726
PID for video ES	726
PID for audio ES (English)	56
PID for audio ES (French)	1022
PID for audio ES (German)	803
PID for subtitles data	585

Figure 7.8 *Programme association table (PAT) and programme map table (PMT)*

the PMT for the selected programme. For instance, if programme 1 (BBC) is selected, then from the PAT in Figure 7.8, PID 306 will be identified and all transport packets with that PID extracted and decoded to produce the PMT for programme 1. The PMT lists the PID of the ESs making up the programme and any other service information related to the programme. The next and final step is for the decoder to extract and decode all transport packets with the listed PIDs to reproduce the PES. For programme 1 (Figure 7.8), that means PID 726 (for the PCR and the video elementary stream), PID 56 (for English language) and PID 585 if subtitles are required. This process naturally takes some time, which explains why changing channels for a digital TV receiver is not as instantaneous as for an analogue TV receiver.

The other two tables, NIT and CAT, are optional. Programme 0 in the PAT is reserved for the NIT, and in Figure 7.8 it points to PID 16. The CAT is retrieved only if the scramble control flag in the PES header is enabled.

Transport stream multiplexing

The transport packets of several programmes are multiplexed as illustrated in Figure 7.9. The simple time-division multiplexing allocates time slots in a regular sequence to participating data streams. For this to function, the data streams must share a time clock and have a common fixed bit rate. But MPEG-2 PESs have a varying bit rate that is determined to a large extent by the amount of residual error produced by the video encoding process. Furthermore, the participating program share the available bit rate of the transport stream as necessary with video facing coding difficulties given a larger share of the available transport stream bit rate. For this reason, *statistical* (also known as *packet*) *multiplexing* is used. First the 188-byte transport packets of each program are sent to their individual buffer with a buffer arbitrator as shown in Figure 7.9. The buffer arbitrator decides which program buffer should have a packet extracted by the statistical multiplexer. Thus if a particular program encounters difficult material with a requirement for increased bit rate, its buffer will fill up and the arbitrator will then allocate more packets to accommodate the interim need for increased bit rate. This is only possible if the other programs are experiencing normal video content, otherwise, the re-quantisation at the video encoder would have to be re-adjusted. Thus there has to be a direct connection between the stat-mux and the relevant video coders by which the level of re-quantising is determine by the multiplexer.

As the process proceeds, the buffers would slowly empty. This allows for the PSI to be inserted in the transport stream. Meanwhile, the buffers are filled again. If the multiplexer finds that having sent enough of everything including PCRs, it is still short of a packet, it will send a null packet to maintain the bit rate at its output. This may be repeated if necessary to give time for the buffers to fill up. This process provides for the bit rate of the transport stream to be fixed independently from the timing of the data

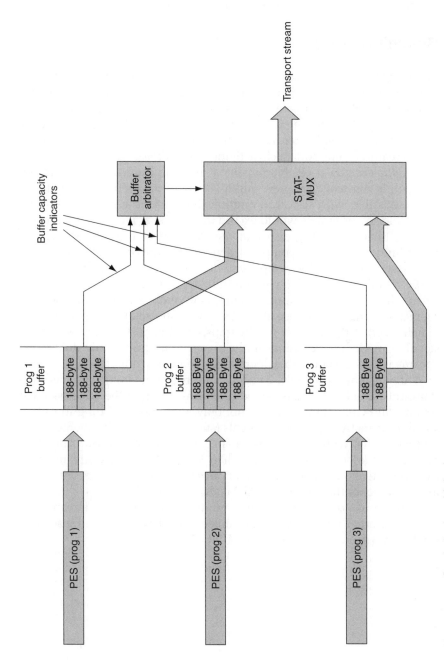

Figure 7.9 *Statistical multiplexing*

being carried. This is important because the transport stream is the bit-stream that is modulated and transmitted and as such, it determines the spectrum of the transmission and that must not vary.

Re-multiplexing

The straight forward broadcasting involves a transport stream carrying a number of programs. However, in several applications, a provider may take in a number of streams from several sources and select programs from among them to be carried by a newly created transport stream. Furthermore, a broadcaster may wish to include an advert within a transport stream. To support such activities, *re-multiplexing*, also known as *transmultiplexing* is used. Nominally, a re-multiplexer does not alter bit rates while constructing a new multiplex out of the input streams. It is of course possible, with statistical multiplexing that the sum of the bit rates of the selected programs would exceed the bit rate of the transport stream. To avoid this, *transcoding* or *re-compression* has to be used. Transcoding is the technique by which a compressed video stream is translated to a lower bit rate strictly within the compressed domain. It can reduce the bit rate of MPEG-2-compressed video without fully decoding and re-encoding a bit-stream. It is in partial compression in that it involves identifying the DCT coefficients and re-quantise them to ensure the totality of the bit rate remains within the limit set for the transport stream. This results in occasional or moderate reductions in the average bit rate of individual video streams with little noticeable degradation in picture quality.

While insertion of MPEG-2-coded advertisements into a channel with stat-multiplexed video requires bit rate transcoding of the advertisement stream, MPEG-4 coding schemes not only allow use of coding tools that are more attractive for advertisements (such as natural video merged with synthetic video with lots of scene changes), they allow for lowering of bit rates to values that are far lower than MPEG-2 stat-mux stream rates. Advertisements could be authored at rates lower than 500 kbps with quality that matches that of the MPEG-2-coded video.

With the use of MPEG-4 coding, multiple advertisement streams can now be inserted into a stat-mux channel instead of a bitrate-transcoded, single advertisement stream using MPEG-2.

8 Channel encoding

The transport stream emerging from the system layer multiplexer forms what is known as a *channel*. It is also known as a *multiplex* carrying a number of different programmes. Before transmission can take place, the packetised transport stream is first scrambled to obtain an even distribution energy across the channel. This is followed by forward error correction (FEC) before the signal is finally modulated and transmitted (Figure 8.1). The channel encoder is agnostic as the type of encoding (MPEG-2, MPEG-4, AVC, ACC, etc.). The delivery of audio and video encoded data is the same regardless of the encoding tools used.

Scrambling

Scrambling is the process of rearranging the order of the data bits. It is not to be confused with encryption, which is the replacement of the original information by an alternative code pattern. Encryption is a secure system and it is used in conditional access applications where restrictions apply. Scrambling is used only for energy dispersal. The problem with plain unscrambled bitstreams is that they are likely to have long series of zeros and ones, which introduce a d.c. component. This results in an uneven distribution of energy making the transmission highly inefficient. If the bitstream can be randomised and the series of 0s and 1s scattered, a more even energy distribution will be obtained. This is the purpose of scrambling. Totally random scattering is not possible as there is no way of descrambling the bits back to their original order at the receiver. However, a pseudo-random scattering with a known pattern can be easily descrambled at the receiving end to regenerate the original order of the data bits. To ensure the start of each transport packet can be recognised, the start byte is not scrambled and the scrambler is disabled for the duration of the start code.

Forward error correction

Digital signals, especially signals with a high level of data compression, require an efficient error protection capability. In digital video broadcasting (DVB), the bit *error rate (BER)* must be better than 10^{-11} or 1 in 100,000 million bits. This is equivalent to less than 1 uncorrected bit in 1 h of transmission. A transmission channel with such a low BER is known as a quasi-error-free (QEF) channel. DVD re-production and cable television are such channels. However, open communication, terrestrial or satellite,

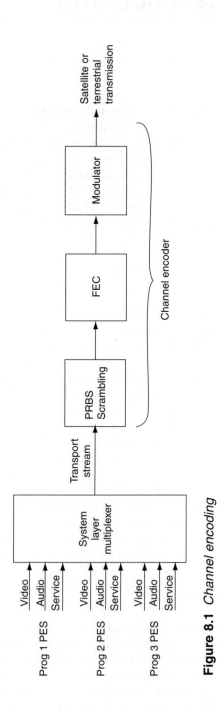

Figure 8.1 *Channel encoding*

are prone to errors caused by a variety of reasons including weather conditions and reflections. To ensure such communication media meet the QEF specification, preventive measures must be taken to ensure that errors introduced by the transmission medium are detected and, where possible, corrected. This is the function of the FEC block.

In general, if an error is detected at the receiving end of a communication system, it can be corrected in two different ways: the recipient can request the original transmitter for a repeat of the transmission, or the recipient can attempt to correct the errors without any further information from the transmitter. Whenever it is a realistic option, communication systems tend to go for retransmission. But if the distances are large, perhaps to contact a space probe, or if real-time signals are involved, such as in audio and video broadcasting, then retransmission is not an option. These cases require error correction techniques.

Error correction is common place in all walks of life including aural communication. It makes use of redundancy in messages. Redundancy exists naturally in all codes including languages. In ordinary English, the u following a q is quite unnecessary and 'at this moment in time' can be shortened to 'at this moment' or even 'now'. Redundant letters or words play a very important role in communication. They allow the recipient to make sense of distorted information. This is how we can make sense of badly spelled emails, a text message, a badly tuned receiver, and so on.

In digital communications, *redundancy* means unnecessary data that occupies precious bandwidth and the purpose of data compressions is precisely to remove them. But without redundancy, there cannot be error correction. For this reason, controlled redundancy bits are added to enable data corrupted in transmission to be corrected at the receiving end.

The most basic technique, parity, provides rudimentary error detection. It involves a single parity bit at the end of a digital word to indicate whether the number of 1s is even or odd (Figure 8.2). There are two types of parity checking. Even parity (Figure 8.2a) is when the complete coded data, including the parity bit, contains an even number of 1s. Odd parity (Figure 8.2b) is when the complete coded data contains an odd number of 1s. At the receiving end, the number of 1s is counted and checked against the parity bit; a difference indicates an error. This simple parity check can only detect an error occurring in a single bit. An error affecting two bits will go undetected. Furthermore, there is no provision for determining which bit is actually faulty. For these reasons, more sophisticated techniques are normally used with higher levels of redundancy. One such technique is FEC employed in digital television (DTV) broadcasting.

Error correction

The introduction of redundancy bits to a package of data increases the data length and with it the number of possible combinations. Consider a 6-bit package consisting of 4 bits of useful data and 2 redundancy bits.

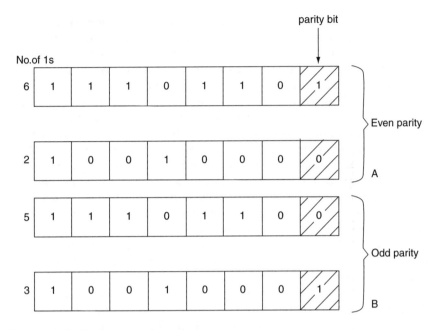

Figure 8.2 *Parity error detection*

The 4 bits of useful data contain $2^4 = 16$ different valid messages. At the receiving end, however, a set of $2^6 = 64$ different messages may be received, of which only a sub-set contains the 16 valid messages. This sub-set is called a *code* and the valid messages are called *code words* or *code vectors* (vectors for short). When a message is received that does not correspond to any of the valid codes, the receiver finds a valid code word 'nearest' to the received message, on the assumption that the nearest is the most likely correct message in the same way as the corrupted word in the sentence 'the fishing boat *saiked* in the morning' is easily deduced to be '*sailed*' being the nearest valid replacement. This technique is known as forward error correction (FEC).

In practice, real codes contain very long strings of binary bits with thousands of valid code words requiring carefully designed coding techniques to produce code words which are well structured into sets and sub-sets. The process of correction is then carried out by advanced and sophisticated mathematics. There are two coding techniques: *block coding* and *convolutional coding*. In block coding, such as Hamming or Reed–Solomon codes, a block of k data digits is encoded by a code generator into a code word of n digits where n is larger than k. The number of redundancy bits is therefore $(n - k)$. The ratio k/n represents the efficiency of the code and is normally known as the *code rate*. In convolutional codes, the coded sequence from the encoder depends not only on the sequence of the incoming block of k bits, but also on the sequence of data bits that preceded it. Unlike block codes, the code word of a convolutional code is not unique to the incoming k bits, but depends on earlier data as well.

For convolutional codes, k and n are usually small, giving small code rates such as $1/2, 2/3, 3/4$ and $7/8$.

Convolutional codes invariably outperform block codes, especially for correcting random and burst errors. One of the more efficient algorithms for decoding convolutional codes was devised by *Viterbi* with especially good results in correcting random channel errors. DTV uses both block coding (Hamming and Reed–Solomon) and convolutional coding.

FEC processing

FEC employed in DTV channel encoding consists of three layers:

- Outer coding
- Interleaving
- Inner coding – for satellite and terrestrial broadcasting only

Outer coding employs the Reed–Solomon code RS (204, 188); this adds 16 bytes to the transport packet, making a total of $188+16=204$ bytes (Figure 8.3). It can correct up to eight erroneous bytes in any single transport packet. If the error is higher than 8 bytes, the packet will be marked erroneous and uncorrectable. A code rate of $1/2$ is normally used, and this has to be set at the receiver to ensure the signals are properly decoded.

Reed–Solomon coding does not provide correction for error bursts, i.e. errors in adjacent bits, hence the need for an interleaving stage. Interleaving ensures that adjacent bits are separated before transmission. If the transmission medium introduces lengthy bursts of errors, they are broken down at the receiving end by the de-interleaver before reaching the outer decoder. For a fuller description of the interleaving technique, refer to Appendix A3.

The inner layer employs convolutional coding to ensure powerful error correction capabilities at the receiving end. Such error correction capabilities are essential for satellite and terrestrial DTV broadcasting where the medium of transmission is 'error-prone'. This layer is not necessary for a QEF medium, such as cable in which there is less than one uncorrected error event per hour of transmission.

Both inner and outer coding involves the addition of redundancy bits which make code words longer. Long code words increase the bit rate,

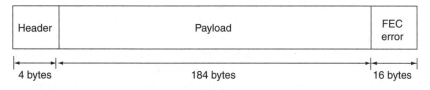

Header	Payload	FEC error
4 bytes	184 bytes	16 bytes

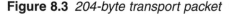

Figure 8.3 *204-byte transport packet*

which in turn increases the potential for errors. These problems may be avoided by shortening the code word, a process known as *puncturing* the code. Puncturing operates by selectively and periodically removing certain coded bits from each code word according to a regular pattern known to the receiver. At the receiver, dummy bits are reinserted to replace the omitted ones, but they are marked as *erasures*, i.e. bits with zero confidence in their accuracy. Consider a code rate of $1/2$ that is punctured by removing 1 bit in 4, a puncturing ratio of $1/4$. The mother code rate of $1/2$ produces 2 coded bits for every 1 uncoded bit and thus 4 coded bits for every 2 uncoded bits. If 1 bit in 4 is punctured, then only 3 coded bits are transmitted for every 2 uncoded bits, which is equivalent to a code rate of $2/3$. In fact, this is exactly how a $2/3$ rate is generated.

Puncturing a code word increases its code rate as the number of redundant bits is reduced. Punctured codes are obviously less powerful than the original unpunctured mother code. However, there is an acceptable trade-off between performance and code rate as the degree of puncturing increases.

Modulation

The final reduction in the bit rate is provided by the use of advanced modulation techniques. Simple frequency modulation in which logic 0 and logic 1 are represented by two different frequencies is highly inefficient in terms of bit rate and bandwidth requirements. Three types of modulation are used in DVB:

- Differential quadrature phase shift keying (DQPSK) for DVB satellite (DVB-S)
- Quadrature amplitude modulation (QAM) for cable
- Orthogonal frequency division multiplexing (OFDM) for DVB terrestrial (DVB-T)

Phase shift keying

A digital signal has only two states, 1 and 0, and when it is used to modulate a carrier, only two states of the carrier amplitude, frequency or phase are necessary to convey the digital information. In terms of bandwidth, the most economical form of modulation is phase modulation, known as phase shift keying (PSK). Here, the carrier frequency remains constant while its phase changes in discrete quantities in accordance with the logic state of the data bit. Binary PSK (BPSK) is a two-phase modulation technique in which the carrier is transmitted with a reference phase to indicate logic 1 and a phase change of 180° to indicate logic 0.

In quadrature phase shift keying (QPSK), also known as 4-phase PSK, the serial bitstream is first converted into a 2-bit parallel format using a

Table 8.1 *DPSK phase angle representation*

Phase angle (°)	Bit combination
45	00
135	01
225	10
315	11

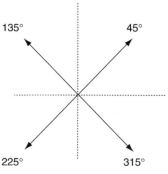

Figure 8.4 *QPSK phasor diagram*

serial-to-parallel converter. The instantaneous states of each pair of bits, known as dibits, can take one of four combinations, namely 00, 01, 10 and 11. For each of these combinations, the carrier is set to a particular phase angle (Table 8.1). The carrier has four phases: 45°, 135°, 225° and 315°. For example, data 00 and 10 are represented by 45° and 225°, respectively. The four phases are obtained from two quadrature (at right angles) carriers having the same frequency (Figure 8.4) I (in-phase) and Q (quadrature). Each phase is used to represent a 2-bit combination known as a *symbol* (Table 8.1). As can be seen, the bit rate is twice the symbol rate (also known as the *baud rate*). It is the symbol rate, i.e. the rate of phase change that determines the bandwidth. For the same bandwidth, twice as much information can thus be sent using QPSK compared with BPSK.

Differential phase shift keying (DPSK) has no specific reference phase. A phase shift occurs only if the current bit is different from the previous bit. The phase reference is therefore the previously transmitted signal phase. The advantage of this technique is that the receiver and the transmitter do not have to maintain an absolute phase reference with which the phase of the received signal is compared. DQPSK combines the advantages of QPSK and DPSK.

Phase shift keying may be improved by increasing the number of carrier phase angles from 4 in the case of QPSK to 8 or 16 in the case of 8-PSK and 16-PSK, respectively. In 8-PSK, the carrier may have one of eight different phase angles (Figure 8.5) with each phasor representing one of eight 3-bit combinations.

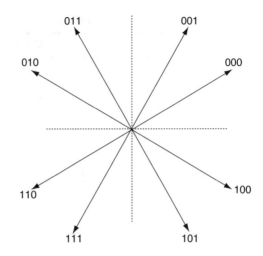

Figure 8.5 *8-PSK phasor diagram*

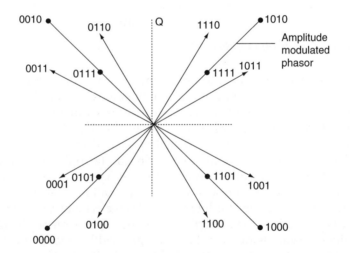

Figure 8.6 *16-QAM phasor diagram*

Quadrature amplitude modulation

QAM is an extension of PSK in that some phasors are changed in amplitude as well as phase to provide increased bit representation. For instance, 16-QAM encoding increases the bit width of the modulation to 4 as shown in Figure 8.6. Twelve different carrier phasors are used, four of which have two amplitudes to provide further 4-bit combinations. Figure 8.7 depicts all possible carrier phase angles and amplitudes; it is known as a *constellation map*.

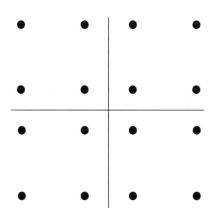

Figure 8.7 *16-QAM constellation map*

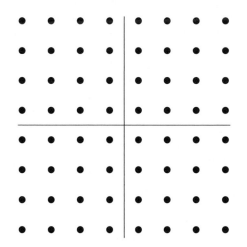

Figure 8.8 *64-QAM constellation map*

A higher order of digital modulation may be employed, for example, in cable, namely 64-QAM encoding, in which each carrier phase/amplitude represents one of 64 possible 6-bit combinations. The constellation map for 64-QAM is shown in Figure 8.8.

Orthogonal frequency division multiplexing

QAM encoding is very effective and very efficient. However, unlike satellite transmission, terrestrial broadcasting suffers from multiple path interference. In terrestrial transmission, besides the direct signal from the transmitter, a receiving aerial may receive one or more signals that have been

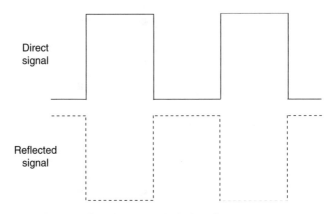

Figure 8.9 *Reflection for short symbol duration*

reflected off buildings, trees or moving objects such as flying airplanes. Reflected signals take a longer time to reach the receiving aerial than direct signals. In analogue TV transmission, the delayed signal will cause another faded picture to appear on the screen, known as a *ghost*. In DTV transmission, delayed signals can cause intersymbol interference and fading, which may result in partial or full picture and sound break-up depending on the amount of delay. If the delay results in a phase delay of 180°, the reflected signal cancels out the direct signal and complete picture and sound failure occurs (Figure 8.9). This can be avoided by using OFDM.

OFDM involves the distribution of a high-rate serial bitstream over a large number of parallel carriers. For each carrier, the bit rate is far below that of the original modulating bitstream. The carriers, 2048 (2K mode) or 8192 (8K mode), are closely and precisely spaced across the available bandwidth. Each carrier is modulated (PSK, 16-QAM or 64-QAM) simultaneously at regular intervals with the set of data bits used to modulate the carriers known as the *OFDM symbol*. Due to the large number of carriers, the duration of the OFDM symbol is considerably larger than the duration of one bit of the modulating bitstream. Consider a bitstream consisting of 1000 bits, with bit duration of 0.1 µs. If the 1000 bits are used to modulate a single carrier (simple PSK), the duration over which each bit remains 'active', i.e. the symbol duration, is 0.1 µs. If, on the other hand, the 1000 bits are used to modulate 500 carriers to form an OFDM symbol, the duration of the OFDM symbol will be $2 \times 0.1 \times 1000 = 50\,\mu s$. With a few thousand carriers and a more efficient modulation, e.g. 6-QAM, the symbol duration could be a few hundred times longer than the duration of the bits in the original bitstream.

With long symbol duration, a reflected signal will only be out of phase with the direct signal for a part of the total duration, shown shaded in Figure 8.10. For the rest of the symbol duration, the reflected signal is actively supporting the direct path signal. To avoid the shaded area, a *guard interval* (also known as a *guard period*) is added at the start of the

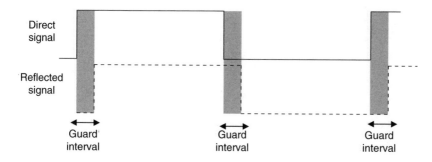

Figure 8.10 *Effect of long symbol duration of reflected signals*

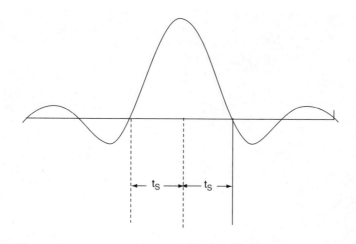

Figure 8.11 *Frequency spectrum of a single OFDM carrier*

symbol during which time the receiver pauses before starting the evalua-
tion of the carriers.

The total symbol duration thus consists of two parts: a *useful duration*
t_u proceeded by a guard period set to a fraction of the useful duration, e.g.
$t_u/4$, $t_u/8$, $t_u/16$. For instance, employing the 8K mode, the useful symbol
duration is set to 896 μs. Using a guard period $t_u/4$ would protect against
echoes with delays as large as 224 μs. The introduction of a guard period,
however, reduces the number of active carriers available for modulation
and hence lowers the data capacity of the system. In the UK, the number
of carriers used is 2048 but the number of active carriers is only 1705.

In OFDM, the carriers are harmonically related with a common and precisely calculated frequency spacing to ensure orthogonality (Figure 8.11).
The spacing is set to the reciprocal of the symbol duration t_u:

$$\text{OFDM frequency spacing} = \frac{1}{t_u}$$

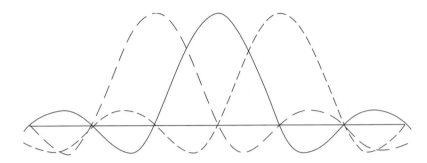

Figure 8.12 *Frequency spectrums of orthogonal carriers*

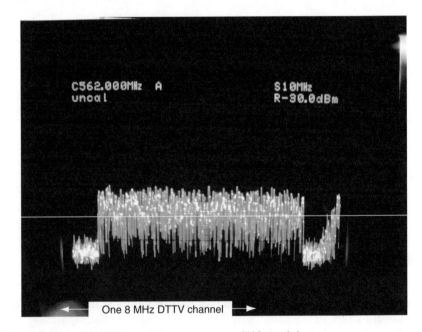

One 8 MHz DTTV channel

Figure 8.13 *OFMD frequency spectrum (2K mode)*

By ensuring orthogonality of the carriers, the demodulator of one carrier is not made aware of the existence of the other carriers, thus avoiding crosstalk. Figure 8.12 shows the effect of orthogonality in which the fundamental frequency of one carrier which, being a sine wave, is the only frequency component, occurs when adjacent carriers spectrums crosses the zero line.

When all the carriers are included, a flat spectrum is obtained (Figure 8.13); this is designed to occupy the same bandwidth as allocated to a conventional analogue television programme, namely 8 MHz for the UK and 6 MHz for the USA.

The flat spectrum of OFDM reduces the effective radiated power (ERP) required of the digital terrestrial television (DTTV) transmitter compared

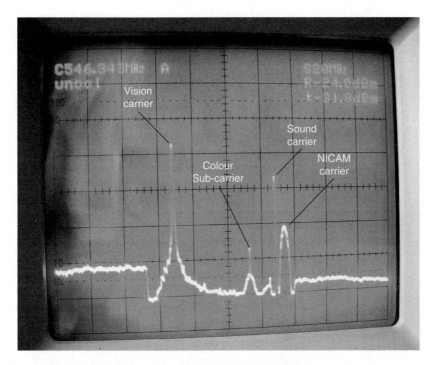

Figure 8.14 *Spectrum of an analogue TV broadcast*

with analogue terrestrial broadcasting, in which the carrier power is concentrated in narrow bands around the vision carrier, colour sub-carrier, FM sound and NICAM carriers (Figure 8.14). In contrast, DTTV transmitted power is more evenly and efficiently spread across the whole spectrum, giving considerable savings in power as shown in Figure 8.15. The complete standard definition terrestrial transmission from Crystal Palace is shown in Figure 8.16. It will be noticed that the digital multiplexes are 15–20 dB below the analogue TV broadcasts. This is necessary while both systems cohabit the same UHF band with adjacent channels to avoid cross-interference. However, when analogue is switched off, this restriction will no longer be necessary. Digital terrestrial broadcasts power may thus be increased making it possible for good un-interrupted reception using a simple indoor aerial.

Single-frequency network

Apart from improving the quality of reception of DTTV, the high tolerance to fading and multipath interference has an added advantage. It allows broadcasting authorities to use a single-frequency network (SFN) throughout the country. A signal from an adjacent transmitter broadcasting an

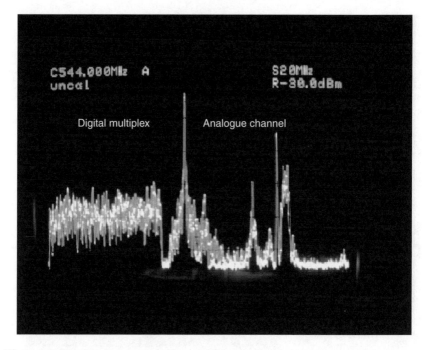

Figure 8.15 *Spectrum showing analogue channel peaky spread compared with the flat response of a digital multiplex*

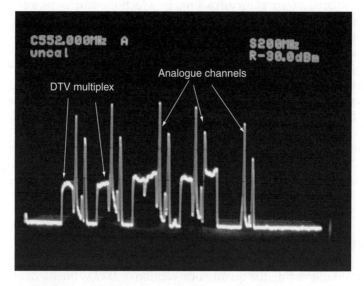

Figure 8.16 *Standard terrestrial television broadcasts from Crystal Palace, London*

identical signal will cause ghosting in analogue TV transmission. In digital terrestrial broadcasting, these signals are indistinguishable from reflected waves, and are therefore treated accordingly. They enhance the direct path signals and improve reception.

Coded OFDM

Besides multipath interference, terrestrial transmission suffers from frequency-dependent noise caused by narrowband interfering signals within the bandwidth. This is why it is necessary to use a more powerful FEC coding system, the C in COFDM, than that used in satellite broadcasting. Noise in the transmission media often causes bits to lose their original logic levels, rather like the erasures introduced by puncturing. At the receiving end, *soft decision* decoding has to establish whether a received bit is actually logic 1 or logic 0 before an irrevocable decision (*hard decision*) is taken on its integrity. The process involves gathering information about the effect of noise on all the carriers in the multiplex. A history of possible sequences and their relative likelihood is thus established. Combined with the FEC punctured codes, this history of sequences and their likelihood, known as *channel state information* (CSI) ensures that errors caused by frequency-selective interference and fading are detected and corrected at the receiving end.

8K/2K COFDM modes

For DTTV, the European DVB system is based on COFDM modulation with either 8K (8192) or 2K (2048) carriers with a symbol duration t_u of 896 and 224 µs, respectively. The effective number of carriers, i.e. the actual number of carriers that may be used for COFDM modulation, is 6817 (8K mode) and 1705 (2K mode). The remaining 343 carriers (2K mode) or 1375 carriers (8K mode) are transmitted without any modulation during the guard period. Apart from removing the effect of reflected signals, the guard interval performs two other important functions: (a) the carriers that are transmitted during the guard interval are used by the decoder for synchronisation purposes and (b) the guard period allows the useful carriers to settle following the abrupt changes in their phases each time they are modulated by a new set of symbol bits. From among the 1705 carriers (2K system), 1512 are used to carry data. The remaining carriers are used to carry the necessary reference signals for accurate decoding to take place at the receiving end. The 2K COFDM mode is currently used in the UK. However, DTTV being a very flexible broadcasting system, the 8K mode may be introduced in the future without making current decoder boxes redundant. Table 8.2 lists the main characteristics of an OFDM signal for a channel with 8 MHz bandwidth.

Table 8.2 *COFDM specification for 8 MHz channels*

Type	8K mode	2K mode
Useful duration t_u (µs)	896	224
Guard interval Δ	$t_u/4$, $t_u/8$, $t_u/16$, $t_u/32$	$t_u/4$, $t_u/8$, $t_u/16$, $t_u/32$
Symbol duration $t_s = t_u + \Delta$		
Number of carriers	8192	2048
Number of active carriers	6817	1705
Carrier spacing $1/t_u$ (Hz)	1116	4464
Spacing between extreme carriers, i.e. bandwidth (MHz)	7.61	7.62

The 2K mode provides optimum performance for mobile receivers whereas the 8K mode is generally preferable for large-scale SFNs. The system may be adapted for other channel bandwidths by merely changing the ratio t_u/n; where n is 2048 for the 2K mode and 8192 for the 8K mode. This ratio is known as the *elementary time element*, T, and $1/T$ is known as the *OFDM system clock*. By changing T, hence $1/T$, the characteristics of the emitted signal (apart from the bandwidth) is maintained. For the 8 MHz, $T = 224/2048$ and $1/T = 2048/224 = 64/7$ or 9.143 MHz. The same figure is obtained for the 8K mode namely $8192/896 = 9.143$. For the 7 MHz channel, $1/T = 2048/256 = 8$ MHz. In a 7-MHz channel, the carrier spacing is 976.6 Hz (8K mode) and 3906 Hz (2K mode). Table 8.3 lists OFDM specifications for channels with 7 MHz bandwidth used in some countries including Australia.

The actual bit rate available depends on the type of modulation of the carriers. 64-QAM would result in a 27 Mbps multiplex whereas 16-QAM would almost halve the bit rate capacity. However, 64-QAM is more error-prone than 16-QAM. QPSK must be used for DTTV reception on fast-moving vehicles, e.g. trains and cars; this reduces the bit rate capacity to 10–12 Mbps but it also reduces the possibility of decoding errors. In the UK, the introduction of DTTV was notorious for picture freeze and pixelisation.

Table 8.3 *COFDM specification for 7 MHz channels*

Type	8K mode	2K mode
Useful duration t_u (µs)	1024	256
Guard interval Δ	$t_u/4$, $t_u/8$, $t_u/16$, $t_u/32$	$t_u/4$, $t_u/8$, $t_u/16$, $t_u/32$
Symbol duration $t_s = t_u + \Delta$		
Number of carriers	8192	2048
Number of active carriers	6818	1706
Carrier spacing $1/t_u$ (Hz)	976.6	3906
Spacing between extreme carriers (MHz)	6.656	6.656

This was mainly as a result of the commercial services using 64-QAM to pack as many programmes as possible within a single multiplex. That bad reputation was only broken with Freeview which changed to 16-QAM.

High definition terrestrial television

In 2006, both the BBC and ITV started HD terrestrial broadcasts on a trial basis from Crystal Palace: channel 27 for the ITV and 31 for the BBC (Figure 8.17). Each HD multiplex occupied a standard 8 MHz bandwidth using 8K mode, 64-QAM, code rate 2/3 and a guard period of 1/32. A lower transmission power (10 dB below SD) was used to avoid any interference with adjacent analogue and digital channels.

Satellite channel encoder

Figure 8.18 shows the main components of a satellite DTV transmission system. The input signal is a multiplexed MPEG-2 transport stream consisting of 188-byte packets. The signal is first randomised with a pseudo-random bit sequencer, for energy dispersal. The Reed–Solomon coding RS (204, 188)

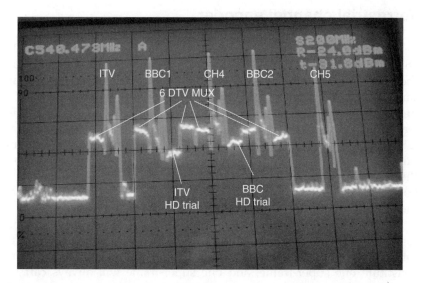

Figure 8.17 *Terrestrial TV broadcasts from Crystal Palace showing analogue channels, SD and HD multiplexes*

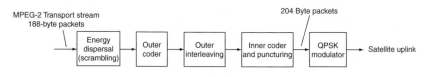

Figure 8.18 *Channel encoder (satellite)*

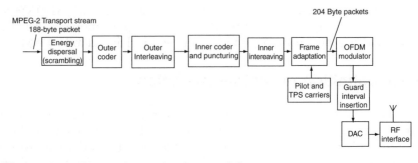

Figure 8.19 *Channel encoder (terrestrial)*

adds 16 bytes of error protection data to the packet, making a total of 204 bytes. There are then interleaved before entering the inner coder. The inner coder employs convolutional coding with a mother code rate of 1/2. The convolutional code is then punctured to produce a code rate of 2/3, 3/4, 5/6 or 7/8 as determined by the service provider. The error-protected packets are now ready for the QPSK modulator before entering the satellite uplink interface for transmission to the transponder.

Terrestrial channel encoder

Figure 8.19 shows the main elements of a DTTV transmission system. The 188-byte multiplexed MPEG-2 transport stream is first randomised and then processed through the inner coder, interleaver and outer coder and punctured in the same way as for a satellite system. However, DTTV broadcasting requires further error protection, provided by the inner inter-leaver, before the packets can be framed, modulated and transmitted via UHF aerial. The inner interleaver is a frequency interleaver which carries out bit interleaving as well as symbol interleaving; here 'symbol' refers to the bits being transmitted by one OFDM carrier during one OFDM symbol period. The error-protected packet must now be framed before the modulating stage.

A DVB-T frame consist of data from 68 consecutive OFDM symbols together with reference *pilots* and *transmission parameter signalling* (TPS) information scattered among the data (Figure 8.20). Four OFDM frames constitute one super-frame, and one super-frame will always carry an integer number of MPEG-2 packets. There are two types of pilots: contin-ual pilots and scattered pilots. Continual pilots are spread at random over each OFDM symbol. They are used to modulate the same carriers on all OFDM symbols. Scattered pilots are spread evenly in time and frequency across the OFDM symbol. The continual pilots are used for synchronisa-tion and phase error estimation whereas the scattered pilots allow the receiver to take account of echoes and other impairments when estimating the channel characteristics. Channel parameters such as the mode (2K or

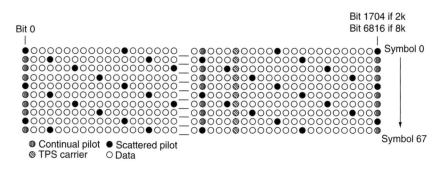

Figure 8.20 *DVB frame (terrestrial)*

8K), guard interval length, modulation type and code rate have to be included within the transmitted bitstream. For this reason, a TPS channel is added to the OFDM signal. The TPS information, 68 bits in length, is carried by specified carriers spread over the entire OFDM frame. One carrier in each symbol is allocated to carry one bit of TPS using simple BPSK.

The OFDM modulation is implemented using an inverse fast Fourier transform (FFT) where the OFDM carriers (1705 or 6817) are modulated by each frame. The modulation of the carriers may be BPSK (used for TPS channel), 16-QAM or 64-QAM as determined by the service provider. The guard interval is added next, followed by a digital-to-analogue converter to prepare the signal for the RF interface.

9 Video re-production

Overview

The process of television re-production involves video and audio processing as well as video formatting and drive for the particular video display unit. Figure 9.1 shows the major sections of the video section of a television receiver set. The video input may be from an analogue or digital source and the latter may be standard definition or high definition. The analogue video may take several formats:

- CVBS signal obtained from an analogue terrestrial tuner or directly from a SCART connection.
- Y and C from an S-video connection.
- Component video Y, P_r, P_b.
- RGB via VGA port.

These analogue video signals are first fed into the next chip which selects the video signal to processes it as determined by the viewer. Modern TV receivers are designed to receive direct audio/video (AV) input from external sources such as video recorders, camcorders, satellite receivers and digital television decoders. They also provide direct AV signals to peripherals such as video recorders and MPEG decoding audio systems. There are several ways for connecting AV devices to each other including the SCART connector, video coupling (S-video, component video, RGB), digital video interface (DVI), HDMI, VGA for PC, universal serial bus (USB), RS232 and FireWire. Refer to Chapter 25 for details.

Video processing includes colour decoding using a comb filter which separates the luminance Y signal from the chrominance components. The resulting RGB signals are then fed into the next stage for video formatting.

The purpose of *video formatting* is to ensure that the incoming video which may be interlaced or progressive, and may have a variety of line and frame frequencies (PAL, NTSC, PC VGA) is converted into the native fixed format or resolution of the display unit. Formatting is followed by the drive circuitry which may be RGB amplifiers in the case of a cathode ray tube (CRT) display, scan and sustain drive in the case of a plasma panel or line (row) and data (column) drivers in the case of an LCD display.

Video formatting is not necessary if the display unit is a CRT. Cathode ray tubes do not have a fixed resolution in which case, analogue RGB is fed directly into the tube via a driver amplifier.

Digital video may be obtained from a standard definition terrestrial DVB decoder consisting of a digital terrestrial tuner, COFDM demodulator and

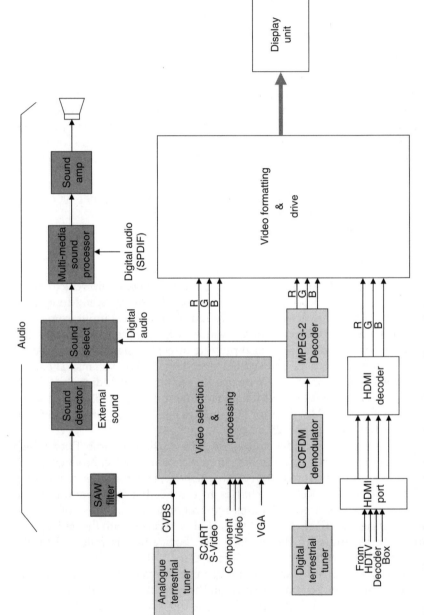

Figure 9.1 *Generalised block diagram of a TV re-production system*

an MPEG-2 decoder. High-definition video would arrive as an uncompressed digitised video via HDMI port to an HDMI decoder or receiver which converts back into digitised R, G, B as illustrated in Figure 9.1. Both digital video signals are fed directly into the video formatting and drive section of the receiver.

Audio from a terrestrial analogue broadcast is tapped off the CVBS signal by a SAW filter and following demodulation by the sound detector, it is fed to a multi-media sound processor, amplified and fed into the speakers. For digital reception, stereo sound forms part of the transport stream produced by the COFDM demodulator. It is MPEG-2 decoded and fed to the sound select switch for further processing and amplification. External sound in the form of phono L/R connection may be fed directly to the sound select switch. However, audio may also be available in a digitised form via a coaxial or a fibre optic (SPDIF) connection in which case it is fed directly to the multi-media sound processor (MSP).

Display units

The purpose of the display unit is to re-produce the original moving pictures in as faithful a way as possible. The faithfulness of the re-production depends on the type of unit used, its properties and mode of operation. The earliest display unit is the traditional CRT which remains the standard by which all other units (plasma, LCD, DLP, etc.) are compared.

The cathode ray tube

A great deal of research and development has gone into the technology of colour picture tubes over the years, and direct-view types are now available with screen diagonals up to 89 cm. Improvements have been made in screen materials, gun technology and energy demand. The colour tube uses all the techniques of its simpler counterpart, and by extension and refinement of these is able to present a full colour picture on a single 'integrated' screen. These colour tubes work on the shadowmask principle originally brought to fruition by Dr A. N. Goldsmith and his research team in the American laboratories of RCA Ltd in 1950. The original device was rather clumsy and cumbersome by today's standards, but it embodied all the fundamental features of the current generation of colour tubes, and was amenable to mass production with all the cost advantages that could bring. The practical realisation of a relatively compact and decidedly cheap domestic colour TV receiver was (and is) dependent on the direct-viewing shadowmask tube concept. Today a number of display devices are available including the LCD and Plasma screen. The CRT however, is standard display device with which all other devices are compared. It remains best the device in terms of picture quality.

In the CRT, high-speed electrons in the form of a beam current are emitted by an electron gun, focused and accelerated by an electron lens and then directed towards a screen which acts as a positively charged anode. The screen which is coated with phosphor gives a visible glow when hit by high-speed electrons. The colour of the emitted light is determined by the type of phosphor used. For monochrome display, only one type of phosphor coating is used. For a colour display, three types of phosphors are used in order to obtain the three primary colours.

Extra high tension

A final anode voltage known as the *extra high tension* (*e.h.t.*) in the region of 15–30 kV is required by tube to attract and accelerate the electrons. This e.h.t. is produced by a voltage multiplier network known as the tripler (Figure 9.2). The a.c. voltage input is obtained from an overwind

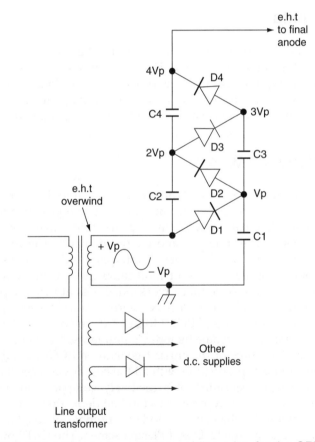

Figure 9.2 *The tripler arrangement providing e.h.t. for the CRT*

on the line output transformer. C1/D1 act as a clamper which charges C1 to the peak of the input voltage, Vp. This is then applied to the second clamping circuit consisting of D2 and C2 charging C2 to 2 × Vp and so. The elements of the tripler are contained in a single well isolated package or capsule known as the Tripler.

Monochrome tube

The monochrome display tube consists of a single electron gun, an anode assembly acting as the electron lens and a viewing surface. The beam passes through a 4-anode assembly (Figure 9.3) which provides acceleration (A1), electrostatic focusing (A2/A3) and final e.h.t. anode (A4). As well as the grid potential the emission of electrons is also dependent upon the potential between the cathode and the first anode. An increase in the potential between these two electrodes causes more electrons to be emitted and vice versa. The actual tube voltages depends on the size of the tube and its design, however the following are typical voltages for a monochrome receiver tube:

Cathode	70 V
Grid	30 V
A1	300–400 V (accelerating anode)
A2	15–20 kV (final anode)
A3	variable up to 500 V (focus anode)
A4	15–20 kV (final anode)

The electron gun and the anode assembly are contained within a vacuumed thick glass envelope. Access to the various electrodes is obtained via

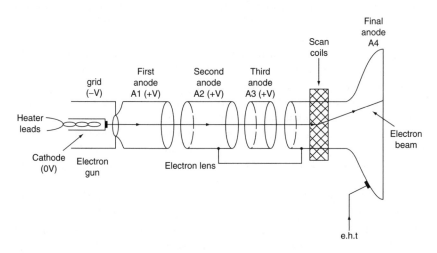

Figure 9.3 *Components parts of a monochrome CRT*

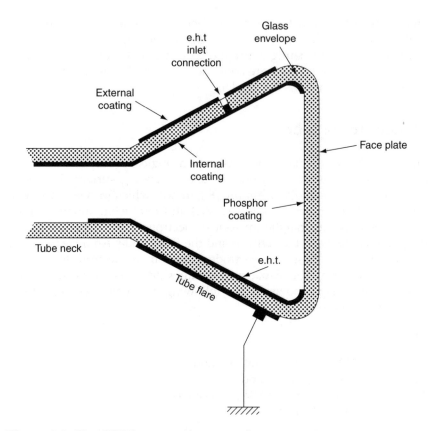

Figure 9.4 *The CRT inner and outer coatings*

pin connections at the back of the neck of the tube with the exception of the final anode which is accessed along the tube flare. The inside and outside of the tube flare are coated with a layer of graphite known as *aquadag* coating (Figure 9.4). The outer coating is connected to chassis and the inner coating is connected to the e.h.t. The glass separation between them forms a reservoir capacitor for the e.h.t supply. This is why it is important that this capacitance is fully discharged before handling the tube otherwise a very violent shock may be experienced.

In order to produce a display the electron beam is deflected in the horizontal and vertical directions. Electromagnetic deflection is employed using two sets of coils (line and field) known as *scan coils* placed along the neck of the tube.

The tube's angle of deflection forms an important specification of the CRT. It refers to the angle through which the beam is deflected as it goes across the screen. The most common deflection angle is 90° although 100° and 110° are widely used for large tube displays. The deflection angle depends on the strength of the magnetic field created by the scan coils, the

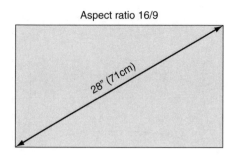

Figure 9.5 *Screen size*

speed of the electron beam which is a function of the e.h.t. and the diameter of the neck of the tube. A narrow neck allows the scan coils to operate in close proximity to the beam and hence exercise greater influence upon it.

Screen size

Another tube specification is the tube size which refers to the diagonal measurement from the lower corner to the opposite upper corner of the screen given in inches or centimetres as illustrated in Figure 9.5. However, the tube size does not translate into the actual display or visible screen size. The television tube is generally partially covered by the external casing of the screen. In addition, the tube cannot project an image to the edges of the full tube size. For instance, a 28 inch (71 cm) tube has a visible screen of 26 inch (66 cm). The precise dimensions of the screen depends on the aspect ratio, e.g. 4:3 or 16:9 (widescreen). The power required by the scan coils is a function of the size and geometry of the tube (screen size and neck diameter) as well as the deflection angle and the e.h.t. applied to the final anode.

Raster geometry

A pure sawtooth scan waveform would produce what is known as 'pincushion' distortion in a flat or nearly flat screen tubes. This is because an equal angular deflections of the beam scans a smaller distance near the centre of a flat screen compared with the distance it scans further away. A similar effect is produced for the vertical scan. When both non-linearities are considered together, the raster produced on a flat screen exhibits the pincushion distortion illustrated in Figure 9.6. Such distortion may be reduced by the design of the scan coils and to a larger extent by 'distorting' the shape of the scanning waveform by introducing a non-linearity as a counter balance to the non-linearity of the display, a process known as S-correction. An old technique is the use of small peripheral permanent

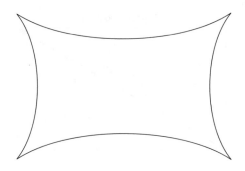

Figure 9.6 *Pin cushioning effect*

magnets placed on the scan coils which may be manually manipulated to obtain the necessary modification of the magnetic field.

Scan velocity modulation

As the electron beam in a CRT display unit retraces the image, line by line across the screen, details and definition may be lost when the scene includes transitions from dark to light, i.e. black vertical edges. Such edges become soft and diffused. Scan *velocity modulation*, or velocity modulation (VM), attempts to overcome this by emphasizing the edges to improve sharpness. This may be desirable with analogue television and low-resolution video from VCRs. However it is not necessary for high-resolution videos sources such as HDTV and DVD and it must be switched off for these applications to prevent the inevitable artifacts from degrading the picture quality.

VM regulates the speed by which the electron beam scans the screen, hence the name scan velocity modulation. The VM circuit examines the luminance signal and where transition from light to dark area is detected; the beam is slowed down before the transition and speeded up during it. When the beam is slowed down it spends more time writing the light area and the phosphor receives more energy and hence increased brightness and vice versa. This results in an enlarged or 'enhanced' black edge. Although this technique remains controversial as far as picture quality is concerned, it has been adopted by many manufacturers and has been promoted as *'picture enhancement'*.

Beam modulation

In order to produce an image on the tube, the brightness of the screen has to be varied as it scans the surface of the tube to recreates the picture information line by line. This is achieved by varying the intensity of the electron beam in accordance with the video signal, a process known as beam modulation. The beam is modulated by varying the potential between the

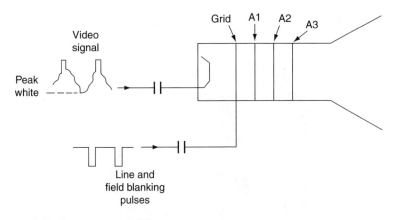

Figure 9.7 *Beam modulation*

cathode and the grid. There are two types of modulation: grid modulation in which the cathode voltage is be held constant while the grid voltage is varied by the video signal and cathode modulation in which the voltage at the grid is fixed and that of the cathode is varied with the video signal. With cathode modulation the video signal is negative-going as shown in Figure 9.7. Peak white is produced when the cathode is at its most negative potential. In practice cathode modulation is normally used because of its greater sensitivity with negative-going blanking pulses applied to the grid to cut the beam off during line and field flybacks.

Colour receiver tubes

CRTs used for colour display have three separate guns, one for each primary colour arranged to bombard a screen which is coated with three different types of phosphors, one for each primary colour. The three phosphors are either grouped in triangular delta (3-colour *triad*) or in-line formation. When a group of triads is caused to glow at proportioned colour intensities, the eye is deceived into perceiving that the primary colours occur at the same point and so it discerns a spot or a dot corresponding to the mix of the three colour intensities.

Typical voltages for colour tubes are considerably higher than their monochrome counterparts with the final anode voltage in the region of 30 kV.

Purity

For correct colour reproduction, the red, green and blue beams must strike only their own particular phosphors and no other. This is known as *purity*. Purity adjustment involves changing the magnetic field formed by the

Figure 9.8 *Tube neck assembly*

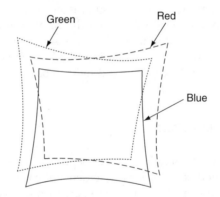

Figure 9.9 *RGB rasters*

scan coils both in terms of strength and direction to move the beams so that they strike the correct phosphor coating of the screen. This is achieved by the use of a pair of two-pole ring magnets placed along the neck of the tube and a rotating disk magnet as shown in Figure 9.8.

Convergence

The three electron beams scanning the screen produce three separate rasters as illustrated in Figure 9.9 thus rendering the problem of raster correction considerably more complicated compared with those encountered with monochrome displays. The three rasters must not only be of the correct rectangular shape with no pincushion distortion, they must also coincide precisely. This is known as *convergence*. There are two types of convergence, static and dynamic. *Static convergence* relate to the central part of the picture. *Dynamic convergence* covers the rest of the screen which involves the establishment of a continuously varying (dynamic) magnetic field to ensure convergence at the outlying areas and corners of the display. As we will see later, this requires electromagnetic waveforms which are a function of both the line and field frequencies.

The shadowmask tube

The first mass-produced colour tube was the delta-gun shadowmask. It has three electron guns mounted at 120° to each other at the neck of the tube each with its own electron lens. The guns are tilted by a small amount towards the central axis of the tube so that their electron beams converge and cross at the shadowmask and pass through carefully positioned holes to strike their correct phosphor dot. A large number of electrons miss their holes and are lost by hitting the mask resulting in what is known as low electron transparency resulting in low efficiency and low brightness. In the *delta* arrangement (Figure 9.10), the three guns are in three different planes making it difficult to achieve accurate conversion as the beams are made to scan the screen. This is the main disadvantage of the delta type. Highly complex and expensive convergence circuits are necessary to overcome the two dimensional distortion of the rasters and to maintain good convergence. Several presets have to be used in a complex sequence

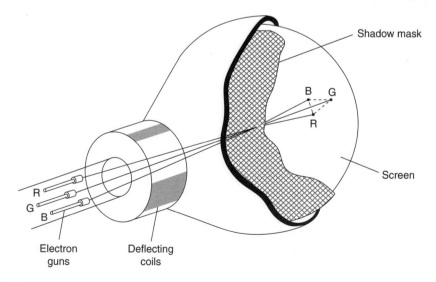

Figure 9.10 *The shadowmask tube*

requiring highly skilled labour. Furthermore, the tendency for convergence to drift means frequent adjustments are necessary. For this reason, delta-gun tubes are no longer used for domestic TV receivers. However, they are still in production for use as monitors for advanced computer displays because of their high definition when fitted with a very fine-pitch shadowmask.

The in-line colour tube

In the in-line shadowmask tube the three guns are placed side by side and the phosphors coating on the screen is in the form of striped triads. Each 3-colour group is arranged to coincide with a longitudinal grill or slot in the shadowmask. Having three beams in the same horizontal plane has two advantages.

First, purity is unaffected by horizontal magnetic fields such as the earth's magnetic field. Secondly, the need for vertical conversion correction disappears because the three beams always travel in the same horizontal plane. Convergence is then reduced to the relatively easy task of deflecting the two outer beams inward slightly to converge with central beam. The first in-line tube was developed by Sony known as the Trinitron. This was followed by the Mullard's *precision-in-line, PIL* (AX series) self-converging tube which eliminated the need for dynamic convergence adjustment altogether.

The trinitron tube

The Trinitron uses a single in-line gun assembly and a single electron lens assembly as shown in Figure 9.11. The phosphors are arranged in vertical

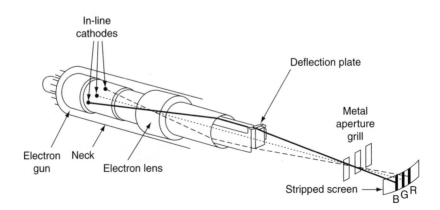

Figure 9.11 *The trinitron tube*

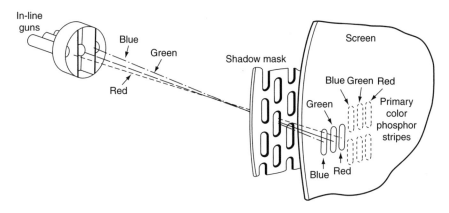

Figure 9.12 *Precision-in-line*

stripes forming three-colour striped triads. Higher electron transparency is achieved as fewer electrons are lost by hitting the mask resulting in improved efficiency and brightness. The single electron gun employs three in-line cathodes. The three beams pass through a complicated anode arrangement which bends the two outer (red and blue) beams so that they seem to be emanating from the same source as the green beam in the middle. The Trinitron suffers from two basic disadvantages. The first is that the construction of the striped mask has very little stiffness in the vertical direction and has to be kept under considerable tension to prevent sagging or buckling. The second disadvantage is the need for some dynamic convergence adjustments especially in wide angle large screens.

The PIL tube

In the precision-in-line, PIL tube, three separate guns are mounted side by side on the same horizontal plane. The phosphors are painted in vertical stripes with the shadowmask consisting of staggered slots (Figure 9.12) providing mechanical rigidity and high electron transparency for improved brightness. The main advantage of the PIL tube is the development of *astigmatic magnetic field* produced by a special deflection yoke designed to produce staggered magnetic field to eliminate the need for dynamic convergence.

10 Plasma panels

Introduction to flat panel displays

Electronic displays have been around almost since the electron was first discovered in 1897. Scientists and engineers have developed a very wide range of electronic display technologies and today, the choice has never been wider.

For years, the most common display technology has been the cathode ray tube or CRT. In the last few years, around 115 million new CRT computer monitors and 130 million new television sets based on CRT technology have been sold annually throughout the world. CRTs have been produced in large volumes for over 50 years and consequently, manufacturers have developed very efficient processes and enjoy great economies of scale. This means that CRT technology delivers a good performance at relatively low cost.

But now there are newer, slimmer, lighter and more versatile products that are true alternatives to the traditional CRT device. Currently, the most popular technologies are:

Plasma display panel, PDP.
Liquid crystal display, LCD.
Digital light processing, DLP.

There are other technologies that are being developed such as organic light-emitting diodes (OLED), field emissive displays and surface-conduction electron-emitter displays (SED).

The CRT is a well tried and tested technology and remains the standard for standard definition picture quality. Nonetheless, flat panel displays (FPDs) have a number of advantages that makes them more desirable than the CRT. These advantages are:

- Fully flat display.
- Large screen formats.
- Thin (40 mm) – suitable for wall hanging.
- Small in size – occupying less desk space.
- Fully digital internal operation.
- Light weight – 1/6th of CRT.
- Unaffected by magnetic fields.
- Fully flicker-free operation.
- Larger viewing area – a 15″ flat panel gives the same viewable screen as a 17″ CRT.
- High resolution.

On the other hand, flat panels have disadvantages, but these are different depending on the type of FPD under consideration. In general, however, FPDs are more complex, more expensive, have restricted viewing angle and use more power than the comparable conventional CRT display.

Viewing angle

Traditional CRT displays may be viewed at virtually any angle without degradation in colour or brightness. The same cannot be said of flat panel displays FPDs. FPDs are specified with a 'viewing angle' defined as the angle (horizontal and vertical) from which the display can be correctly seen without discolouring or brightness degradation. Most FDPs have a viewing angle of 160°V and H. That means that a picture may be viewed at 80° (160°/2) from a line drawn vertically from the centre of the screen. Given that at 90° the viewer is facing the edge of panel, such a viewing angle is a very tall order. Some manufacturers claim a viewing angle as high as 170°. That being said, a viewer may find that images at the extreme ends of the specified viewing angle do not provide comfortable viewing or good picture quality.

Matrix format

All flat panels consist of a number of picture elements, known as *pixels* arranged in a matrix format of rows and columns as illustrated in Figure 10.1. For colour production, each pixel is sub-divided into three sub-pixels or cells. The construction of a colour flat panel will be as illustrated in Figure 10.2 with barrier ribs separating the pixel cells.

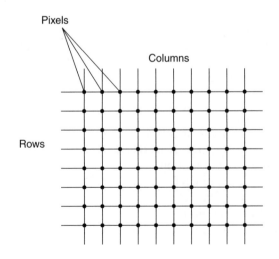

Figure 10.1 *Matrix format*

Resolution

The resolution of an FPD is defined as the number of pixels that make up the panel. It is specified as $X \times Y$ where X and Y are the number of pixels in the horizontal and vertical directions respectively. Although a flat panel may be manufactured with any resolution, the following is the accepted standard resolutions:

VGA (video graphic array): 640×480
SVGA (supper VGA): 800×600
XVGA or XGA (extended VGA): 1024×768
SXGA (super extended graphic array): 1280×1024
UXGA (ultra extended graphic array): 1600×1200

The total number of pixels in a panel is the product $X \times Y$. For example, an SVGA panel has total number of pixels $= 800 \times 600 = 480,000$. It should be noted here that the number of cells is $3 \times 800 \times 600 = 1,440,000$. It follows that while the number of rows or pixel lines $= 600$, the number of columns $= 3 \times 800 = 2400$.

Pixel specifications

The size of a pixel is normally defined by its horizontal pitch which is the distance between the middle of one cell and the next cell of the same colour as illustrated in Figure 10.2. A pixel also has a vertical pitch dimension. Pixel size depends on screen size as well as resolution. Higher resolutions for same screen size necessitate a smaller pixel size and vice versa.

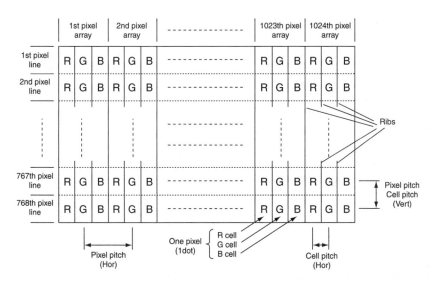

Figure 10.2 *Pixel cell construction of a flat panel display*

Typical values for a PDP are 1080 μm (*H*) and 810 μm (*V*) for a 42 in. VGA panel. For high resolution PDPs, a typical pixel pitch could be as low as 800 μm (*H*) and 500 μm (*V*).

The size of a cell is normally defined by its horizontal pitch or width which is the distance between the middle point of adjacent cells. A cell also has a vertical pitch dimension which is the same as the pixel pitch (vertical). The cell pitch depends on screen size as well as resolution. A typical cell width for a PDP is between 360 and 300 μm.

The final specification of a pixel is the rib width and height (or vertical) dimensions, typically 80 and and 150 μm respectively for a plasma panel. For deep cell construction, the rib height is doubled.

Resolution: flat panel versus CRT

It is appropriate here to compare the resolution of a flat panel with that of a CRT display. CRTs are completely analog devices, and unlike FPDs, they do not have discretely addressable pixels that define a *native resolution*. A CRT resolution is defined as the maximum number of points or dots that can be displayed without overlap. A colour CRT has a matrix of red, green and blue phosphor dots. A phosphor dot is not a discrete pixel. The sweeping electron beam does not uniquely turn each individual dot on and off, as is the case with discrete pixel types of displays. Rather, as the beam scans across the face of the tube, the areas of phosphor that the beam strikes glow with an intensity proportional to the beam current's instantaneous amplitude. The size of the phosphor area illuminated by the beam depends on the focus of the electron beam and the distance between adjacent dots of the same colour. If the electron beam is sufficiently well-focused, the tube will be able to resolve an area smaller than a single dot which is in fact just a tiny area of continuous phosphor. If the electron beam's focus is sufficiently diffuse, it may not just illuminate a single dot, but may also simultaneously illuminate a whole or part of an adjacent dot. The dot pitch plays a significant role in determining a colour CRT's resolution capability, because it determines the distance between adjacent dots of the same colour. If we consider just blue dots, for example, one blue dot is not directly adjacent to another blue dot. There are green and red dots that are to some degree positioned between the two red dots. The result is a gap between adjacent blue dots, and the size of this gap influences the attainable resolution. The farther apart the blue dots, for example, the less blue resolution the screen is capable of displaying, and likewise for the red and green dots. The smallest dot pitch found in a typical top-quality monitor tube is about 0.22 mm, or about 4.45 dots per millimetre, which is about 115 dots per inch. Such a display with a width of 36 in. has about 4156 dots in a horizontal line. It is safe to say that it is possible for such a CRT monitor to display the resolution of HDTV, either 1280 × 720 or 1980 × 1020, if the proper conditions of dot pitch, electron beam focus and scanning speed are met.

To summarise, the resolution capability of a given colour CRT depends on a number of factors, the most important being the dot pitch, how tightly the electron beam is focused and the electron beam's scanning speed. Most manufacturers do not list the dot pitch ratings anymore which are typically between 0.12 and 0.28 mm with an average of 0.25 mm. CRTs are, within limits, flexible in their resolution capabilities and can handle a video signal of any resolution. This is not the case with flat panel FPDs containing discrete pixels. If you want to display a resolution other than the native pixel resolution on a display containing discrete pixels, the image must be re-scaled.

Plasma operation

The principle of operation of a plasma pixel is closely related to that of the simple neon light. It has long been known that certain mixtures of gases such as neon, helium and xenon when subjected to a sufficiently strong electric voltage will break down into 'plasma' and what is known as *plasma discharge* occurs. Plasma is an electrically neutral, highly ionised substance consisting of electrons, positive ions and neutral particles. Being electrically neutral, it contains equal quantities of electrons and ions and is, by definition, a good conductor. Apart from conducting a current, plasma discharge converts a part of the electric energy into electromagnetic waves including *ultra violet (UV)* and visible light.

The first drawback of this simple plasma unit is the fact that the light emission from such a pixel is the familiar orange glow of neon signs. To overcome this, and produce colour, appropriate phosphors which emit different colours when excited is used to cover the inside of the pixel. Ultraviolet rays generated by the discharge are then used to impact onto the phosphor coating to emit colour light as shown in Figure 10.3. To optimise the UV emission, the gas mixture is modified with the addition of xenon.

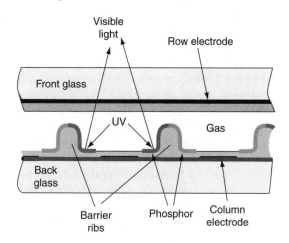

Figure 10.3 *Double (matrix) substrate plasma pixel cell*

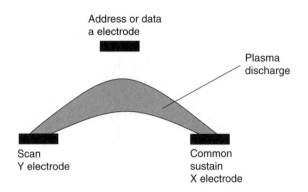

Figure 10.4 *Coplanar discharge*

This process is used as the basis for a matrix of tiny pixel cells formed by an array of row and column electrodes placed within two glass substrates. In this type of construction, known as the matrix or *double substrate plasma panel*, the two electrodes are deposited in strips on two separate parallel glass plates with a discharge cell placed at each intersection.

This early design was modified to simplify manufacture, reduce cost and increase brightness. This is how the *co-planar* panel was born (Figure 10.4) in which the two electrodes, *scan* (Y) and *common sustain* (X or C) are placed on the same plane rather than facing each. The discharge thus occurs between two electrodes which are parallel to each other and are deposited on the same glass plate. A third electrode, the address or data electrode is added to ensure the discharge takes place perpendicular to the parallel electrodes and to determine which cell will suffer a discharge.

Figure 10.5 shows the construction of a coplanar pixel cell. The cells are separated by barrier ribs to prevent electrical and optical interaction (*crosstalk*) between the cells. The inside walls of the ribs are covered with a layer of phosphor. The electrodes are deposited on the front and rear glass plates: scan and sustain electrodes on the front and the address electrodes on the rear. The dielectric layer above the electrodes which is typically 20–25 µm thick protects the electrodes and provides a capacitive reactance in series with the cell which apart from acting as a current limiter, is crucial to the formation of the wall charge which acts as a temporary memory storing *one-bit video* data. This is followed by the MgO (Magnesium Oxide) layer to protect the dielectric. Its large secondary electron emission coefficient when bombarded by neon or helium ions helps to maintain a low plasma breakdown voltage. The final layer is the phosphor layer responsible for light emission.

The panel is constructed by first mounting the address electrodes on the rear glass (Figure 10.6). A protective coating of a dielectric is then added. Ribs are formed as dividers and colour phosphors are then added to the cells. Scan and Sustain transparent electrodes are mounted on the front glass and coated with a protective dielectric. Finally an MgO overcoat is

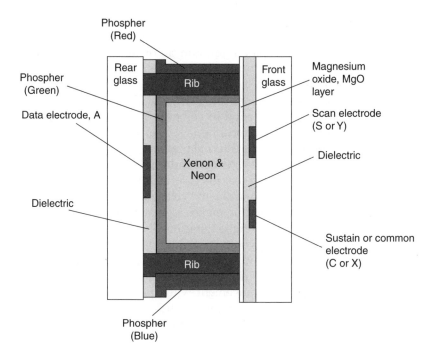

Figure 10.5 *Coplanar pixel cell construction*

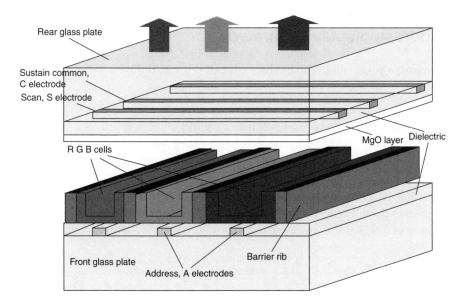

Figure 10.6 *Plasma panel construction. Note assymetrical construction*

applied on the front glass for protection of electrodes from the discharge. The scan and sustain electrodes occupy a large area on the front glass plate and therefore must be transparent. Indium tin oxide, ITO is used for that purpose. Due to the resistivity of ITO, a small metal electrode, a bus electrode, is generally deposited on the edge of each coplanar electrode to maintain a constant voltage along the electrode. The two plates are then sealed and plasma gas, e.g. Xenon in Neon or Helium are pumped at a pressure lower than normal atmospheric sea level. In this way, the two glass plates are constantly under external pressure thus sustaining the seal.

From this basic structure, Red, Green and Blue colour are produced. It will be noticed from Figure 10.6 that the pixel cells have different sizes with the blue cell being the largest and the red cell smallest. The reason for this *asymmetric cell structure* is that blue makes a greater contribution to colour temperature and when blue is brighter, it is possible to use brighter red and green so that the panel as a whole is brighter. However, the asymmetric arrangement has the major disadvantage of increasing manufacturing costs and reducing lifetime. For this reason, some manufacturers have opted to the symmetrical phosphor arrangement with appropriate changes to the video signal to counter the varied colour contribution of primary colours.

The life time of a display is mainly limited by the sputtering of the MgO layer. Although the secondary emission properties of MgO do not change significantly even after hundreds of nanometers are removed from its surface by sputtering, its life time is limited, typically 50,000 or 60,000 h at average brightness.

Address display separated

Driving a pixel cell involves three phases (Figure 10.7):

- *set* or *initialise* (also known as erase),
- *address* or *write*,
- *sustain* or *discharge* (also known as display).

The set phase

The cell is set by removing any residual charge that may remain from a previous drive cycle. This is accomplished by applying a step pulse in the region of 300–400 V, between the scan (S) and common sustain (C) electrodes

Figure 10.7 *Three-phase pixel cell drive cycle*

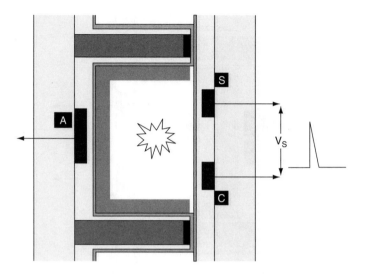

Figure 10.8 *Set or erase pulse applied between scan S and common C electrodes*

(Figure 10.8). The step voltage causes a small start discharge to take place and the gas in the cell becomes ionised. This start discharge is not a fully fledged discharge and very low light emission occurs.

The address phase

Once the cell has been cleared and set, the next stage is to determine if the cell is required to be 'turned ON', i.e. required to produce light emission. This depends on the picture content represented by the video information fed into the address (A) electrode. At this stage, the scan (S) electrode is taken to a negative potential by a scan pulse. If the data electrode is at logic 1, then the cell is selected for emission. The potential between the data and scan electrode causes a small discharge to take place in the cell. The discharge is quickly quenched as the dielectric surrounding the electrodes charges up. This charge is known as the *'wall charge'* which primes the cell for a fully fledged plasma discharge at the next phase (Figure 10.9). The pixel cell that does not have a wall charge will not produce light at the next phase. This technique in which the active cells are selected by forming a wall charge at the address stage is known as *'selective write'* addressing.

The sustain phase

The final stage is to discharge the cell and produce light emission. This is accomplished by applying a high frequency pulse (in the region of 200 kHz) known as the *sustain pulse* between the scan S and the common sustain

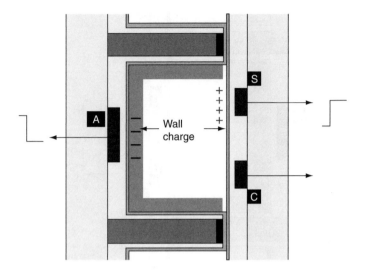

Figure 10.9 *Address phase and wall charge formation*

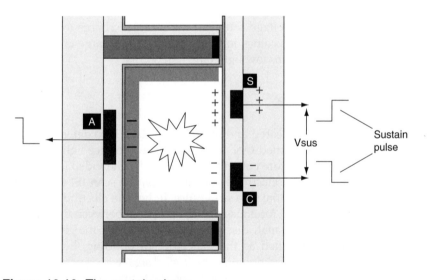

Figure 10.10 *The sustain phase*

C electrodes. The amplitude of the sustain pulse is smaller than the break-down voltage of the plasma. However, its polarity is such as to add to the existing wall charge causing plasma discharge to take place and with it the emission of UV which upon bombarding the phosphor coating release the appropriate colour glow (Figure 10.10). This discharge will also be quenched by the dielectric creating a new wall charge. This time, however, the wall charge is in the reverse polarity to the previous one. If now a sustain pulse with an opposite polarity is applied to the scan-sustain electrodes,

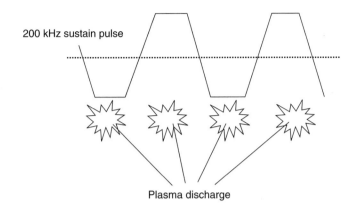

200 kHz sustain pulse

Plasma discharge

Figure 10.11 *Discharge takes place at each positive and negative edges of the sustain pulse*

it will add to the newly created wall charge to produce another full discharge which in turn reverses the wall charge and so on. Discharge will thus occur at each positive and negative halves of the sustain pulse as illustrated in Figure 10.11. The brightness of the pixel cell is determined by the number of sustain pulses applied to the electrodes. If the cell was not selected for emission during the address phase, there would not be a wall charge and a pixel cell discharge will not occur even when a sustain pulse is applied.

From the above, it can be seen that the set–address–discharge process have three distinct and separate phases. This technique is known as *address display separated* (*ADS*) in which pixel cells are 'addressed' individually and then in a 'separate' phase the pixels are 'displayed' by a simultaneous discharge.

Driving the panel

The process of driving a plasma panel follows the sequence of ADS described above. First, the set voltage is applied simultaneously to all the pixels and their individual cells. Next, the pixels are addressed one line at a time. Each line is selected in turn and the pixel cells along that line are addressed individually by the digitised video data fed down the address electrode bus. If a particular cell in that line is to be selected for discharge, then the dielectric in that cell is charged and a wall charge is formed. Once the pixel cells have been addressed and the appropriate cells primed by a wall charge, the next line is selected and the cells in that line are addressed and primed as dictated by the next set of video data clocked down the address columns and so on.

When all the lines have been scanned and all the pixel cells addressed and primed as necessary, the sustain phase is then executed by applying a series of sustain pulses to all S electrodes and a similar but anti-phase

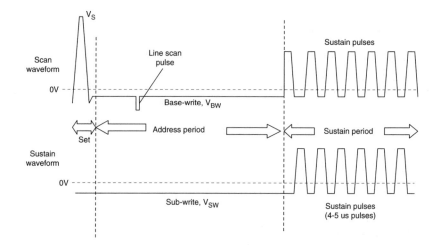

Figure 10.12 *Scan and sustain drive waveforms*

sustain pulses to the common C electrodes. The scan and sustain wave-forms will thus have the distinctive shapes shown in Figure 10.12. A set voltage pulse V_S is applied to all scan electrodes simultaneously to clear and set all pixel cells. This is then followed by the address or write period in which, each line or row of pixels is selected in turn by a negative going select pulse. The cells along the selected line are then addressed and primed by one-bit video data fed down the address bus, hence the name *one-bit plane* technique. A video data bit of 1 will select the cell for emission and a wall charge will be formed. A data bit of 0 will not select the cell and wall charge will not be formed. After the line has been addressed, its scan voltage is kept at a negative voltage known as *base-write* voltage V_{BW} for the remaining duration of the address period. This is necessary in order to avoid pre-mature discharge of the cells, known as *self-erasing* discharge which will remove the wall charge. As for the sustain electrode, it is kept at a small negative voltage known as *sub-write* voltage V_{SW} during the address period. At the end of the address period, out-of-phases 4–5 µs pulses are fed into the sustain and scan electrodes for the simultaneous discharge and display of all selected pixel cells.

Figure 10.13b shows the waveforms associated with addressing three pixel cells, K_N, L_N and M_N on three consecutive lines, Y_K, Y_L and Y_M respectively. All three cells fall along the same address line A_n (Figure 10.13a). At the set phase, a set pulse is fed to all scan lines. The address phase involves selecting each line separately by a negative going pulse. Pixel line Y_K is selected by Y_K select pulse and at the same time, data for pixel cell K_n appears on address line A_n to prime the cell for discharge during the sustain phase. Pixel line Y_L is then selected by Y_L select and simultaneously data L_n appears on address bus A_n to prime the cell. Similarly for pixel line Y_M except that in this case the data for cell M_n is zero and the cell is not

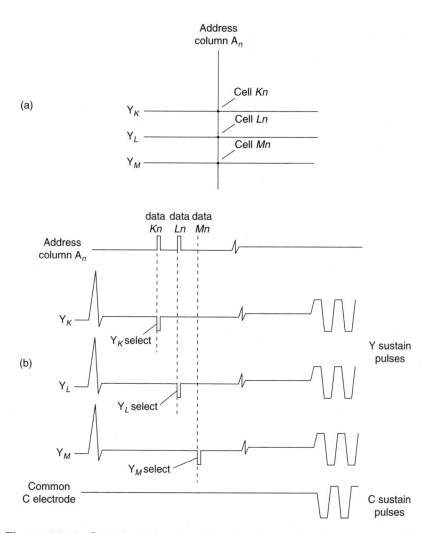

Figure 10.13 *Sustain and scan pulses for three cells along one data line*

primed for discharge. When the scan phase is completed, anti-phase sustain pulses are applied to the Y and C electrodes.

Scanning: sequential and interlaced

One of the major drawbacks of CRT displays is flicker. There are two types of flicker that affect a CRT display: inter-field caused by interlacing and the less significant intra-field flicker. Both are caused by electron beam scanning. When the beam scans a line, the phosphor glows and as soon as the beam moves along, the phosphor begins to fade. By the time the beam

Table 10.1 *CRT–PDP comparison*

	CRT	PDP
Scan	Interlace	Progressive
Emission	Dot sequence	Simultaneous
Phosphor excitation	Electron beam	Ultra violet rays
Brightness control	Beam current	Discharge (sustain) period
Power consumption	100–150 W	200–400 W

reaches the end of a field, the phosphor at the start of the field would have almost completely faded, causing intra-field flicker to occur. Similarly for intra-field flicker, by the time the beam reaches the end of a line, the phosphor at the start would have faded.

In plasma panels, the process of line-by-line scanning is separate from the actual process of light emission. The latter takes place during the sustain period exciting and discharging all pixels simultaneously. For this reason, flicker disappears completely. It will also be noticed that the ADS plasma addressing technique is sequential and since television broadcasting involves interlaced pictures, it is necessary for *interlace-to-progressive (I/P) conversion* to take place before video data are fed to the video driving board. Table 10.1 shows a comparison between a CRT and a PDP.

Sub-field coding

As was stated earlier, the brightness of a pixel cell depends on the number of sustain pulses and therefore the duration of the sustain period. A long sustain period produces more plasma discharges and therefore higher brightness levels and vice versa. A single drive cycle of set–address–sustain produces one level of brightness for all selected cells. To produce different levels of brightness for different cells within a single picture frame, in other words, to produce a greyscale, we need more than one drive cycle in which each drive cycle known as a *sub-field (SF)* is given a brightness weighting determined by the length of its sustain period. A SF with a short discharge period would result in low brightness and vice versa. If two SFs are used, then $2^2 = 4$ different greyscale levels can be obtained, whereas 6 SFs would result in $2^6 = 64$ different levels. And since each drive cycle or SF requires one bit of video, then the number of bits used to describe the video signal determines the number of SFs. For example the standard eight-bit colour signal would generate eight SFs per frame and with it $2^8 = 256$ greyscale levels.

Table 10.2 shows a binary weighting stretching from lowest order SF1 valued at 1 to the highest order SF7 valued at 128. Figure 10.4 shows the weighting of each SF in terms of its period. Each pixel cell is selected for one, two or more SF to the maximum depending on the required brightness of the cell. For low brightness, the cell is selected and primed for low SFs

Table 10.2 *Weighting values for an 8-SF frame*

SF1	$2^0 = 1$	SF5	$2^4 = 16$
SF2	$2^1 = 2$	SF6	$2^5 = 32$
SF3	$2^2 = 4$	SF7	$2^6 = 64$
SF4	$2^3 = 8$	SF8	$2^7 = 128$

while high brightness is achieved by priming the cell for the high order SFs. The total brightness of each pixel cell in a frame is the summation of the brightness produced by each SF. Minimum brightness, i.e. black is obtained when the cell is not selected in any SF and the highest brightness is obtained when a cell is selected for all SFs giving a brightness value of 256.

Plasma panel brightness

The complete set of SFs has to be processed within the duration of one complete TV picture frame which for the UK's 625-line system is 20 ms and for the USA's 525-line system is 16.7 ms. In any SF, the sustain or discharge time progressively increases in binary steps. The address period, however, is constant throughout. This is because the address period depends on the number of lines and the duration of the write pulse. Given a typical write pulse of 2 μs, then for a VGA panel with 480 lines and 8 SF (256 greyscale levels), the address period = 2 μs × 8 SF × 480 = 7.68 ms.

With a PAL TV field period of 20 ms, the available sustain period per frame = 20 – 7.68 = 12.32 ms. The percentage of the TV field available to sustain the cells = 12.32/20 = 61.6%. The equivalent calculations for NTSC is 17.6 – 7.68 = 9.92 ms giving a percentage sustain of 9.92/17.6 = 56.4%. These are low percentages which result in low panel brightness, something that early plasma panels were notorious for. One of the first methods used to increase the time available for the sustain period is the dual scanning.

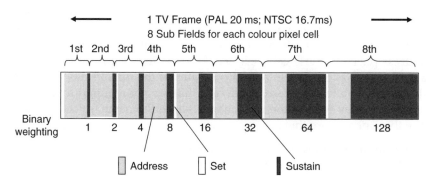

Figure 10.14 *SF weighting in terms of its sustain period*

Dual scan

With dual scan the screen is divided into two halves: upper and lower. The two halves are initialised and sustained simultaneously, but scanned as two separate screens. Pairs of lines, one from each half are scanned simultaneously thus reducing the time for the address period by almost a half and consequently virtually doubling the time available for the sustain period and with it the maximum brightness of the screen. However, the simultaneous scanning of plasma lines requires two independent panel drive circuits, with the consequent increase in cost.

Selective erase addressing

For high definition television with 1000 or more lines, even dual scanning is insufficient to provide the necessary sustain period for good quality picture. Given a write pulse of $2\,\mu s$, then for an XSGA panel (1280 × 1024),

$$\text{address period} = 2\,\mu s \times 8\ SF \times 1024 = 16.38\ ms$$

This gives a total sustain period = 20 − 16.38 = 3.62 ms with the equivalent for NTSC of 17.6 − 16.38 = 0.32 ms. The percentage sustain period = 3.62/20 = 3.68% for PAL and 0.32/17.6 = 1.9% for NTSC. Neither value is any where adequate for good quality picture reproduction.

One way of increasing the sustain time is to reduce the duration of the address phase. For HD panels, selective write addressing requires the address cycle time per line to be reduced to $1.0\,\mu s$ or less. That means addressing frequencies approaching 1 MHz. Such high frequencies would increase the power requirements and with it the cost of the address drivers. Furthermore, regardless of how fast the line addressing cycle is, the address pulse itself cannot be very narrow because time is needed to form the wall charge if a cell is to be selected. Another way to reduce the pulse width is to raise the addressing voltage, but this often results in higher power consumption. For this reason, a different technique known as 'selective erase' is used.

The *'selective erase'* (also known as *'erase wall-charge'*) addressing reverses the process of creating the wall charge. In the 'selective write' the cells are cleared in the initialisation set phase and a wall is formed during the address phase in those cells selected for emission. With 'selective erase', a wall charge is formed in all the cells during the initialisation phase. In the following address phase, those cells that are not required to ignite are 'de-selected' by erasing the wall charge, hence the name 'selective erase'. The three phases are thus:

Initialisation and wall charge formation,
addressing (erasing wall charge),
sustaining.

The advantage of this is that it takes a relatively short time to discharge a cell and remove the wall charge compared with charging a cell to form the wall charge thus making it possible to use narrower line select pulses. With a typical address (or select) pulse width of 1.2 µs, a SXGA (1280 × 1024) panel with 256 greyscale levels (8 SFs), and the address period is

$$1.2 \, \mu s \times 8 \, SF \times 1024 = 9.83 \, ms$$

The resulting total sustain period per TV frame = 20 − 9.83 = 10.17 ms for PAL and 16.7 − 9.83 = 6.84 ms for NTSC. This gives a percentage sustain period of 10.17/20 = 50.8% for PAL and 6.84/17.6 = 40.9% for NTSC.

Sustain-during-erase

Even with the use of the selective erase technique, the amount of time available for discharging the cells continue to impose a limitation on the brightness and contrast of the picture of an HD panel. To overcome these limitations, a new scanning technique has been developed known as *'scan-during-sustain'*, *SDS*. As the name suggests, with SDS, scanning and sustaining are carried out simultaneously.

In the SDS technique, the row electrodes are grouped into a number of blocks. At any one time, two of the blocks known as the *scan blocks* are addressed while the other blocks, known as *sustain blocks* are discharged. The process is then repeated with two different blocks being addressed while the remaining blocks are sustained and so on. Thus scan pulses are applied to the scan blocks simultaneously with the sustain pulses to the sustain blocks. For example, an SXGA panel which contains 1024 row electrode pairs and $1024 \times 3 = 3840$ data D electrode is divided into eight blocks: B1 − B8. Each block contains 128 pairs of S and C electrodes. In each block, the scan electrodes are driven separately while the common (sustain) electrodes are connected and driven together as shown in Figure 10.15. The odd numbered blocks use the left side of the panel for S and the other side for C electrode connections. The reverse is true for the even numbered blocks. With this arrangement, a single type of drive board can be used for both left and right sides of the panel.

Greyscale and colours

In digitised video applications, the number of colours that may be displayed depends on the number of greyscale levels available and that of course is determined by the number of bits allocated to each colour. In broadcast application 8 bits are allocated to each primary colour making a total of 24 bits for the RGB video signal, hence the name *24-bit video*. Each colour thus have 2^8 greyscale levels and hence a pixel with three colour

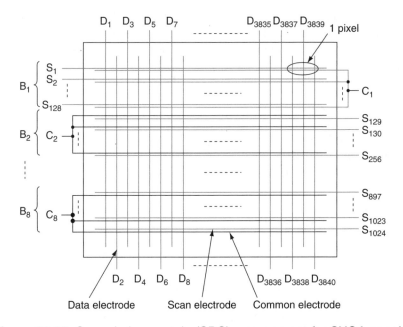

Figure 10.15 *Scan-during-sustain (SDS) arrangement for SXGA panel*

cells can produce $256 \times 256 \times 256$ (or 2^{24}) $= 16.778$ million colours. A 10-bit video results in $2^{10} \times 2^{10} \times 2^{10} = 2^{30} = 1.07$ billion colours and so on. In general, the number of colours $= 2^n$ where n is the total number of bits allocated to the video which for a 24-bit video is 24 bits. However, with the introduction of SFs, the number of greyscale levels and with it the number of colours is now dependent on the number of SFs which may be different than the original bit allocation at the analogue-to-digital converter stage. In modern plasma panels, 10, 11, 12 and even 13 SFs are used which allows manufacturers to claim astronomical number of colours. For instance, 10 SFs would produce $2^{10} = 1024$ greyscale levels and $1024 \times 1024 \times 1024$ (or 2^{30}) $= 1.07$ billion colours. As for 12 SFs, the figure is 67.8 billion. Increasing the number of SFs does not improve the inherent quality of the picture as the amount of video information remains unchanged. It merely allows it to be presented differently, which as we will see later, does enhance the video experience.

False contours

The basic plasma technology described so far would produce an acceptable moving pictures but its quality leaves a lot to be desired as many of those who purchased early plasma television discovered to their dismay. The basic PDP suffers from two major shortcomings: *dynamic false contours* and *black-level luminance*.

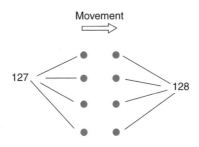

Figure 10.16 *Adjacent pixels with almost equal brightness*

False contours are a serious impairment of picture quality that appears on a moving picture. This artefact is the result of the SF coding techniques used to control the brightness of the pixel cells. As described earlier, TV picture frames are broken up into a number of SFs, conventionally, 8 binary-weighted SFs are used. Consider an image in which the brightness of two adjacent column of pixels is 127 and 128 respectively (Figure 10.16). The set of pixels on the right reach their brightness level of 127 by being selected and primed to discharge for the first seven of the 8 SFs while the second set to the left fires for the 8th SF only. For a still image there are no problems and the picture information would be re-produced without artefacts. However, if there is movement from left to right, then what is known as false contours may be observed depending on the direction and speed of the movement. Figure 10.17 shows a schematic time chart representing two such adjacent pixels. The left pixel has a level of 127 moving towards a second to its right that has level of 128. Because the human eye follows the motion of an image, the movement of the pixel causes the visual sensation to be integrated along the arrow shown in the figure. The seven SFs of the pixel on the left and the 8th SF of the second pixel are added up by the eye resulting in a brightness intensity of 127 + 128 = 255 with a boundary of 127 on the left and 128 on the right. The effect for the human eye is a bright spot and in the case of few such pixel combination, a line contour. This artefact is most pronounced when an image of middle and almost uniform intensity such as a Caucasian skin or blue sky moves across the screen resulting in the smudging illustrated in Figure 10.18.

Several techniques have been developed by PDP manufacturers to minimise the effect of false contour, namely *time compression, SF splitting* and *non-binary weighting* of SFs.

Time compression

The time-compression technique as the name suggests, reduces the light emission period of the SFs by introducing a time interval between one set of SFs and the next and consequently squeezing the SFs as shown in Figure 10.19. The time interval reduces the possibility of the eye adding up

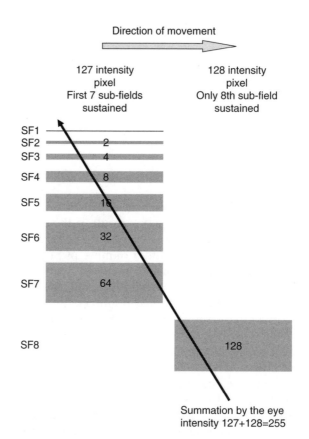

Direction of movement

127 intensity
pixel
First 7 sub-fields
sustained

128 intensity
pixel
Only 8th sub-field
sustained

SF1
SF2 ———————— 2
SF3 ———————— 4
SF4   8
SF5   16
SF6   32
SF7   64
SF8   128

Summation by the eye
intensity 127+128=255

Figure 10.17 *False contours visualised by human eye along the arrow*

Figure 10.18 *The effect of a severe form of false contours*

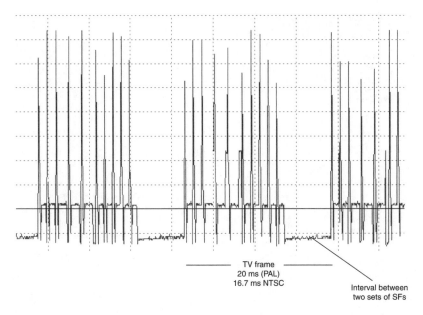

Figure 10.19 *Time compressed sustain waveform*

the intensities of consecutive SFs. In this way, the area that may suffer false contours can be greatly reduced by a factor depending on the compression ratio.

Adaptive time compression

The drawback with time-compressed SF drive is the reduction in panel brightness as a result of reducing the total sustain period of each SF. This is especially significant for darker images. To overcome this, adaptive time-compressed drive is used in which the compression ratio is changed with varying picture brightness. For a bright image, a high compression ratio is used and vice versa for dark images. In this way, images with a low *average picture level* (*APL*) have a longer sustain period and increased level of brightness. Figures 10.20 and 10.21 show the scan and sustain waveforms for low and high APL pictures respectively.

Sub-field splitting

In addition to time-compression, one or two highly weighted SFs may be split into two lower-weighted SFs. With the conventional binary coding, the maximum intensity weighting is provided by SF8 with 128. If this SF was split into two SFs, each with a weighting of 64, then if a false contour described earlier is to take place, the visualised intensity would be greatly

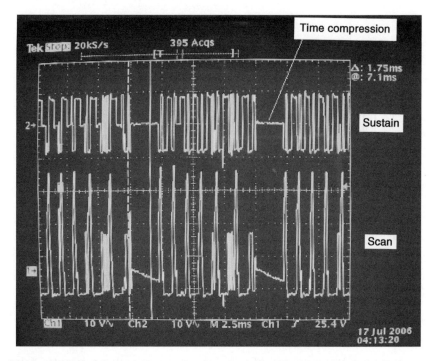

Figure 10.20 *Adaptive time compression. Sustain and scan waveforms showing time compression for a picture with high brightness*

reduced to 127 + 64 = 191. In practice, using this SF splitting, nine SFs are employed with a weighting distribution shown in Figure 10.22. SFs corresponding to 128 and 64 are divided into four SFs of weighting 48 each and positioned at either end of the SF set. In the figure, the SF with weighting 1 is neglected to reduce the number of SFs and maintain the total level of brightness. The effect of removing the LSB SF is compensated for by an error diffusion circuit.

Non-binary sub-field coding

Non-binary coding is similar to SF splitting in that it removes the very high-weighted SF8 and introduces a larger number of lower-weighted SFs that can be selected for any greyscale level to reduce the effect of false contours. Figure 10.23 shows a typical non-binary 12-SF drive system.

Dynamic brightness control

As started earlier one of the major drawbacks of early plasma screens has been its low level of brightness compared with a CRT equivalent. Dynamic brightness control which Panasonic calls Adaptive Brightness Intensifier

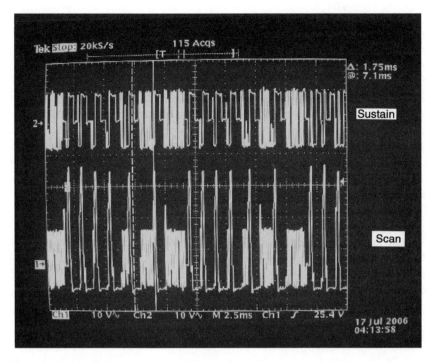

Figure 10.21 *Adaptive time compression. Sustain and scan waveforms for a low average picture brightness (not absence of time compression)*

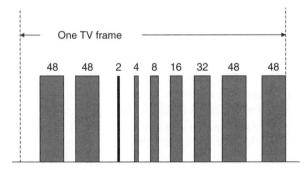

Figure 10.22 *Sub-field splitting in which the maximum SF value is 48*

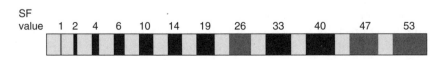

Figure 10.23 *Twelve non-binary sub-field drive*

(AI) and Pioneer calls i-CLEAR attempts to rectify this. The average picture level, APL is used to adjust the number of SFs. The standard number is 11 for normal brightness; it is reduced to 10 for dark scenes (low APL) and increased to 12 for bright scenes (high APL). The use of a larger number of SFs allows for finer weighting distribution and hence increased greyscale levels which improve the picture contrast range. In darker images, reduced number of SFs provides more time for the discharge period which improves peak brightness making dark scenes more visible. The reduction in greyscale levels in a low luminance picture is not discernable by the human eye. Furthermore, variable SF coding helps to maintain power consumption within small variations which improves efficiency and reduces cooling requirements making it possible to dispense with the use of cooling fans.

Other techniques employed by various manufacturers to improve panel brightness and contrast include box-shaped and *deep cell structures* to enlarge the effective phosphor surface and the use of *'direct colour filter'* to reduce ambient light reflection. Ambient light reflection may also be reduced using *'black stripes'* applied to the front glass, a technique developed by Pioneer.

Black level drive

One of the most important factors in the perception of natural images on a television screen is the contrast. Contrast is the difference between black level and maximum brightness. It may be improved by either increasing peak brightness or decreasing the black level. Light emission in a plasma panel is produced whenever a discharge takes place. While plasma discharges take place mainly during the sustain period, there are other small discharges that take place outside the sustain phase. The most significant of these is that caused by the high-voltage high-rising initialisation pulse at the start of each set–address–sustain cycle. The effect of this discharge is a fairly bright radiation known as *'priming light'* emanating from the phosphor which makes the blacks slightly grey making it difficult to reproduce dark scenes. As this is repeated for each SF, some ten to twelve times per frame, the accumulated effect is to lift the black level and decrease the contrast ratio. The effect of the set pulse may be reduced by employing the ramp-shaped pulse shown in Figure 10.24. When a ramp is applied to the sustain-scan electrodes, a very weak discharge occurs for a longer period generating fewer UV rays and hence less phosphor radiation.

The shape of the set pulse in terms of its gradients and voltage levels have to be precisely set according to the manufacturer's specifications to within very tight limits. One such specification is illustrated in Figure 10.25.

The shape of the ramp pulse may be adjusted with two or more variable pots as indicated by the manufacturer. Slight changes in the shape

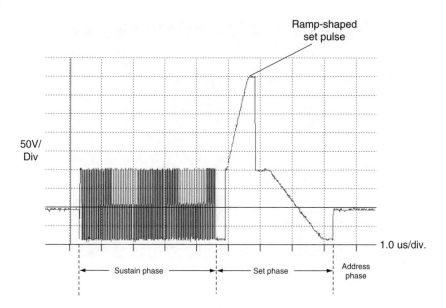

Figure 10.24 *Ramp-shaped set pulse*

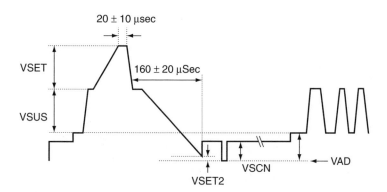

Figure 10.25 *Set pulse ramp specifications*

of the ramp set pulse would result in colour noise on part or the entire displayed image depending on the type and severity of the change. A typical symptom of a badly adjusted set ramp pulse is shown in Figure 10.26.

Another technique called *'Real Black drive'* used by some manufacturers dispenses with the use of large initialisation pulse at the start of all SF except the first one. The first SF starts with the normal high-amplitude initialisation pulse while the remaining SFs use smaller initialisation pulses as shown in Figure 10.27.

Figure 10.26 *Effect of maladjusted ramp set pulse*

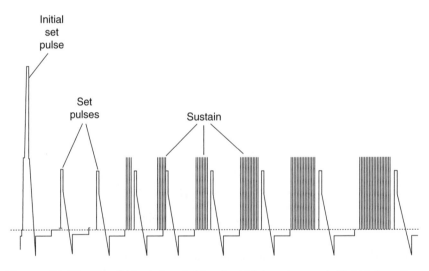

Figure 10.27 *'Real' black level drive in which only one initial high volt-age set pulse is used for each set of sub-fields*

Sub-field generation

Figure 10.28 shows a block diagram incorporating the adaptive or dynamic brightness control technique. Analogue RGB signals are digitised by the A/D converter and following interlace-to-progressive (I/P) conversion, the digitised 24-bit video is fed to the scalar. The function of the scalar is to convert the format of the incoming video to the native resolution of the panel.

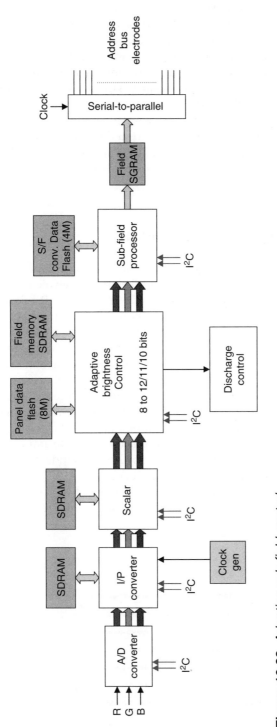

Figure 10.28 *Adaptive sub-field control*

The converted video is then fed to the adaptive brightness control chip which determines the number of SF to be used, frame by frame and passes this information to the SF processor. It also synchronises the discharge of the pixel cells to control the sustain pulses and the duration of the discharge in line with the selected SF coding. A Flash memory holds the necessary software and a field memory stores the field contents. The SF processor produces the video information in the required bit stream to be stored in a Static Graphic RAM (SGRAM) memory chip before going into the serial-to-parallel converter to be clocked into the data electrode bus SF by SF. All processing chips are controlled via a serial control bus, normally an I²C from the microprocessor. For details of serial control bus I²C, se Appendix 2.

Alternate lighting of surfaces

In the conventional PDP construction, each pixel cell depends on two electrodes: scan and sustain. Discharge takes place between these two electrodes when a sustain pulse is applied across them. It follows that each line of pixels requires two strips of electrodes. As the panels resolution increases, the number of pixels per line and the number of lines increase demanding a much finer-pitched pixel cells. Furthermore, the number of electrode drivers has to rise which increases production costs. Another drawback of the conventional structure is the gap between electrode pairs that cannot be utilised for luminance radiation. While this structure is adequate for VGA-resolution PDPs, it is certainly impractical and costly for high-definition panels. Several developments have taken place in the construction of the pixel cells and the plasma panel to improve their high luminance efficiency and contrast ratio. These include *alternate lighting of surfaces (ALiS)* developed by Fujitsu-Hitachi.

With ALiS, the Y electrode is shared between two X electrodes as illustrated in Figure 10.29. The electrodes are arranged in equal intervals and

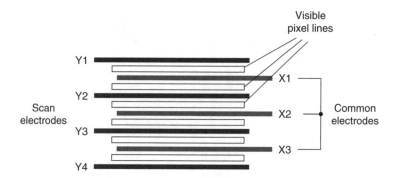

Figure 10.29 *ALi's electrode-sharing construction*

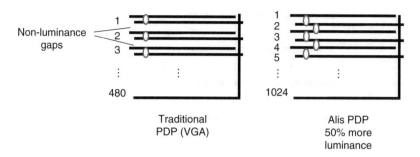

Figure 10.30 *Not provided*

the space between them is used as display lines thus doubling the resolution for the same number of electrodes. The non-luminance area between pairs of electrodes in the conventional PDP has been removed as shown in Figure 10.30, raising the aperture rate by 65% and improving luminance by 50.

Since each Y electrode is shared between two X electrodes, in order to ensure that each pixel line is independently addressed and discharged, interlaced addressing must be used. First the 'odd' field is addressed, line by line by energising pairs of electrodes in the following order: X1–Y1; X2–Y2; X3–Y3 and so on (Figure 10.29). This is followed by the sustain stage in which all pixel cells that have been selected in the 'odd' field are discharged. The same process is then repeated for the 'even' field by energising the following pairs of electrodes: X2–Y1; X3–Y2; X4–Y3 and so on followed by sustaining the even filed. The whole process is then repeated for each individual SF. The interlaced driving of the PDP removes the need for interlace-to-progressive (I/P) converter. With ALiS, the number of electrodes required for a given number of pixel lines = ½ no. of lines + 1.

Compared with a conventional PDP, ALiS has higher resolution, brighter picture, lower cost, lower noise, longer life as the lighting duty of each cell is reduced by half as a result of interlacing and requiring no I/P conversion.

Enhanced-ALiS

Like all interlaced picture re-production, ALiS suffers from intra-frame line flicker. Although flicker with ALiS is far less than that experienced with a CRT, there still remains an element of it which is removed with the enhanced-ALiS (e-ALiS) developed by Fujitsu. Such displays are frequently known as A1 panels. The most outstanding characteristic of the new e-ALiS display type is that they once more offer progressive driving.

For ALiS technology to be used with progressive driving, it is necessary to control two pixel lines both simultaneously and independently using one electrode only. This is made possible by the introduction of horizontal barriers to separate the cells into individual units that can be pre-charged independently from each other.

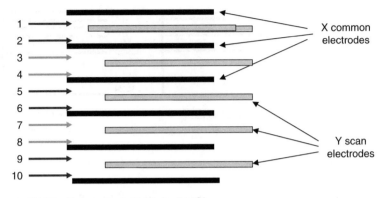

Figure 10.31 *Enhanced-ALiS (e-ALiS)*

4-phase sub-field drive

With e-ALiS, the display is broken up into pairs of visible lines allocated to two groups: Group 1 comprising lines 1–2, 5–6, 9–11 and so on and group 2 comprising lines 3–4, 7–8, 11–12 and so on as shown in Figure 10.31. A 4-phase drive cycle is used:

Erase – Pre-condition (or reset) – Address – Discharge

The sequence begins with a simultaneous X–Y erase pulse to all lines and pixels (Figure 10.32). This is followed by the pre-conditioning phase to initialise lines in group 1 to be ready for the addressing phase. Pre-conditioning is achieved by applying a common reset pulse (Reset 1) to the Y electrodes while at the same time applying a pulse to the X electrodes involved in group one, line by line. A small discharge takes place which creates a small pre-conditioning wall charge. This is followed by the address phase in which the pre-conditioned pixel cells are addressed line-by-line and primed to be 'ON' or 'OFF'. Next, group 2 (line-pairs 3–4, 7–8, 11–12, etc.) are pre-conditioned and addressed in a similar way. Once both groups of line-pairs are addressed and the cells are selected for discharge, a sustain pulse is applied to discharge all selected cells simultaneously. The process is then repeated for the next SF and so on. Thus, while interlacing is used in the address phase, the discharge or light emission is simultaneously eliminating line flicker completely.

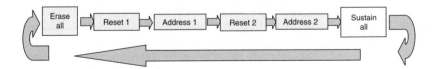

Figure 10.32 *4-phase SF drive sequence for e-ALiS*

Pioneer's Waffle rib and T-shaped structured panel

A number of developments have taken place and continue to take place in the structure of a plasma panel to improve its brightness and contrast. Box-shaped cells are commonly used to provide a larger effective phosphor surface and hence higher brightness. Pioneer developed the Waffle-structured Rib curbing like the surface of a waffle and the T-shaped electrode structure. The deep cells are encapsulated with horizontal ribs so emitted light won't escape into neighbouring cells. Further improvement was produced with the addition of 'Crystal Emissive Layer' to increase luminous efficiency. Figure 10.33 shows the basic layers of a modern Pioneer plasma display panel.

Plasma panel faults

The first step in fault finding on PDP panels is to ascertain if the fault is a panel malfunction, in which case it has to be replaced, or if the cause of the fault is outside the panel in which case a repair is possible. Some of the symptoms point clearly to a panel fault, others may be ambiguous. One of the more obvious symptoms of a faulty panel is Pixel defects. This may be a single pixel, several pixels or cluster pf pixel failure as illustrated in Figure 10.34. Manufacturers allow for a certain number of pixel defects before the panel is rendered obsolete. The display panel is divided into a central area and side areas and the number of bad pixels permitted depends on the area.

Other classic symptoms pointing to faulty PDP panel are shown in Figure 10.35. A horizontal line across the screen (Figure 10.35a) may be caused by a faulty panel or a faulty tape carrier in which case, the panel has to be replaced. However, it could also be caused by a bad connection between scan driver board and the panel. A single vertical line shown in Figure 10.35b is caused by a faulty panel. Multiple vertical lines shown in Figure 10.35c are caused by a faulty tape carrier. In either case, the panel has to be changed.

Drive faults

The classic symptom for a source (or column) drive is a vertical band across the screen as shown in Figure 10.36. A horizontal band would indicate a faulty line scan drive.

Image burn

Image burn is a result of a residual image remaining on a panel after the same still picture has been displayed for some time. Image burn is caused when the phosphor of some pixels is continuously bombarded with UV

Front glass substrate
Dielectric layer
T-shaped electrodes
Protective layer
Crystal emissive layer
Black stripe
Auxilary electrode
Green phosphor
Blue phosphor
Red phosphor
Deep waffle RIO structure
Address protective layer
Address electrode
Rear glass substrate

Figure 10.33 *Deep waffle-shaped cells with T-shaped electrodes*

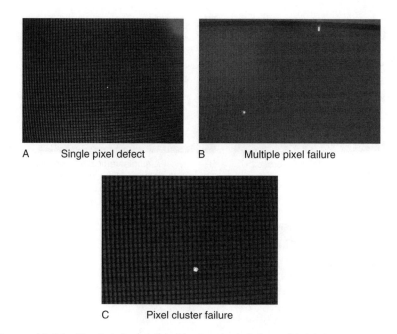

| A | Single pixel defect | B | Multiple pixel failure |

C   Pixel cluster failure

Figure 10.34 *Pixel defects. (a) Single pixel defect. (b) Multiple pixel failure. (c) Pixel cluster failure*

radiation due to displaying a permanent image such as one or two lines of text for a long time. The pixels involved in displaying the image use their phosphor to radiate visible light continuously causing the phosphor to darken resulting in a fall in the brightness of light emission. Meanwhile those pixels not involved in the displayed image retain their strength. This differential in phosphor strength is revealed as a burned image.

A residual image or burn may be permanent or temporary A temporary burn is where the affected pixel cells have been subjected to full level emission for a relatively short time, as short as half an hour. Such temporary burn is reversible. Permanent image burn occurs after the temporary burn phase when the constant high emission of pixel cells has caused the phosphor to be permanently damaged. This type of burn is irreversible. Image burn may be avoided by ensuring an actively moving picture at all time, regular change of video display and not displaying bright images for too long. Of particular importance is the burn caused by displaying a 4:3 picture on a wide screen panel. Leaving such a picture on for a long time could cause a burn at the picture edges. Furthermore, 4:3 curtain lines may appear when the display goes back to wide screen format. Later panels have been equipped with side curtain colour and level adjustments to avoid these burns. Temporary burns may

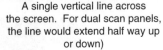

A   A single horizontal line across
the screen

B   A single vertical line across
the screen. For dual scan panels,
the line would extend half way up
or down)

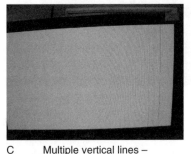

C   Multiple vertical lines –
faulty Tape carrier

Figure 10.35 *Symptoms of a faulty panel. (a) A single horizontal line across the screen. (b) A single vertical line across the screen. For dual scan panels, the line would extend half way up or down. (c) Multiple vertical lines – faulty tape carrier*

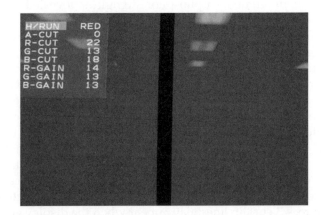

Figure 10.36 *Source (or column) drive fault*

be removed by applying a natural moving picture for as long a time as possible, in the order of few days, to reset the pixel's response, activate the scroll bar from the user setup for few hours, apply a totally white picture a number of minutes depending on the length of time the fixed picture has been displayed.

11 Liquid crystal display (LCD)

Liquid crystal display (LCD) units used for the purposes of moving picture reproduction are some of the more popular flat panel displays. Like all flat panel displays, LCDs employ a matrix structure in which the active element, in this case a liquid crystal (LC), forming the pixel cell is located at the intersection of two electrode buses.

So, what is a liquid crystal? An LC is neither crystal nor liquid. It exhibits liquid-like as well as crystal-like properties. This feature is a result of the LC's comparatively elongated molecules and their structure. Though an LC is a natural material, the liquid crystal which is used for LC displays is a multi-component mixture that is artificially created by blending of biphenyl, cyclohexane, ester and the like.

Polarisation

Light is a transverse electromagnetic (EM) wave composed of an electric and a magnetic field. The two fields are at right angles to each other travelling at the speed of light. In an EM wave, the electric field defines the designation of the wave in terms of its polarisation: if the electric field is vertical, the wave is said to be *vertically polarised* or *'p-polarised'* (Figure 11.1) and conversely if the electric field is horizontal, the wave is said to be *horizontally polarised* or *'s-polarised'*. Natural light from the sun or any other light source such as a lamp is unpolarised. It contains both vertical and horizontal polarisation. Light may become polarised if the vertical or horizontal polarisation is reduced or removed completely by, for instance a polarising filter, the type used in Polaroid sunglasses. If the horizontal polarisation is removed, the light would be vertically polarised and vice versa. This simple principle is used to control the brightness of an LC cell.

Principles of operation of LC cell

By themselves, the molecules in an LC are arranged in a loose order. However, when they come into contact with a finely grooved surface, the molecules line up in parallel along grooves of alignment layer as shown in Figure 11.2.

Furthermore, the application of an electric field across the LC causes a change in the molecular structure. This change affects the optical properties of the crystal in the way light is reflected off it or passes through it.

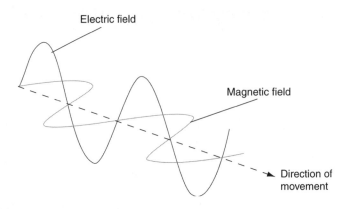

Figure 11.1 *Vertically polarised electromagnetic (EM) wave*

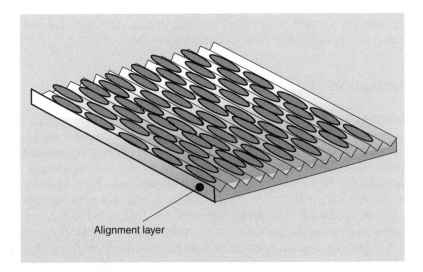

Figure 11.2 *Alignment of LC molecules along tiny grooves of glass plate*

The most popular type of LC is the *Twist-Nematic* (*TN*). In this mode, the grooves in the two plates are at right angles as shown in Figure 11.3. The molecules along the upper plate point in direction 'A' and those along the lower plate in direction 'B' thus forcing the molecules of LC to arrange themselves in a helical form. The helix has the effect of twisting the EM wave (light) passing through it by 90°. Thus, if a vertically polarised light is forced through such a crystal, it will suffer a 90° twist and become horizontally polarised. However, what is special about LCs is that if an electric field is applied across it (Figure 11.4), the helical structure begins to break down and with it the polarisation of light resulting in a smaller twist than the natural 90°. The voltage level determines the extent to which

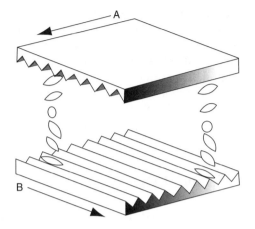

Figure 11.3 *Twist-Nematic (TN) liquid crystal (LC) with a 90° twist*

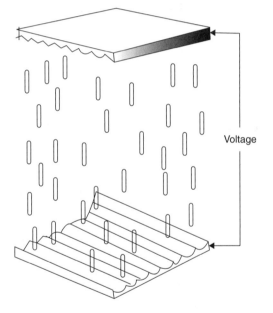

Figure 11.4 *TN crystal molecules re-arranged removing the twist when a voltage is applied*

breakdown occurs and the amount of twist. If a high enough voltage, in the region of 8 V, is applied, the twist is eliminated completely.

Reflective and transmissive

An LC does not produce light, so the technology is *'non-emissive'* and therefore does not give off a glow like a cathode ray tube (CRT) or a plasma panel does. An external form of light is therefore necessary which may be provided in two ways for two types of LC displays: *reflective* and *transmissive*. In the reflective type, the change in the molecular structure controls the reflected light while in the transmissive type it controls the light passing through it. The former is dependent on an external or ambient light for its brightness while the transmissive type has its own backlight and is not dependent on ambient light. For this reason, the transmissive type is the more popular of the two. In either case, the voltage across the LC controls its luminance.

The TN transmissive LCD

Consider two differently polarised filter plates placed opposite each other with a backlight unit as shown in Figure 11.5. Plate A allows only vertically polarised light through while plate B permits only horizontally polarised light. The effect of the two glass plates is to block the unpolarised light emanating from the backlight completely. Now, consider the same arrangement

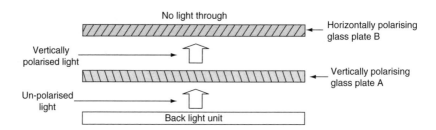

Figure 11.5 *Light is blocked by two opposite polarised filters*

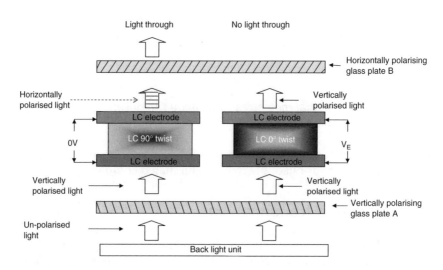

Figure 11.6 *Amount of light controlled by voltage across LC cell*

with an LC placed between the two polarised glass plates as illustrated in Figure 11.6. The unpolarised light from the backlight passes through plate A, becomes vertically polarised and goes through the LC which, without any voltage across it, forces a 90° twist changing its polarisation from vertical to horizontal which pass through the second plate B without any hindrance. If now a voltage is applied across the LC, the 90° twist would be removed and light would be blocked. If a smaller voltage is applied, a twist angle less than 90° is introduced by the LC and low-intensity light would appear at the other end. Since the applied voltage determines the twist angle of the LC, it follows that the voltage now controls the intensity of light appearing at the other end and a greyscale may thus be obtained by varying the voltage applied across the LC.

Normally white and normally black

The LC display may be used in two different modes: *normally white* (or bright) and *normally black* (or dark). The former allows the backlight through while the latter blocks the backlight when the voltage across the LC cell is zero. The arrangement in Figure 11.6 is that for the more popular normally white LCD. A normally black LCD would have only one polarising plate.

There are several types of TNLC cells depending on the angular twist the molecules are subjected to. In the simple TN type, the molecules are twisted by 90° resulting in a drop in contrast when used with large screens. The *Super Twist-Nematic (STN)* has its molecules twisted from 180 to 260° to improve the contrast ratio. Finally, the *Film Super Twist-Nematic (FSTN)* twists the molecules by 360°. This is used for very high quality black and white LCDs.

Passive- and active-matrix LCDs

There are two matrix LCD technologies: passive-matrix LCD (PMLCD) and active-matrix LCD (AMLCD). In the PMLCD, pixels are addressed directly with no switching devices involved in the process as illustrated in Figure 11.7). The effective voltage applied to the LC must average the signal voltage pulses over several frame times, which results in a slow response time greater than 150 ms and a reduction of the maximum contrast ratio. The addressing of a PMLCD also produces a kind of cross-talk resulting in blurred images because non-selected pixels are driven through a secondary signal-voltage path. This places a limit to the number of pixels that may be used in a display and with it a limit on the maximum resolution.

In the AMLCDs, on the other hand, a switching device is used to apply the voltage across the LC (Figure 11.8) and hence a better response time becomes possible. In contrast to PMLCDs, the active type, AMLCDs has no inherent limitation in the number of pixels, and they present fewer cross-talk problems.

There are several kinds of AMLCD depending on the type of switching device used. Most use transistors made of deposited thin films, which are accordingly called *thin-film transistors* (*TFTs*). The most common TFT semiconductor material is made of *amorphous silicon* (*a-Si*). a-Si TFTs are amenable to large-area fabrication using glass substrates in a low temperature (300–400°C).

An alternative TFT technology, polycrystalline silicon, normally known as *polysilicon* or *p-Si* is costly to produce and especially difficult to fabricate when manufacturing large-area displays. Nearly, all TFT LCDs are made from a-Si because of the technology's economy and maturity, but the electron mobility of a p-Si TFT is 100 times better than that of an a-Si TFT. This makes the p-Si TFT a good candidate for a TFT array containing integrated drivers, which is likely to be an attractive choice for small, high definition displays such as view finders and projection displays.

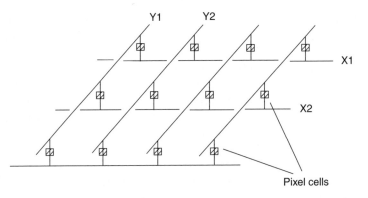

Figure 11.7 *Passive-matrix LCD (PMLCD)*

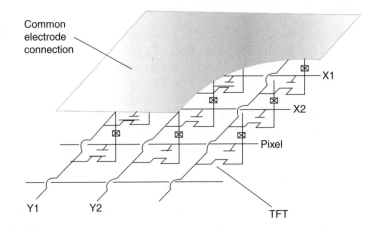

Figure 11.8 *Active-matrix LCD (AMLCD)*

TFT cell drive

In the TFT LCD, switching transistors are provided for each pixel cell as shown in Figure 11.9. One side of each LC cells is connected to its own individual TFT while the other side is connected to a common electrode which is made of transparent *indium tin oxide* (*ITO*) material. This is necessary to ensure high aperture ratio. *Aperture* ratio is the ratio of the transparent area to the opaque area of the panel. A cross-section of a TFT is shown in Figure 11.10.

Unlike the CRT in which the phosphor persistence allows for continued luminance of the picture even after the electron beam has moved to scan other lines, in flat display applications, no such persistence exists and

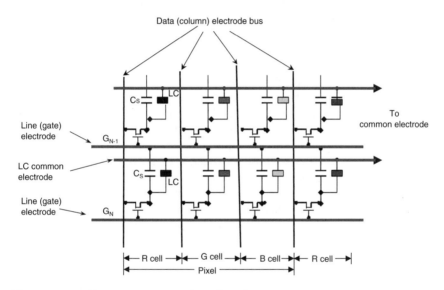

Figure 11.9 *Equivalent circuit for TFT-LC display*

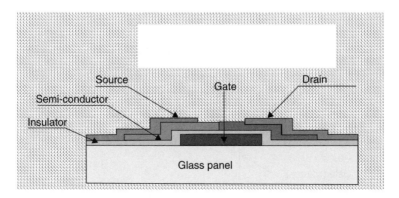

Figure 11.10 *TFT cross-section*

refreshing of pixels to produce natural moving pictures becomes difficult as the number of pixels increases. Hence, the need for a pixel cell 'memory'. A charge on a storage capacitor C_S is used for this purpose as illustrated in Figure 11.9. Each cell consists of three sub-pixels (RGB) normally referred to as cells. Each cell contains an LC driven by a TFT acting as a switch. The LC is placed within two electrodes. One electrode is connected to the TFT's source electrode and the other goes to a common electrode.

The TFT-LCD panel is scanned line by line. Each line is selected in turn by a V_{SEL} pulse to the *line* (or *gate*) *electrode* bus. Once a line is selected, the pixel cells along that line can be addressed and their luminance levels set by a voltage applied via a source driver to their corresponding data (also known as *source* or *column*) *electrode*. The source driver supplies the desired voltage level known as the *greyscale voltage* representing the pixel value, i.e. the luminance of the pixel cell. The storage capacitor C_S is charged and this charge maintains the luminance level of the pixel cell while the other lines are being scanned. When all the lines have been scanned and all the pixel cells addressed, the process is repeated for the next frame and the picture is refreshed.

Figure 11.11 shows the operation of an TFT-LC cell where G_N is the currently selected gate line and G_{N-1} is the immediately preceding gate line. The TFT gate is connected to the line (or gate) electrode bus, also known as the gate bus and the drain is connected to the data (or column) bus, also known as the source bus. Storage capacitor C_S is connected between the current gate line G_N and the immediately preceding gate line (G_{N-1}). For this reason, C_S is known as C_S-*on-gate*. It forms the drain load for the TFT. The TFT turns fully on when its gate voltage is 20 V and turns off when its gate goes to at least -5 V. To select the pixel cell, a 20 V pulse, V_{SEL}, is

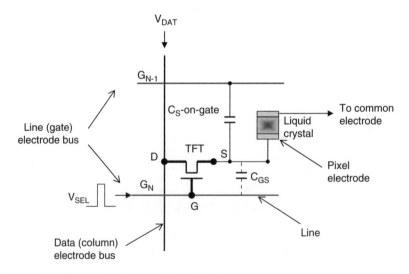

Figure 11.11 *TFT cell equivalent circuit*

applied to the gate. At the same time, data in the form of an analogue positive voltage V_{DAT} is applied to the drain. For peak white, V_{DAT} is 0 V while for pitch black V_{DAT} is a maximum of about 8 V. With the TFT on, the source and drain are shorted and V_{DAT} is applied across the LC. The storage capacitor, C_S-on-gate charges up and this charge is sustained even when the TFT is turned off. This is then repeated for the next line and so on. The main function of C_S is to maintain the voltage across the LC until the next line select voltage is applied when the picture is refreshed. A large C_S can improve the voltage holding ratio of the pixel cell and improve the contrast and flicker. However, a large C_S results in higher TFT load and lower aperture ratio. In determining the value of the storage capacitance, account must be taken of the stray capacitance between the TFT's gate and source, G_{GS} which is effectively in parallel with C_S.

Response time

Response time is one of the few areas remaining where the performance of a traditional CRT still holds an advantage over LCD displays. CRTs have nearly instantaneous pixel response times, but LCDs tend to be much slower. The result is the user might see smearing, motion blur or other visual artefacts when there is movement on the screen.

A pixel's response time is the time it takes a pixel to change state. If it is a *rise-and-fall* response, then it is a measure of the time it takes a pixel to change state from black-to-white-to-black as illustrated in Figure 11.12. More specifically, it represents the pixel ability to change from 10% 'on' to 90% 'on' and then back from 10% 'off' to 90% 'off' again. Originally, this was the standard way of reporting response times of LCD TVs and computer monitors, and was normally listed as a *TrTf* (*time rising, time falling*) measurement. Some manufactures started using a *grey-to-grey* (*GtG*) measurement for LCD response times which is different from TrTf. There are as yet no standards and manufacturers can state any figure that suits them.

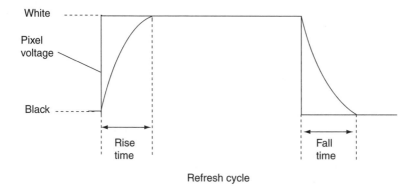

Figure 11.12 *'Time rising, time falling' (TrTf) response of an LC*

One factor that affects the response time is the viscosity of the LC material. This means it takes a finite time to re-orientate its molecules in response to a changed electric field. A second factor is that the capacitance of the LC material is affected by the molecule re-alignment which changes the TFT load and with it the brightness to which the cell ultimately settles.

A good response time starts at around 25 ms with some LCD TV manufacturers claiming a response time as fast as 16 ms or less. Short response times are required for fast moving images such as games. New techniques have been developed to reduce the response time. Such techniques include the use of lower viscosity LC material. Reducing the cell gap thickness is another technique which results in fewer LC material to re-orientate giving a response time as little as 8 ms. Thinner cells make production more difficult with lower yields and hence more expensive. Another technique apply a drive signal for a brief duration in order to give the pixel cells a 'jump start' and then reducing it to the required level as illustrated in Figure 11.13. This technique known as *amplified impulse* provides grey-to-grey transition to be completed up to five times faster than a typical LC display.

For TV images, two techniques have been developed, both attempt to hide the cell transition time. '*Backlight strobing*' involves flickering the backlight off momentarily and the '*black frame insertion*' introduces a black frame during the LC transition. Backlight strobing also helps improve motion blur caused by the '*sample-and-hold*' effect in which an image when held on the screen for the duration of a frame-time, blurs the retina as the eye tracks the motion from one frame to the next. By comparison, when an electron beam sweeps the surface of a cathode ray tube, it lights any given part of the screen only for a miniscule fraction of the frame time. It's a bit like comparing film or video footage shot with low- and high-shutter speeds. This type of motion-blur has come about as manufacturers moved from the traditional resistor type digital-to-analogue converter (DAC) to the much more compact sample-and-hold type. Motion blur originating

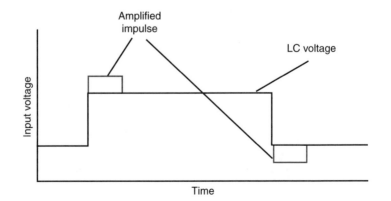

Figure 11.13 *Amplified impulse LC cell drive technique*

from sample-and-hold in the display can become less of an issue as the frame rate is increased.

Polarity inversion

In LC cells, it is the magnitude of the applied voltage which determines the amount of light transmission. Such voltage may be d.c. or a.c. Applying a voltage of the same (d.c.) polarity to an LC cell would cause electroplating of one electrode resulting in what is known as *'d.c. stress'* causing deterioration in image quality. To prevent polarisation (and rapid permanent damage) of the LC material, the polarity of the cell voltage is reversed, a process known as *polarity inversion*. Polarity inversion may be implemented in three different ways: *frame inversion*, *line* (or horizontal) *inversion* and *dot inversion* (Figure 11.14). It will be noticed that line inversion incorporates frame inversion as well since a positive line in one frame becomes negative in the following frame and vice versa.

Unfortunately, it is very difficult to get exactly the same voltage on the cell in both polarities, so the pixel-cell brightness will tend to flicker. This flicker is most noticeable with frame inversion in which the polarity of the whole screen is inverted once every frame resulting in a 25-Hz and 30-Hz flicker for PAL and NTSC, respectively. Flicker may be reduced by having the polarity of adjacent lines using line inversion thus cancelling out the flicker. Better results may be obtained with dot inversion. In this way, the flicker can be made imperceptible for most 'natural' images.

Polarity inversion is carried out by inverting both the pixel cell electrode V_P and voltage at the common electrode, V_{COM} frame-by-frame, line-by-line

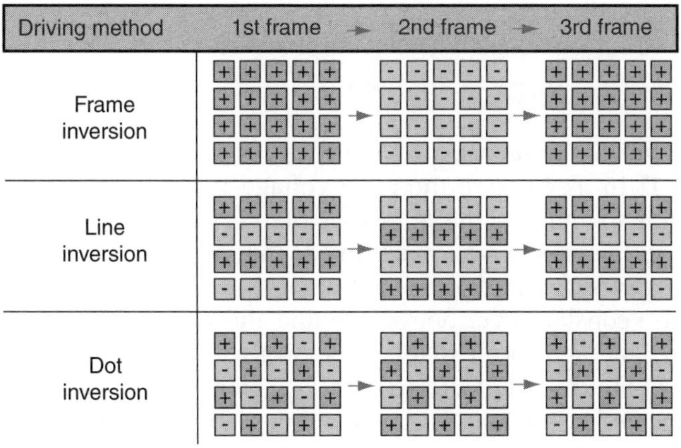

Figure 11.14 *Polarity inversion*

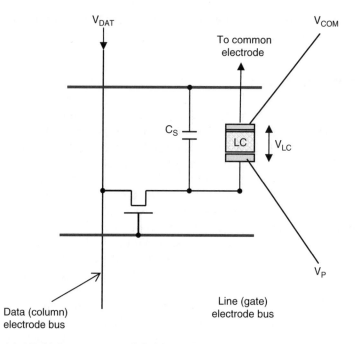

Figure 11.15 *Voltage across LC, $V_{LC} = V_P - V_{COM}$*

or dot-by-dot. Referring to Figure 11.15, the voltage across the LC cell, V_{LC} is the difference between V_P and V_{COM}:

$$V_{LC} = V_P - V_{COM}$$

When both V_P and V_{COM} are inverted,

$$V'_{LC} = -V_P - (-V_{COM})$$

$$V'_{LC} = -V_P + V_{COM}$$

$$V'_{LC} = -(V_P - V_{COM})$$

The line inversion sequence for three lines of a plain white screen is shown in Figure 11.16. For line n, the pixel voltage V_P for all the pixels on the line is high to remove the 90° twist of the LC and remains constant throughout 'line n' since all pixels are at the same luminance level. They are then inverted for the following 'line $n + 1$' and so on. The pixel voltage V_{COM} is also constant over one whole line and inverted for the following line. For a 5-step greyscale display, the pixel values change along the line as shown in Figure 11.17 starting with V_{white} for peak white and $-V_{black}$ for black at the end of the line. This is then inverted and repeated for the next line and so on. V_{COM}, on the other hand, is constant over one line and inverted over the next. The LC voltage V_{LC}, being the difference between

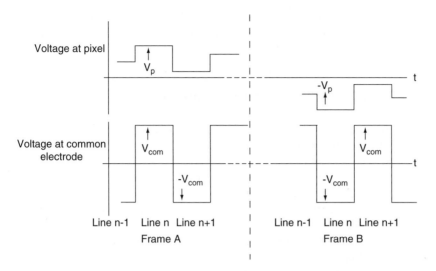

Figure 11.16 *Line inversion pixel voltage sequence for plain white display*

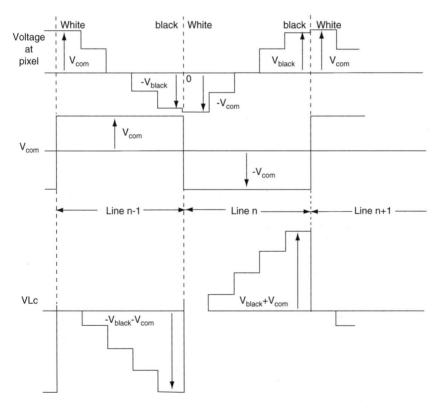

Figure 11.17 *Line inversion waveform for a 5-step greyscale display*

V_P and V_{COM} is that shown in Figure 11.17 inverted every line. For dot inversion, the pixel voltage and the common electrode voltage will invert on successive dots.

Greyscale and colour generation

In general, there are two ways of generating greyscale: digital and analogue. The digital method includes the sub-field coding known as '*time ratio*' technique employed in plasma panels. The other digital technique is the '*area ratio*' in which the pixel cell is divided into several smaller areas, e.g. 6, 8 or 10. Each area is energised independently. The sum total of the luminance of the separate areas is the luminance of the pixel cell. The *analogue method* is by far the simplest technique by which luminance is determined by the instantaneous value of the voltage fed into the data bus and this is the method used in LCD panels.

With the video data being in digital format, the analogue technique requires a digital-to-analogue DAC. Traditionally, a resistor-DAC was used employing a resistor ladder to generate a set of binary-weighted reference voltages, which may be combined to produce an instantaneous voltage that represents the value of the signal. Given eight different reference voltages, $2^8 = 256$ different greyscale voltage values may be produced. In the case of LCD panel drives, two separate sets of reference voltages are needed because of the asymmetrical gamma characteristics encountered during polarity inversion. The resistor chains may of course be incorporated within the DAC chip itself which can also provide the necessary gamma correction. Current LCD panels use linear DACs and sample-and-hold architecture in which the DACs are shared by the column video data thus reducing their number and with it the chip size.

To produce colour, each pixel is divided into three R, G and B sub-pixels or cells in the same way as a plasma pixel. While this arrangement can produce a greyscale, it cannot generate a colour image. To do that, three filters, one for each of the three primary colours are provided on a *masked filter* substrate. The RGB elements of this colour filter line up one-to-one with the pixel cells on the TFT-array substrate. Because the cells are too small to distinguish independently, the RGB elements appear to the human eye as a mixture of the three colours. Practically, any colour can thus be produced by mixing these three primary colours. Pixel cells may be arranged in three formats: *vertical stripe*, *horizontal stripe* and the *triad* or *delta* as illustrated in Figure 11.18. The most popular with manufacturers is the vertical stripe which is sometimes described as $3\,m \times n$ format. Thus, for an SVGA-resolution LCD panel (resolution 800×600), the total number of pixel cells $= 800 \times 3 \times 600 = 1,440,000$ with the number of columns $= 800 \times 3 = 2400$.

The number of colours provided by the LC display is determined in the same way as that for plasma panels namely by the number of combinations

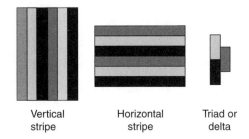

| Vertical | Horizontal | Triad or |
| stripe | stripe | delta |

Figure 11.18 *Colour filter formats*

of R, G and B that may be produced given a particular greyscale levels. In general, the number of colours $= 2^n$ where n is the total number of bits allocated to the video which for a 24-bit video is 24 bits. Given this, the number of different colours $= 2^{24} = 16.78$ million. Alternatively, the same figure may be produced by multiplying the greyscales levels of each colour, namely $256 \times 256 \times 256 = 16.78$ million.

Panel drive

In a TFT-LCD panel, each pixel cell may be addressed by selecting two electrodes: the line (or gate) electrode and the data (or source) electrode with respective line and data drives as illustrated in Figure 11.19. 24-bit video is fed into the LCD controller which addresses each line in turn by a line select voltage which turns all the TFTs on that line on. At the same time, the controller feeds the corresponding RGB video data for the pixel cells of the selected line to the greyscale generator. The greyscale generator consists of a DAC and a sample-and-hold circuit. It converts the digital greyscale value of each cell into an equivalent analogue voltage that when fed into the LC electrode produces the correct luminance. The sample-and-hold ensures the analogue value remains steady long enough for the cell to respond to any change in value. Before going into the cell electrode, the voltage is inverted by a line sync for line inversion (shown) or by a dot frequency pulse for dot inversion. Similarly, the common electrode voltage V_{COM} is inverted by a line sync pulse of a dot-frequency pulse. The backlight which forms a part of the panel assembly is fed by a sine wave in the region of 2000 V from an external d.c.–a.c. converter.

The analogue cell values are fed into the panel using a *tape carrier package (TCP)* in which an LSI chip is installed on a thermostabile film and sealed with plastic as illustrated in Figure 11.20. The chip may be a driver or a shift register or both delivering not an inconsiderable power to the cells, power that must be dissipated using a heat sink. The heat sink normally forms part of the panel frame. A faulty TCP IC would result in

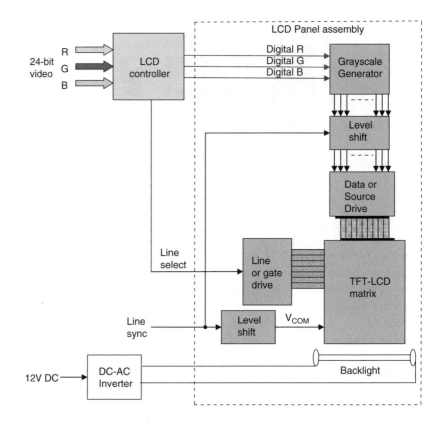

Figure 11.19 *TFT-LCD panel drive*

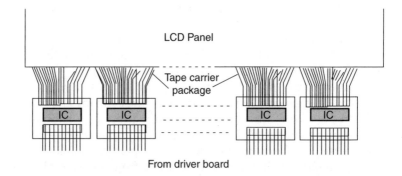

Figure 11.20 *Tape carrier package connection to the LCD panel*

vertical lines along the section of the screen which is fed by the particular TCP as shown in Figure 11.21. This is an un-repairable fault and the whole panel must be changed. The same symptom would be observed if the tape carrier is damaged in any way.

Figure 11.21 *Symptom of a faulty tape carrier package (TCP) chip or tape carrier itself*

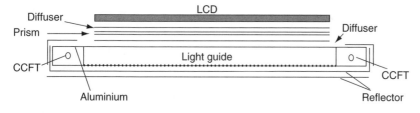

Figure 11.22 *Backlight assembly (light guide type)*

The backlight assembly

There are two types of backlight formats: the *guided type* for screen of 20 in. or less and the *direct type* for larger screen sizes. Both use *cold cathode fluorescent tubes* (CCFT) for their low energy consumption and low cost. The guided type is slim and compact but suffers from complicated structure and low light efficiency. By contrast, the direct type is thick in structure but simpler in structure with high efficiency.

The assembly of the guided type consists of one CCFT on either side of the screen with a *light guide*, and a *reflector* behind the light together with one or more *microprisms* and one or more diffusers in front of it (Figure 11.22). The light guide is based on a methacrylate material and it is used to guide the light through the layers. The reflector is located in the back of panel and it is used to improve the reflection of light. It is made of a material called polyethylene (PET). The diffuser layer has two functions, to diffuse and to collimate the light to make it parallel and uniform. Using the same material PET as the reflector, the diffuser improves brightness by 20%. The light from the diffuser is collimated by the next microprism layer which improves brightness by 40%.

The direct light assembly has more than two lamps as shown in Figure 11.23. The same layers are used as for the guided light type performing the same functions. Direct light diffuser are used to diffuse the light and to avoid seeing the backlight. A transparent indium tin oxide (ITO) sheet connected to ground is used to filter out the noise produced by the lamps.

When servicing a backlight assembly, care must be taken to ensure that the layers are replaced in the correct order as well as orientation otherwise a permanent pattern may appear on the screen. Other precautions that must be observed when servicing the LCD panel include keeping the surfaces clean and avoiding scratching their surfaces, keeping the LCD dry as water could cause electrical shorts and corrosion, avoiding swift temperature changes as dew and ice could cause non-conformance and malfunction, avoid electrostatic discharges by ensuring proper body earth before handing the LCD and do not operate the LCD for a long time with the same pattern as this would cause image persistence which may result in permanent damage.

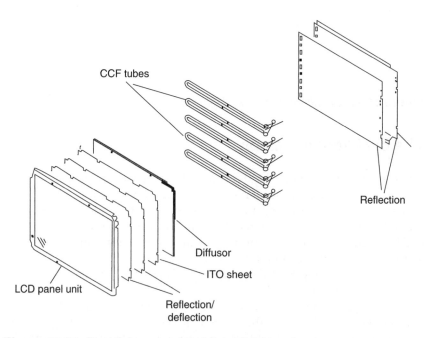

Figure 11.23 *Backlight assembly (direct light type)*

CCFT parameters

The following are the main parameters of a cold cathode fluorescent tube that have to be considered when designing the driving d.c.–a.c. inverter or in replacing the tube:

- *Starting voltage* (typical values 2000–3000 V_{peak}): Also known as the discharge voltage, the starting voltage is the minimum voltage required to ignite, i.e. start the tube. The starting voltage is usually 50% higher than its operating voltage. The starting voltage is the primary parameter which determines the 'end of life' for the tube. The older is the tube, the higher is its starting voltage.
- *Operating voltage* (typical values 2500–3500 V_{p-p}): This is the voltage across the tube when it has been lit. It is a key parameter in the design of the d.c.–a.c. inverter.
- *Tube current*: The current through the tube determines, to a large extent, its brightness. It also indirectly determines the tube's useful life. In general, the tube's life is the square of its current. If the current increases by 20% above its normal value, its life span decreases by 40%. Higher than normal current also results in excessive heat.
- *Frequency* (typical value: 40–60 kHz): Frequency generally has no effect on the brightness of the tube, its efficiency or its useful life. However, it does have an impact on the compatibility between the tube, the display itself and the graphic information displayed by the tube.
- *Waveform*: An undistorted current and voltage sinusoidal waveforms are required to avoid radiated electric noise that may impact on the system and surrounding environment introduced by a distorted sine wave. Although the a.c–d.c. inverters produce pure sine waves, the dynamic nature of the tube distorts both the voltage and the current waveforms.
- *Impedance*: A high impedance is presented by the CCFT assembly which is in the region of 50–70 kΩ.

Tube brightness control

The CCFT requires a sine wave with an amplitude of few thousand volts and a frequency in the region of 50–70 kHz. This is provided by a d.c.–a.c. converter which is essence an oscillator. While maximum brightness may be obtained by turning the tube fully on, in most application, there is a need to reduce the lamp's brightness. There are two basic methods for dimming the CCFT. The first simply reduces the tube current either directly or indirectly by reducing the voltage applied to it. The second method maintains a constant current but turns the lamp on and of to control its brightness (Figure 11.24). If the inverter is turned on for longer periods than it is off, a brighter light is produced and vice versa. This technique employs a *pulse-width modulated* (PWM) waveform to turn the inverter oscillator on and off. A typical tube drive and control signals are shown in Figure 11.25.

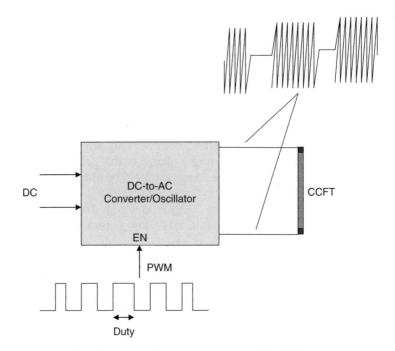

Figure 11.24 *Cold cathode fluorescent tube (CCFT) brightness control*

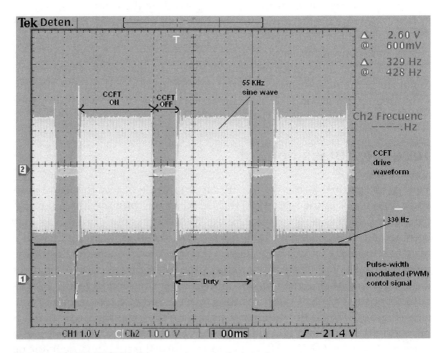

Figure 11.25 *CCFT drive and control signals*

The frequency of the PWM waveform has to be chosen carefully to avoid interaction with the frame rate. Typical values are 270 and 330 Hz and in the case of PC monitors, it is varied with the frame rate itself to avoid interference with the graphics.

The d.c.–a.c. inverter

Essentially, the d.c.–a.c. inverter is a tuned collector oscillator (Figure 11.26). When power is switched on, the transistor conducts feeding energy into inductor L. When the inductor saturates, current ceases and the back e.m.f. forces the current to reverse. Energy in L is now transferred to C. When C is fully charged, charging current ceases causing an opposite back e.m.f. across the inductor and the capacitor discharges into L with its energy transferring back to L until the inductor saturates and so on. The output across the secondary of the transformer is a pure sine wave.

The basic elements of a d.c.–a.c. inverter are illustrated in Figure 11.27. Capacitor C_P is the primary tuned capacitor which resonates with the inductance of the primary winding of transformer T1. Capacitor C_S is connected in series with the tube to ensure constant current operation. At the frequency of operation, the impedance of the tube assembly together with the ballast capacitor is very high making the inverter act as a constant current source. The output is a sine wave with a slight distortion caused by the reactance of the tube. Because of the very high impedance, measuring devices such as a DVM or a CRO would load the output so much as to render the readings almost meaningless. A

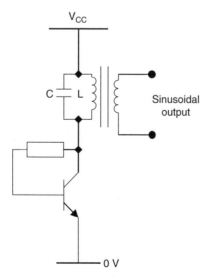

Figure 11.26 *Tune-collector oscillator as a d.c.–a.c. inverter*

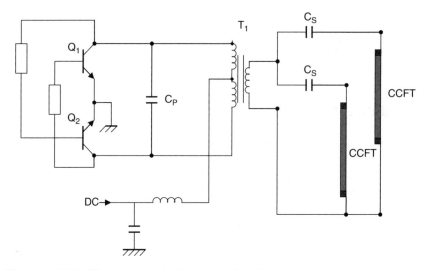

Figure 11.27 *The essential elements of a d.c.–a.c. inverter*

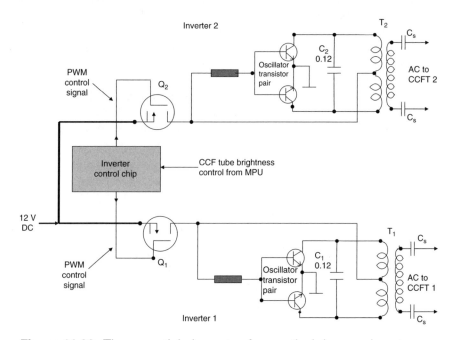

Figure 11.28 *The essential elements of a practical d.c.–a.c. inverter*

current probe should be used to observe the shape of the waveform on an oscilloscope.

Figure 11.28 shows practical backlight inverter driving two CCFTs together with the control chip as used by a Panasonic 15-in. LCD receiver.

The control chip provides the PWM signal to drive two separate inverters, one for each CCFT. The control chip itself is controlled by a signal from the microprocessor. For inverter 1, a centre-tapped step-up transformer is used to feed $700\,V_{\text{rms}}$ to drive CCFT1. The tuned circuit is formed by C1 and the primary of the transformer and a pair of transistors is used to oscillate back-to-back. The d.c. power to the oscillator is obtained from switching transistor Q1. While Q1 is on, oscillation takes place and the tube lights up. However when Q1 is off, the oscillator turns off and with it the tube itself. Switching transistor Q1 is controlled by a PWM signal from the control chip. The width of the pulse controls the ON/OFF ratio of Q1 and with it the brightness of the tube. For inverter 2, Q2 is the switching transistor and C2 is the tuning capacitor. The tuning capacitors may be recognised by their non-nominal values, in this case $0.12\,\mu\text{F}$. Capacitors marked C_s are the series capacitors that ensures the lamp presents a very high impedance to the inverter. It is normal to include over voltage/over current protection as well as a panel enable from the microprocessor controller.

Lamp error detection

Invariably, LC displays go into standby if one or more lamps fail to light up. This may be caused by a actual malfunction lamp or a faulty inverter circuit. A typical lamp error detection circuit is shown in Figure 11.29. When the lamp is functioning normally, lamp current flows through R1 and turns D1 on feeding a positive voltage to the inverter. The output from the inverter is negative which turns D2 off producing an zero error voltage. If the lamp fails to light up, current through R1 ceases, D1 turns off, inverter output is positive and D2 turns on resulting in a positive ($\sim 4\,V$)

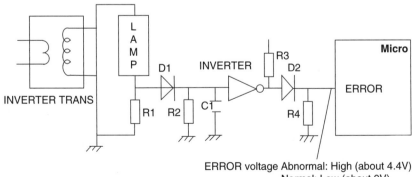

ERROR voltage Abnormal: High (about 4.4V)
Normal: Low (about 0V)

Figure 11.29 *Lamp error detection circuit*

error voltage which will interrupt the microcontroller and the set will be forced into standby. Smoothing is provided by R2C1.

Adaptive transmissive scaling

Light, which is emitted from the LCD panel, is a function of two parameters: the intensity of the backlight and the *transmissiveness* of the LC cells. The latter is the amount of polarising twist the LC imposes on the light passing through it. Therefore, by carefully adjusting these two parameters, one can achieve the same perception of brightness at different values of the backlight intensity and the LCD twist. Since the changes in energy consumption of the backlight lamp are higher than that of the LCD cells, energy may be saved by simply dimming the backlight and increasing the LC cells' transmissiveness. This is known as *brightness enhancement* which may be carried out dynamically, frame-by-frame. Apart from power savings, a dramatic improvement in brightness and contrast ratio as well as black reproduction is achieved.

Figure 11.30 shows a block diagram for an adaptive transmissive scalar. The video information is assessed by the brightness detector to obtain three-frame parameters: average picture level (APL) and maximum and minimum luminance. These parameters are then used to establish the optimum combinations of pixel voltage and the backlight brightness. Signals are sent to the video gain control and the lamp current control as shown.

Transmissive scaling invariably introduces distortion which may be minimised by the use of a histogram. A histogram is used to calculate or estimate the distortion produced by transmissive scaling and if it is above a specified threshold, both the backlight and the LC pixel values are readjusted to bring the distortion down to an acceptable level.

LCD panel faults

The first step in fault finding on LCD panels is to ascertain if the fault is a panel malfunction, in which case it has to be replaced, or if the cause of the fault is outside the panel in which case a repair is possible. Some of the symptoms point clearly to a panel fault, others may be ambiguous. One of the main symptoms of a faulty panel is pixel defect. This may be a single pixel, several pixels or cluster of pixel failure in the same way as pixel failures in a plasma panel shown in Figure 10.34 in the previous chapter. Again, in the same way as plasma displays, manufacturers allow for a certain number of pixel defects before the panel is rendered obsolete. The display panel is divided into a number of areas and the number of bad pixels permitted depends on the area. Other classic symptoms pointing to faulty LCD panel are shown in Figure 11.31. A horizontal line across the screen may also be caused by a bad connection between gate driver board and the panel. The 'negative picture' effect in Figure 11.32 is also a result of panel failure.

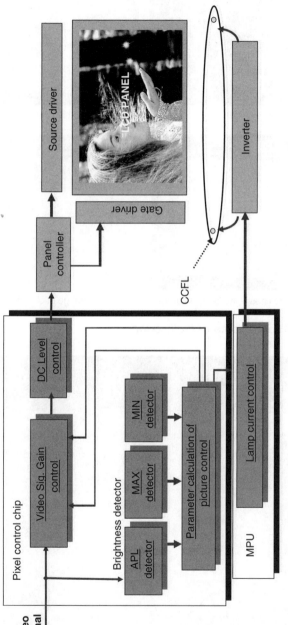

Figure 11.30 *Adaptive transmissive scaling*

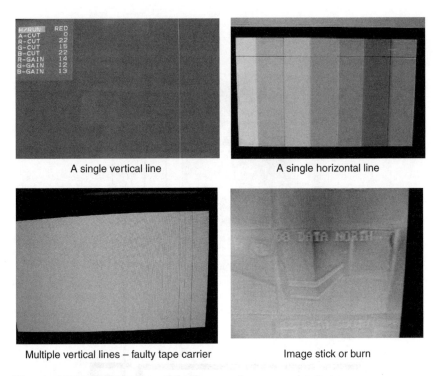

| A single vertical line | A single horizontal line |
| Multiple vertical lines – faulty tape carrier | Image stick or burn |

Figure 11.31 *Symptoms of faulty panel*

Figure 11.32 *'Negative picture' effect due to panel malfunction*

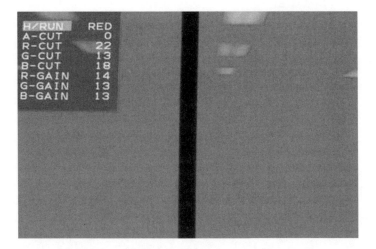

Figure 11.33 *Source or column drive fault*

Figure 11.34 *Symptom of a line gate driver fault*

Drive faults

The classic symptom for a source (or column) drive is a vertical band across the screen as shown in Figure 11.33. A horizontal band (Figure 11.34) would indicate a faulty line scan drive.

12 DLP and SED

Plasma and LCD television receivers have long enjoyed the singular billing as the 'Big Thing' in consumer electronics. Recently, two new display technologies have begun to make some headlines of their own, namely *digital light processing (DLP)* and *surface-conduction electron-emitter display (SED)* technologies. DLP is a video projection technology that has recently been available as rear projection large-screen domestic television receivers. The technology took decades to become a viable technology for the ever expanding consumer market. It was developed by Larry Hornbeck at Texas Instruments (TI) in the 1970s, perfected in the 80s and finally introduced to the public in 1996. Since then, DLP TV has started making waves in the fixed-pixel display market by meeting the surging demand for less expensive – though no less capable – large-screen TVs.

SED uses surface-conduction electron emission to excite a phosphor coating of individual pixels, the same basic concept of the traditional cathode ray tube (CRT).

Principles of DLP display

Central to the DLP display is the *digital micromirror device (DMD)* developed by TI. The DMD is a thumbnail-size semiconductor light switch. It consists of an array of millions of microscopic-size mirrors, each mounted on a hinge structure so that it can be individually tilted back and forth. Figure 12.1 shows the basic components of a simple DLP system composed of a light source and a projection lens. Light from the lamp is reflected off the micromirrors and directed towards the projector lens if the mirror is tilted in one direction and away from the lens and towards a light absorber if tilted the opposite way. In the diagram, the two end mirrors are tilted so that the reflected light goes through the lens to be projected on an external surface as two bright square dots. These two mirrors are said to be 'on'. The middle mirror is tilted in the opposite direction and its reflection avoids the lens to be absorbed by a light absorber. It is said to be 'off'. The absence of light reflection from the middle mirror appears as a dark square dot on the screen. The top view of the three micromirrors is shown in Figure 12.2.

Now imagine thousands of these tiny micromirrors, arranged in a matrix. The result is an image with a resolution determined by the number of micromirrors with each micromirror corresponding to one pixel. Thus a VGA-resolution image requires a DMD matrix of 640×480 and an XGA a matrix of 1024×768 and so on. To produce a moving image, the

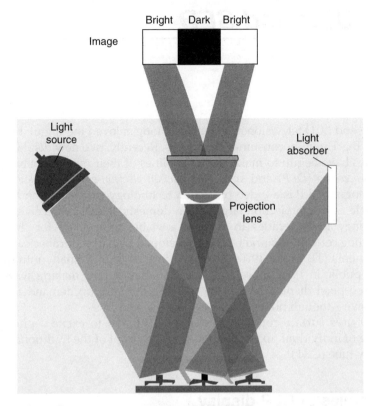

Figure 12.1 *Component parts of a simple DLP*

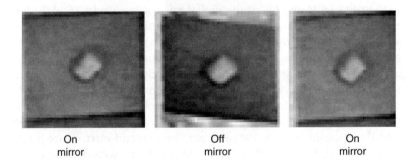

| On mirror | Off mirror | On mirror |

Figure 12.2 *Top view of the three micromirrors shown in Figure 12.1*

micromirrors have to be addressed and set to one position or the other and refreshed at every frame.

Greyscale generation

The simple operation described above does not provide for shades of grey as the operation of the micromirrors is purely digital. A mirror can only be turned 'on' for full brightness or 'off' for no brightness or black. To produce

Table 12.1 *Bit plane weighting for an 8-bit pixel value*

Pixel value bit	Portion of frame period
Bit 0 (LSB)	1/255
Bit 1	2/255
Bit 2	4/255
Bit 3	8/255
Bit 4	16/255
Bit 5	32/255
Bit 6	64/255
Bit 7 (MSB)	128/255

greyscale light has to be modulated using a technique known as *binary-weighted pulse width modulation* (PWM). The binary-weighted PWM is a time ratio technique not dissimilar to that used for plasma panels, in that the frame period is divided into a number of binary-weighted intervals, also known as *bit intervals*. The weighting for each interval reflects the binary code of the pixel value, a technique known as *bit plane weighting* in which the pixel's least significant bit (LSB) consumes $1/(2^n-1)$ and the LSB +1 bit consumes double that and so on where n is the number of bits used to describe the luminance or value of the pixel. The most significant bit (MSB) consumes $(n-1)/(2^n-1)$ of the frame interval. The human eye integrates the pulsed light to an average intensity. The greyscale perceived is proportional to the proportion of time the mirror is 'on' during the frame refresh cycle. Table 12.1 lists the bit plane weighting of an 8-bit pixel value. As can be seen, the MSB addresses (i.e. tilts and holds) the mirror in position for 'half' (128/255) the frame period, 10.04 ms for PAL and 8.38 ms for NTSC. The MSB-1 bit addresses the mirror for a 'quarter' (64/255) of the frame period and so on up to the LSB which addresses the mirror for the smallest bit interval, namely 1/255th.

A pixel value bit of 1 would turn the mirror 'on' and a 0 would tilt the mirror in the opposite 'off' position. For maximum light intensity, the pixel value would be 11111111 (all 1s) which sets the micromirror 'on' throughout the frame period. For minimum intensity or black, all bits would be 0 and the micromirror would be 'off' throughout the frame. Other combinations of 0s and 1s would produce different shades of grey, a total of $2^8 = 256$.

Figure 12.3 shows a typical binary-weighted PWM frame period for an 8-bit pixel value showing the bit intervals as well as the 'on' and 'off' sequence for two different light intensities, 60 and 35.5%.

Bit splitting

Current DLP systems are either 24-bit colour (8 bits or 256 grey levels per primary colour) or 30-bit colour (10 bits or 1024 grey levels per primary colour). In the simple binary PWM addressing, spatial and temporal

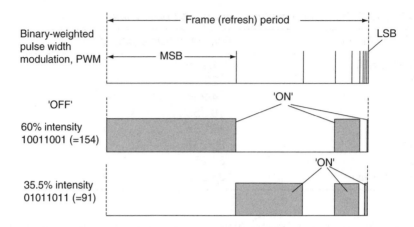

Figure 12.3 *Binary-weighted pulse width modulation*

artefacts can be produced because of imperfect integration of the pulsed light by the viewer's eye. These artefacts can be reduced to negligible levels by what is known as *'bit-splitting'*. In this technique, the longer duration bits are subdivided into shorter durations, and these split bits are distributed throughout the video frame time. DLP displays combine PWM and bit splitting to produce analogue-quality projection systems.

DMD structure

The DMD chip consists of a *monolithically integrated micro-electronic mechanical system (MEMS)* superstructure fabricated over a CMOS SRAM memory as illustrated in Figure 12.4. The MEMS superstructure consists of three layers: the micromirror itself, the *yoke* and *hinge* layer and the base referred to as *metal-3*. Each mirror is associated with an individual SRAM memory cell, the content of which determines the tilt of the mirror: 'on' or 'off'. A schematic drawing of the construction of an individual MEMS mirror superstructure is shown in Figure 12.5.

The layers of the superstructure are separated by a microscopic air gap. The air gaps allow the structure to rotate freely about two compliant torsion hinges. The mirror is rigidly connected to an underlying yoke. The yoke, in turn, is connected by two thin torsion hinges to the underlying substrate. The address electrodes are connected to the complementary sides of the underlying SRAM cell while the yoke and mirror are connected to a bias bus fabricated at the metal-3 layer. The bias bus interconnects the yoke and mirror to a bond pad at the chip perimeter. An off-chip driver supplies the bias waveform necessary for proper digital operation.

The DMD mirrors are 16 µm square and made of aluminium for maximum reflectivity. They are arrayed 17 µm apart leaving a gap between

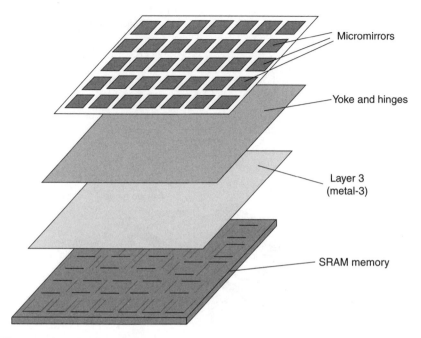

Figure 12.4 *DMD structure*

them of only 1 µm to form a matrix having a high aperture ratio known as *fill factor* of about 90%. The high fill factor produces high luminance efficiency. Furthermore, the high fill factor results in a 'seamless' image avoiding the pixelisation or *'screen door effect'* experienced with large-screen LCD panels and projectors.

DMD operation

At the beginning of each frame, the value of each pixel is fed, bit by bit into the corresponding SRAM cell starting with the MBS. All mirrors remain in the MSB state for half of a frame time. The next MSB is then loaded and held for one quarter and the next bit for one eighth of a frame time and so on until all the bits have been loaded. The process is then repeated for the next frame and so on.

The logic state of a bit in a SRAM cell is fed to an embedded driver to create an electrostatic torque between the mirrors and address electrodes. This works against the restoring torque of the hinges producing a rotation of the yoke and mirror in the positive or negative directions depending on whether the content of the cell is a 1 or a 0. The mirror and yoke rotate until the yoke comes to rest (or lands) against mechanical stops that are at the same potential as the yoke. Because geometry determines the rotation angle, as opposed to a balance of electrostatic torques, the rotation angle is precisely determined.

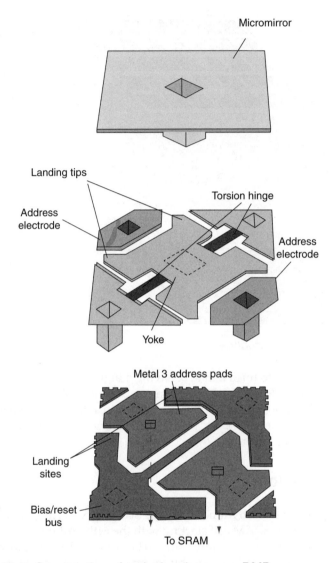

Figure 12.5 *Construction of a single mirror on a DMD*

Once the memory array has been updated, all the mirrors in the array are released simultaneously and allowed to move to their new positions by a bias address voltage. They stay latched in that position by applying a higher bias latch voltage. This prevents the mirrors from responding to changes in the memory while the memory is being written with updated video data. While the mirrors are latched, the memory cells are updated with new data.

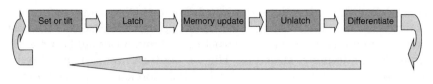

Figure 12.6 *Addressing the DMD*

DMD addressing cycle

The DMD addressing scheme takes advantage of the 'mechanical latching' feature of the DMD. If a bias (latch) voltage in excess of the address voltage is applied to the array of DMD mirrors after the '1' or '0' address voltages are set, the mirrors will stay latched (tilted) in the selected state even if the address voltage changes. Only if the bias (latch) voltage is removed will the mirrors be free to respond to any changes in the address voltage.

The complete address cycle is shown in Figure 12.6. The mirrors are *set* or tilted into position determined by the contents of their corresponding cells and stay in that position for the bit interval. To keep them into that position, a bias (*latch*) voltage is applied. While the mirrors of the array are latched, the underlying memory array is *refreshed* or *updated* by the next bit of the pixel value for the next bit interval. At the end of the bit interval, the latch bias is turned off to release mirrors allowing them to rotate if necessary. However, the stickiness of the mirrors keeps them in position. The next phase, the *differentiation phase*, is to identify which mirrors are to remain in the same state and which are to cross over to a new state for the next bit interval. The latter are then released by applying a retarding field to the yokes and mirrors. Having done that, the rotationally separated mirrors are set by applying an address bias voltage and rotating them to their new states. The others remain in their previous states. A bias latch voltage is then applied and so on.

The bias voltage has three important functions: first, it produces a bistable condition to minimise the address voltage requirement. In this manner, large rotation angles can be achieved with conventional 5 V CMOS. Second, it electromechanically latches the mirrors so that they cannot respond to changes in the address voltage until the mirrors are reset. The third function is to reset the pixels so that they can reliably break free of surface adhesive forces and begin to rotate to their new address states.

Typical mirror tilt is $\pm 10°$ from a plane parallel to the underlying silicon substrate, so the illumination beam is incident on the DMD 20° from the perpendicular to the plane of the silicon. The contrast ratio of a display is limited to a large extent by light which, regardless of the 'on' or 'off' position of the mirrors, is diffracted from the DMD, enters the projection lens. Such light may be diffracted from the edges of the mirrors, torsion beams, posts and the underlying circuit structure. Decreasing the amount of diffracted light improves the black level and thus increases the contrast ratio. To do this requires a mirror structure that minimises light diffraction from the underlying structure. Higher contrast ratios may also be achieved by limiting the aperture of the projection lens, thus discriminating preferentially in favour of reflected rather than diffracted light. However, this results in a reduction in the light throughput of the system.

Multiplexed addressing

The traditional addressing technique described above suffers from two drawbacks. The first is a result of the manner in which the SRAM chip is refreshed and the other is the relatively low throughput of the manufacturing process. In the traditional addressing technique, the MSB of a new frame must be loaded into the DMD SRAM during the LSB bit interval of the preceding frame as shown in Figure 12.7. As the LSB bit interval is the shortest interval in a frame, the bit rate associated with loading the MSB is at a peak. The lowest bit rate is when the MSB-1 data is loaded during the long MSB bit interval. The DMD device and its drivers must therefore have high bandwidths to cope with the highest bit rate even though the data rate is essentially zero throughout the 50% of the frame time represented by the MSB time period and the average bit rate is relatively low. The second drawback is the difficulty in achieving defect-free SRAM substrate reducing the manufacturing throughput. DMD throughput would be improved if the SRAM cell count is reduced below the pixel counts, i.e. if one cell can service more than one pixel. To increase throughput and improve the bandwidth requirements multiplex addressing was developed.

In the traditional addressing technique, the mirrors share a single bias voltage which is applied to all the mirrors simultaneously. The mirrors are latched, released and rotated simultaneously. In multiplex addressing, the DMD mirror array is divided into a number of separate sections, normally 16, each with its own separate source of bias voltage. The mirrors are addressed section by section. In this way, 16 mirrors, one from each section can be addressed by the same SRAM cell thus reducing the number of SRAM cells required.

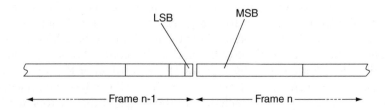

Figure 12.7 *Traditional addressing*

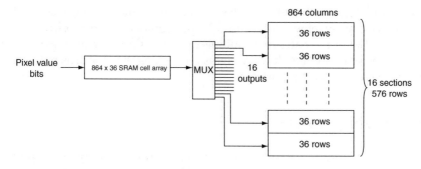

Figure 12.8 *Multiplex addressing*

As an illustration, consider a DMD with a resolution of 864×576 divided into 16 horizontal sections of 864×36 as shown in Figure 12.8. With multiplex addressing, a SRAM cell array of only 864×36 would be required to address each section in turn. The memory would then be refreshed 16 times to update just one bit of the frame. This addressing scheme requires a change in the simple bit plane sequence described above. Figure 12.9 gives an illustration of multiplexed addressing sequence for a system with a 5-bit (0–4) pixel depth and a DMD with four horizontal sections labelled A, B, C and D. It will be noticed that the pixel bits are not loaded in the normal binary sequence and that they appear at a different time throughout the video frame in each of the four sections of the device. With the X-axis divided into LSB bit intervals, it can be seen from the diagram in Figure 12.9 that updating bit A4 is carried out in one LSB interval. Similar is the case for B3, C2 and D3. They are followed by B0 which has three LSB intervals to be updated and so on. One section or another of the device is updated with one bit or another almost constantly. The peak data rate for the DMD and its drivers is thus reduced almost to the average data rate. Furthermore, the DMD memory requirement has been reduced to a quarter of the pixel count.

This multiplexing, or time sharing, of memory among groups of pixels within the display implies that particular bit weightings appear on different

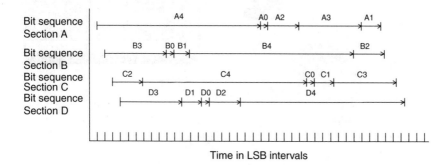

Figure 12.9 *Multiplexed addressing bit sequence*

groups of lines within the display at different times during each video frame. This arrangement introduces the possibility of fixed-pattern artefact associated with the pattern of multiplexed lines. Careful rearrangement of the temporal location of the bit weights goes a long way towards mitigating the visual effects of these artefacts.

Generating colour

Colour DLP optical systems have been designed in a variety of configurations distinguished by the number of DMD chips (one, two or three) used. The *one-chip* and *two-chip* systems rely on a rotating colour disk to time multiplex the colours.

The one-chip configuration is used for lower brightness applications and is the most compact. Two-chip systems yield higher brightness performance but are primarily intended to compensate for the colour deficiencies resulting from spectrally imbalanced lamps (e.g. the red deficiency in many metalhalide lamps). For the highest brightness applications, three-chip systems are required.

In the one-chip configuration (Figure 12.10), white light is focused onto a colour wheel filter system. The colour wheel spins causing a sequence of red, green and blue light to shine onto the DMD mirrors to produce the red, green and blue frames, bit interval by bit interval. The DMD SRAM memory cells are thus fed with red, green and blue pixel values, bit by bit in the same sequence as the spinning colour. The eye integrates the sequential images and a full colour image is seen. The data as well as the mirrors' transition rates are increased by three times as the DMD cells are

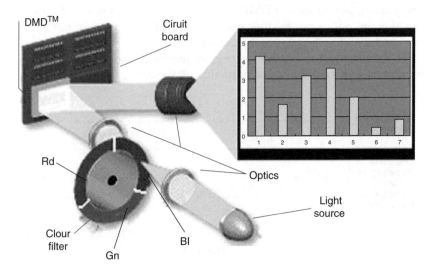

Figure 12.10 *A single-chip DLP system*

refreshed by the full pixel depth three times for each video frame. This was made possible with improvements in the optical and mechanical switching times of the DMD. Single-chip DLP systems made small size projectors a possibility. The drawback is a reduced brightness level compared with the three-chip type.

The rainbow effect

If there is one single issue that people point to as a weakness in DLP, it is that the use of a spinning colour wheel to modulate the image has the potential to produce a unique visible artefact on the screen commonly referred to as the *'rainbow effect'* (Figure 12.11). This is simply due to colours separating out in distinct red, green and blue because of the sequential colour updating from the wheel (three-chip DLP projectors have no colour wheels, and thus do not manifest this artefact). Basically, as the colour wheel spins the image on the screen is either red, or green, or blue at any given instant in time, and the technology relies upon your eyes not being able to detect the rapid changes from one to the other. Unfortunately some people can see it. Not only can some see the colours break out, but the rapid sequencing of colour is thought to be the culprit in reported cases of eyestrain and headaches. Since LCD projectors always deliver a constant red, green and blue image simultaneously, viewers of LCD projectors do not report these problems.

How big of a deal is this? Well, it is different for different people. Most people cannot detect colour separation artefacts at all. However, for some

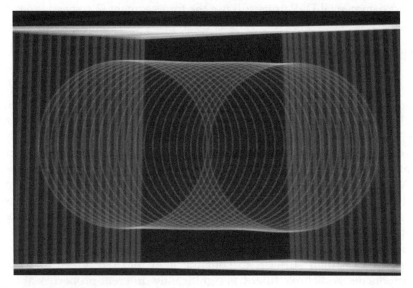

Figure 12.11 *The rainbow effect*

who can see the rainbow effect, it is so distracting that it renders the picture literally unwatchable. Others report being able to see the rainbow artefacts on occasion, but find that they are not particularly annoying and do not inhibit the enjoyment of the viewing experience.

TI and the vendors who build DLP-based projectors have made strides in addressing this problem. The first-generation DLP projectors incorporated a colour wheel that rotated 60 times per second, which can be designated as 60 Hz, or 3600 RPM. So with one red, green and blue panel in the wheel, updates on each colour happened 60 times per second. This baseline 60 Hz rotation speed in the first-generation products is known as a '1×' rotation speed.

Upon release of the first-generation machines, it became apparent that quite a few people were seeing rainbow artefacts. So in the second-generation DLP products the colour wheel rotation speed was doubled to 2× at 120 Hz or 7200 RPM. The doubling of the colour refresh rate reduced the time between colour updates, and so reduced or eliminated the visibility of colour separation artefacts for most people.

Today, as noted above, many DLP projectors being built for the home theatre market incorporate a six-segment colour wheel which has two sets of red, green and blue filters. This wheel still spins at 120 Hz or 7200 RPM, but because red, green and blue are refreshed twice in every rotation rather than once, the industry refers to this as a 4× rotation speed. This further doubling of the refresh rate has again reduced the number of people who can detect them.

For the large majority of users the six-segment, 4× speed wheels have solved the problem for home theatre or video products. Meanwhile, due to the higher lumen output requirements for business presentation use, most commercial DLP units still use the four-segment, 2× speed wheels.

The three-chip DLP

In the three-chip configuration, three separate DMDs are used, one for each colour in the arrangement shown in Figure 12.12. Light from a metal halide or xenon lamp is collected by a condenser lens. The light must now be separated into its three primary components red, green and blue. This is carried out by a set of colour-splitting and colour-combining prisms. Furthermore, these light waves must be directed at 20° relative to their DMD chip. This must be accomplished in a way that eliminates mechanical interference between the illuminating and projecting waves. This task is performed by a *'total internal reflection'* (*TIR*) prism which is interposed between the projection lens and the DMD *colour-splitting/colour-combining* prisms. The colour-splitting/colour-combining prisms use *dichroic filters* deposited on their surfaces to split the light into red, green and blue components. A dichroic filter has significantly different properties at two different wavelengths that can be used to selectively pass light of a small

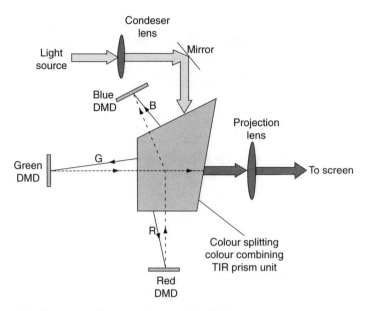

Figure 12.12 *An outline of a three-chip DLP system*

range of colours only. The red and blue prisms require an additional reflection from a TIR surface of the prism in order to direct the light at the correct angle to the red and blue DMDs. Light reflected from the 'on' mirrors of the three DMDs is directed back through the prisms and the colour components are recombined. The combined light then passes through the TIR prism and into the projection lens.

Principles of SED panel

The SED relies on electrons striking a phosphor-coated screen to produce light. This is the same technology as that used by the traditional CRT. The difference is that while the CRT has an electron emitter or gun that scans the phosphor-coated surface of the screen, the SED panel has an array of electron emitters, one for each pixel as illustrated in Figure 12.13.

The display consists of two flat piece of glass, sealed with a vacuum in between. One of them is covered with electron emitters, while the other is covered with phosphor. The vacuum in between the glasses is only half an inch thick, which allows for extremely thin monitors. Each electron emitter is matched up with a pixel on the screen.

The electron emitters, at the heart of the SED, are characterised by an extremely narrow gap measuring only a few nanometres in width known as a *nanogap*. It is formed between two electrodes on the electron-emitting layer. When a voltage of approximately 10 V is applied to the nanogaps, what is known as *tunnelling effect* takes place with electrons emitted from

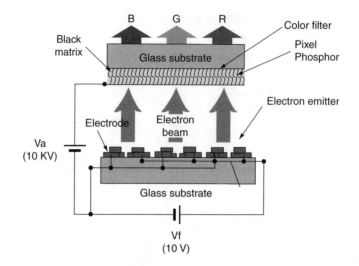

Figure 12.13 *The SED panel construction*

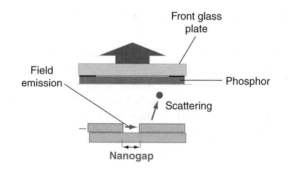

Figure 12.14 *Electron tunnelling*

one side of the nanogap as shown in Figure 12.14. Some of these electrons are scattered at the opposite electrode and accelerated by the roughly 10 kV applied between the front and back glass substrates, causing light to be emitted when they collide with the phosphor-coated glass substrate.

SEDs have the same advantages in terms of fast video response and high contrast. Unlike CRTs, however, they do not require electron beam deflection, so they can produce high-definition images with low distortion for high overall image quality.

Cannon, the SED manufacturer claims lower power consumption than equivalent CRT, LCD and plasma panels. SEDs combine a slim form factor and contrast ratios of LCDs and plasma displays with the superior viewing angles, black levels and pixel response time of CRTs.

13 Television receivers CRT-type

In the previous chapters, we described the various types of display devices from the traditional CRT to the plasma, LCD and the latest DLP rear-projection. The next few chapters will consider the technology involved in receiving and processing video signals in preparation for display.

There are several ways in which a video signal may be received for processing. These include the traditional analogue TV broadcasts as well as the current digital television. Video signals may also be received from a variety of video equipments such as DVD players, VCRs and set-top-boxes via a variety of connections such as SCART, S-video and HDMI. This chapter will cover terrestrial analogue television using a CRT display unit.

The analogue TV receiver

The basic functional units of an analogue terrestrial television receiver are shown in Figure 13.1.

The tuner selects the UHF carrier frequency for the TV channel as chosen by the user and converts it to an *intermediate frequency* (IF) of 39.5 MHz. The modulated IF is then amplified through several stages of amplification and demodulated to obtain the original composite video, blanking and sync (CVBS) signal. The CVBS signal is then separated into its three component parts: video, sound and sync.

The *video selection switch* (sometimes inappropriately called video processor) selects one of the video signal for further processing as chosen by the user. The video in the form of CVBS from the video selector is fed to the *colour decoder* to reproduce the original RGB (or YUV) signals which following the RGB matrix, are fed to the CRT via individual RGB amplifiers. The 6-MHz *sound inter-carrier* is taken off at the video detector stage using a *surface acoustic wave* (SAW) filter. The FM sound signal is detected, amplified and fed into the loudspeaker via the *sound select* chip. The sync pulses are clipped from the video information at the video output stage, separated into line and field, and taken to the appropriate timebase. After amplification, line and field pulses are used to deflect the electron beam in the horizontal and vertical directions via a pair of scan coils.

Automatic gain control (a.g.c.) is employed to ensure that the output of the IF stage remains steady irrespective of changes in the strength of the received signal. The a.g.c. performs three basic functions in a TV receiver.

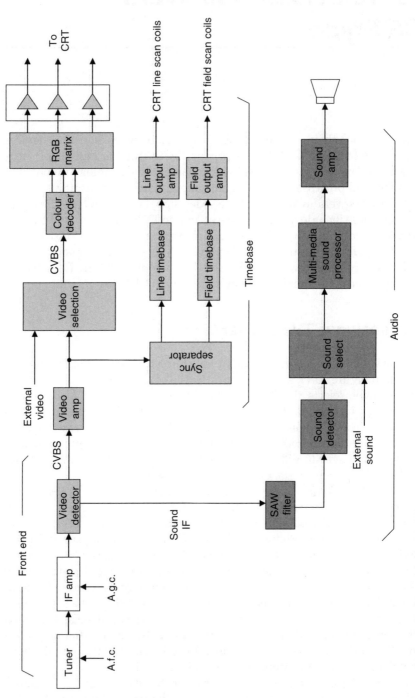

Figure 13.1 *Analogue terrestrial TV receiver*

First, it enables the switching over from a strong channel to a weaker one, or vice versa, without having to adjust the receiver. Second, it avoids overloading the RF/IF amplifying stages which would introduce severe distortion. Third, it attempts to reduce the flutter caused by reflections of transmitted signals from moving objects such as airplanes. The effectiveness of flutter reduction depends on the time constant of the a.g.c. circuit. *Automatic frequency control (a.f.c.)* is sometimes used to keep the IF stable at 39.5 MHz.

Apart from providing the drive for the line scan coils, the *line output* stage also provides the extra high tension (e.h.t.) for the c.r.t. by the use of an overwind at the line output transformer (LOPT) as was mentioned in Chapter 9. The line output stage is also used to provide other stabilised d.c. supplies for the receiver.

The front end

The component parts of the front end of an analogue terrestrial television are shown in Figure 13.2. It consists of the tuner and the IF amplifier stage. The function of the tuner is to select a TV channel frequency, amplify it and convert it into an IF for further amplification by the IF stage. The tuner must be capable of selecting any channel from bands IV and V, and it must provide sufficient RF amplification with good *signal-to-noise ratio (SNR)* and minimal frequency drift. One or more stages of RF amplification are therefore necessary before the *mixer-oscillator* stage. A high-pass filter is normally used at the input to the RF amplifier to produce a correctly shaped response curve. The mixer-oscillator changes the tuned RF to a common IF of 39.5 MHz. The tuner unit is built inside a metal case to screen it from outside RF interference. Further screening is also provided between various stages of the unit by metal walls. These inner walls which form part of the tuned circuits prevent unwanted coupling between one

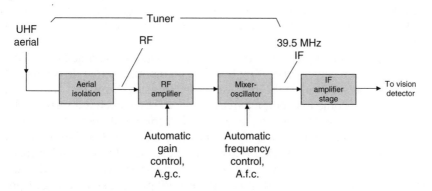

Figure 13.2 *Terrestrial TV front end*

Figure 13.3 *Typical front end unit*

compartment and another. But when a coupling capacitance is needed, a small hole or slot may be cut in the dividing wall.

Both the RF amplifier and RF oscillator use LC tuned circuits. Given the resonant frequency of a tuned circuit $f = 1/2\pi\sqrt{LC}$ and the very high carrier frequencies (370–862 MHz) to which the tuned circuits must resonate, the values of L and C has to be so small as to make it impractical to use normal inductors and capacitors. Instead, transmission lines, normally a quarter of a wavelength long, known as *lecher lines* are employed in combination with a tuning capacitor. Coupling between stages is achieved by simple loops of wire or by tapping the lecher line. Figure 13.3 shows a typical front end unit incorporating the tuner and the IF stage, all within a single metal case.

The aerial is usually connected to the tuner via an isolating circuit. This is essential in receivers operating from the mains supply that do not use an isolating transformer. If the receiver is fed directly from the mains, the chassis can become live; and if a live chassis were connected directly to the aerial, it would make the aerial live as well. A commonly used aerial isolation circuit is shown in Figure 13.4. Capacitors C1 and C2 are large enough to give adequate coupling for radio frequencies, but of high enough impedance at the mains frequency of 50 Hz (or 60 Hz) to effectively isolate the aerial from the mains supply. Resistors R1 and R2 prevent static charge from building up on the aerial.

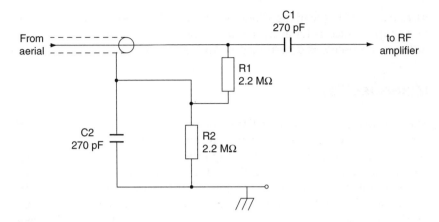

Figure 13.4 *Aerial isolation circuit*

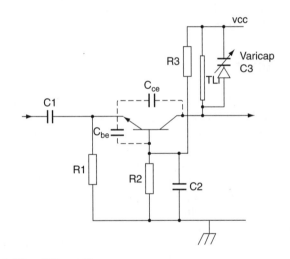

Figure 13.5 *The RF oscillator*

RF oscillator

At the UHF range of frequencies, feedback in the oscillator is obtained by mutual inductance, employing lecher lines to form a Hartley or a Colpitts oscillator. Use may also be made of the inter-electrode capacitors of the transistor. The circuit in Figure 13.5 shows a common-base Colpitts oscillator in which inter-electrode capacitors C_{ce} (between collector and emitter) and C_{be} (between emitter and base) provide the necessary feedback for sustained oscillation. C_{ce} in series with C_{be} effectively fall across the output developed between the collector and base. Part of this output, that part across C_{be}, is fed back into the input between the emitter and base. R1 is the emitter resistor, R2/R3 is the base bias chain with C2 as the base or bias decoupling capacitor and TL1 is the output lecher line resonating with *varicap diode* C3. A varicap diode is a reversed-biased diode which looks like a capacitance as far as the circuit is concerned. The value of the capacitance it represents is determined by the reverse-bias voltage.

Mixer-oscillator

The purpose of the mixer-oscillator stage is to change the incoming UHF carrier frequency to a common IF, a technique known as *super-heterodyne* (*superhet* for short). Frequency changing may be achieved either through multiplication or addition. Addition is the preferred method for TV tuners. A single transistor is made to oscillate at frequency f_o, which is 39.5 MHz above the selected channel frequency f_c. The non-linear part of the transistor characteristic is then used to produce the sum $(f_o + f_c)$ and the difference $(f_o - f_c)$ of the two frequencies together with the two original frequencies, f_o and f_c. A tuned circuit at the output is then made to select the frequency difference, $f_o - f_c = 39.5$ MHz.

Complete tuner

The various tuned circuits in the tuner unit must keep in step with each other as different frequencies are selected, and this calls for good matching in manufacture. The variable tuning is carried out by means of fixed Lecher bars in conjunction with varicap diodes. Tuning control, then, is carried out by varying the d.c. potential applied to the varicap diode. A swing of 30 V would scan the whole UHF television band. The tuning voltage source may be as simple as a single potentiometer or as complex as a frequency-synthesis, self-seeking ensemble. A self-seeking system, when initiated, sweeps up the TV transmission band(s) by itself, stopping each time it encounters a station for storing (manually or automatically) to a non-volatile memory, NVRAM. Frequency-synthesis offers a self-seek and memory facility along with a 'direct addressing' feature in which a required channel number (21–68 for UHF) can be requested by the viewer and automatically tuned. This involves a very stable crystal oscillator in a *phase-locked loop* (PLL) embracing the tuner's local oscillator.

The phase-locked loop

The PLL is commonly used in TV receivers to ensure stability of the IF output without the use of an a.f.c. Available in integrated circuit packages, the PLL is today widely used in a variety of electronic applications, including chrominance decoding. The principle of operation of the PLL is illustrated in Figure 13.6. It consists of a phase discriminator or detector, low-pass loop filter and a *voltage-controlled oscillator* (VCO). Without an input signal to the phase discriminator, the VCO is free running at its own natural frequency, f_o. When a signal arrives, the phase discriminator compares the input frequency f_1 with that generated by the VCO. A difference results in a d.c. output which after filtering is fed back into the VCO to change its frequency. The process continues until the two frequencies are equal and the PLL is said to be locked.

Figure 13.7 shows a block diagram for a typical UHF tuner used by modern TV manufacturers. As well as a UHF aerial connection, the tuner

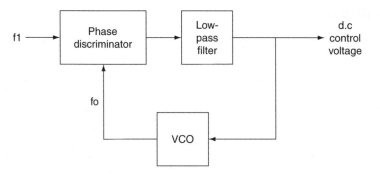

Figure 13.6 *Phase-locked loop (PLL)*

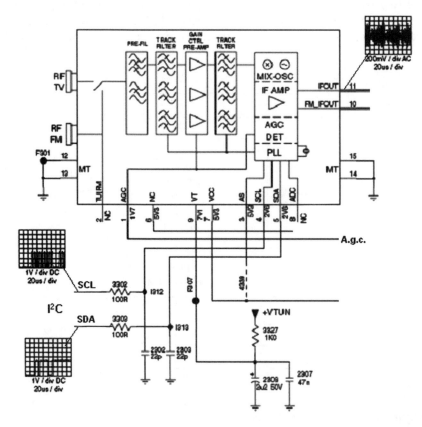

Figure 13.7 *UHF tuner using PLL*

provides a facility for FM radio reception. TV IF output is provided at pin 11 and the FM IF output is available at pin 10. The a.g.c. is fed into pin 1 and tuning is controlled by an I²C serial bus (SCL and SDA) from the microcontroller. For detailed information on I²C bus operation, refer to Appendix 2.

Synthesised tuning

Synthesis is the process of combining or adding up incremental amounts to obtain a certain quantity of, say, a voltage or a frequency. In the *frequency synthesised tuner (FST)*, the tuning voltage is obtained from a programmable PLL composed of the tuner's local oscillator, a controlled pre-scaler, a phase discriminator and a low-pass filter (Figure 13.8). A sample of the tuner's local oscillator frequency is fed back to the phase discriminator via a controlled divider (divide by N), known as a pre-scaler. The value of N is set by the channel select input which determines the frequency f_1 going into the phase discriminator. The discriminator then compares f_1 with reference frequency f_r. A d.c. output is produced that reflects the difference between the two frequencies. After filtering, this d.c. voltage is used to tune the tuner to the selected channel. Synthesised PLL-controlled tuners have extremely stable output with practically no drift, thus removing the need

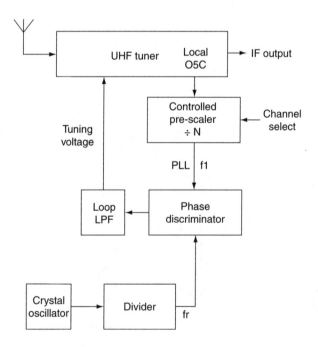

Figure 13.8 *Synthesised tuning*

for an a.f.c. circuit. The stability of the tuner output depends to a large extent on the stability of the reference frequency. For this reason, a frequency divider is used to divide the frequency of a crystal-controlled reference oscillator by a large factor to improve its stability.

The IF stage

Recall that the IF is derived at the mixer-oscillator stage of the tuner. The local oscillator is made to oscillate at a frequency which is 39.5 MHz higher than the selected carrier frequency. The IF is then obtained by selecting the difference between the carrier and the oscillator frequencies. For example, if the TV receiver is tuned to a channel frequency of 511.25 MHz (BBC 1 transmission from Crystal Palace) then the oscillator must be tuned to frequency $f_o = 511.25 + 39.5 = 550.75$ MHz. The relative position of the various frequencies of a modulated UHF carrier may be observed on a spectrum analyser such as that illustrated in Figure 13.9 for a modulated 511.25 MHz carrier. The 8 MHz bandwidth extends from f_{min} to f_{max} where

$$f_{min} = 511.2 - 1.75 = 509.5 \text{ MHz and}$$
$$f_{max} = 511.25 + 6.25 = 517.5 \text{ MHz}.$$

The sound carrier is 6 MHz above the vision carrier, so it has a frequency of $511.25 + 6.00 = 517.25$ MHz.

After the mixer-oscillator stage, the vision carrier is replaced by a vision IF of 39.5 MHz with RF oscillator frequency of $511.25 + 39.5 = 550.75$ MHz. The sound carrier is translated into a sound IF of

$$\text{oscillator frequency} - \text{sound carrier} = 550.75 - 517.25$$
$$= 33.5 \text{ MHz}$$

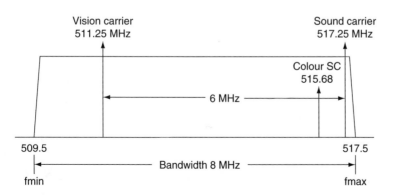

Figure 13.9 *Spectrum of a modulated 511.25-MHz carrier*

The sound IF is now 6 MHz *below* the vision IF Similarly, all other frequencies will reverse their position relative to the vision carrier when they are converted to their equivalent on the IF spectrum.

The IF response curve

Apart from providing sufficient IF amplification to drive the detector, the IF stage is required to shape the frequency response of the received signal to that shown in Figure 13.10. The IF response curve has four purposes:

1. To reject the vision IF of the adjacent higher channel. The adjacent vision IF falls 8 MHz below the vision IF at $39.5 - 8 = 31.5$ MHz.
2. To reject the sound IF of the adjacent lower channel. The adjacent sound IF falls 8 MHz above the sound inter-carrier at $33.5 + 8 = 41.5$ MHz.
3. To provide 26 dB attenuation at 33.5 MHz. This is necessary to prevent any interference caused by a beat between the sound and vision IFs. A small step or ledge is provided as shown to accommodate the FM deviation of the sound inter-carrier. The FM step prevents amplitude modulation of the sound carrier; otherwise it would be detected by the vision demodulator, causing a pattern to appear on the screen and a buzz on the sound, a symptom known as *sound on vision*.

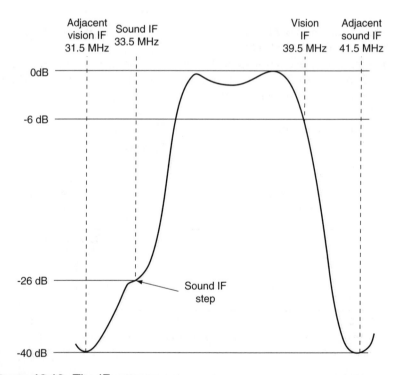

Figure 13.10 *The IF response curve*

4. To provide a steady fall in amplitude from 38 to 41 MHz at the vision IF end. This is necessary because the vestigial sideband transmission gives increased emphasis to these frequencies.

These four functions become very critical in colour TV reception. The 4.43 MHz chrominance sub-carrier falls at the higher end of the video spectrum, and when this is converted to an IF it becomes $39.5 - 4.43 = 35.07$ MHz, only 1.57 MHz away from the 33.5 MHz sound IF. It follows that, in order to retain the full chrominance information and its correct relationship to the luminance information, the response curve on the one hand must not be allowed to fall too early at this end, thus restricting the chrominance information, and on the other hand must provide sufficient rejection of the sound IF. Failure to do this produces cross-modulation between the 4.43 MHz chrominance sub-carrier and the 6 MHz sound inter-carrier. This cross-modulation appears as a 1.57 MHz $(6.00 - 4.43)$ pattern on the screen, known as *herring-bone* pattern.

The vision detector

In the amplitude-modulated waveform, the information is contained in the change of amplitude of the peak of the carrier waveform. By joining the tips of the carrier, an envelope is obtained which represents the original modulating information. The purpose of an a.m. demodulator is therefore simply to retrieve that envelope while removing the high-frequency carrier wave. This may be carried out by the simple rectifier diode detector with a smoothing low pass filter shown in Figure 13.11. For adequate reproduction of the video information, the time constant of the detector C1R1 must be shorter than the period of one cycle of the highest video frequency, 5.5 MHz, and longer than the duration of one cycle of the 39.5 MHz IF. It follows that the time constant has to be somewhere in between $1/5.5 = 0.18\,\mu s$ and $1/39.5 = 0.025\,\mu s$. For this reason, $0.1\,\mu s$ is normally aimed for. A time constant longer than $0.1\,\mu s$ would reduce the high-frequency response of the detector. The output which is the charge across C1

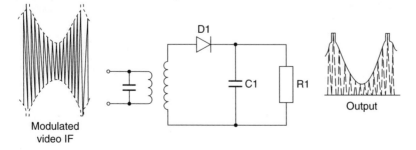

Modulated
video IF

Figure 13.11 *Diode a.m. detector*

is then no longer able to follow the fast changes in brightness which are represented by high video frequencies. A very short time constant, on the other hand, would retain a high proportion of the IF in the form of a ripple.

Synchronous demodulator

The rectifier is in essence a switch which closes during one half-cycle of the carrier wave and opens during the other half. From this point of view, any switching device may be used, provided it allows the carrier through for one half-cycle only. Since we are only interested in the amplitude of the peak of the carrier, the switch need only be open for the duration immediately before and immediately after the positive (or negative) peak of the carrier. In truth, the modulated carrier is sampled once every cycle of the carrier, a sampling rate equal to the carrier frequency itself. This is the principle of the *synchronous demodulator*.

The sampling pulses are obtained by the use of a limiter which removes the envelope and leaves a clipped 39.5 MHz carrier only as shown in Figure 13.12. The switching or sampling pulses, which have the same frequency and the same phase as the IF, are used to control a sampling gate that switches on at the peaks of the modulated carrier. The peak levels are then used to charge a capacitor which, given the correct time constant, will reproduce the original modulating signal. An important characteristic of the synchronous demodulator is that it will only demodulate those a.m. waveforms which have a carrier that is equal in frequency to and is in phase with the sampling pulses. In this way, synchronous demodulators remove both stray modulation caused by noise and beat frequencies between the sound IF on one hand and the adjacent vision carrier and adjacent sound inter-carrier on the other.

The output of the synchronous demodulator may be improved by doubling the sampling rate. Two switching square waves in antiphase to each other are used to operate two separate gates. The two gates are also fed

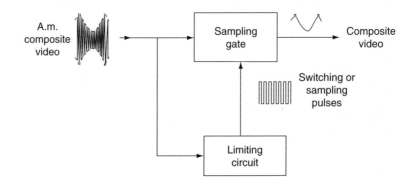

Figure 13.12 *Synchronous detector*

with out-of-phase IF signals. The effect is to produce a signal which appears to have passed through a full-wave rectifier. The output contains a carrier component which is twice the frequency of the original carrier, making it easy to filter out. Synchronous demodulators are too complex and expensive for construction from discrete components but lend them easily to design on a silicon chip as part of an IF or video integrated circuit.

Tuner IC package

An IC package incorporating a tuner, IF stage and vision detector is shown in block-diagram form in Figure 13.13. It uses no conventional tuned circuits and thus needs no tuning control voltage: all its tuning instructions come from the TV's control microcomputer via the SDA and SCL lines of the I²C control.

This IC tunes from 50 to 860 MHz and uses a double-superhet technique. The RF amplifier stage consists of a wideband gain-controlled low-noise amplifier suitable for use with both aerial and cable input signals. It feeds mixer 1 where the first frequency conversion takes place; in fact it is an up-conversion to a first IF frequency well above 1 GHz. This passes through an external filter, a simple inexpensive two-pole ceramic resonator with a bandwidth of about 15 MHz: this defines the initial passband and provides image-frequency rejection. In conjunction with on-chip

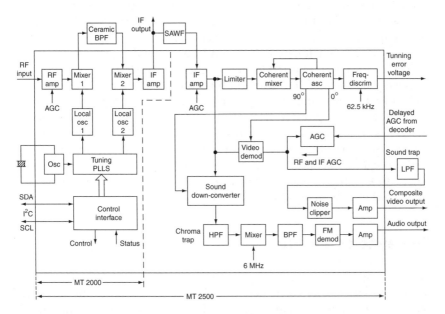

Figure 13.13 *Vision detector IC package*

image-rejecting mixer design, image-signal suppression of 65 dB is achieved over the entire tuning range. The signal then undergoes a second, down-conversion in mixer 2, whose local oscillator 2 runs at a frequency such that the second IF frequency is 39.5 MHz (vision). The second mixer stage is again a special image-rejecting type to provide suppression of 65 dB to the image signal produced in the first mixer stage. The two local oscillators are fully integrated into the IC, and generate the required frequencies with reference to a single external crystal. Tuning resolution is 62.5 kHz, giving 128 steps in the (typical) 8 MHz channel width. The system uses a complex frequency-synthesis circuit whose components – varicap diodes, voltage-controlled oscillators, phase/frequency detectors, programmable dividers and charge pumps – are all on the chip. The IF amplifier is gain-controlled in the same way as for a conventional tuner/IF ensemble to optimise the noise performance and minimise cross-modulation when large input signals are present. This device has an a.g.c. range of 96 dB, a noise figure of 8 dB, image rejection of more than 57 dB, and cross-modulation performance of less than 1 dB in the presence of 30 mV input signal.

The 'tuner-only' version of the IC in Figure 13.13 ends at the dotted line to the left of diagram centre. The 'complete-receiver' IC incorporates further procession which will now be described; either chip can be used for digital TV by taking the IF output signal (top centre) to a suitable demodulator.

Continuing to the right, the 39.5 MHz IF signal is 'shaped' in a SAW filter for passage to a further IF amplifier, after which analogue demodulation takes place. This involves a PLL locked to the vision carrier frequency. The demodulated video signal passes through an external trap (right of diagram) to take out the sound carrier, then back through a chip-internal noise clipper to remove impulse interference.

Down-converted inter-carrier sound has an FM carrier frequency of 6 MHz for the UK, and is fed through a chroma trap to remove colour sub-carriers, then an on-chip self-tuned filter on its way to the FM demodulator. After filtering and de-emphasis, the baseband signal is ready for amplification and passage to the speaker(s). Take-off of a Nicam stereo signal feed is also possible with this IC. The tuner chip is controlled by the industry-standard I²C serial control bus which allows interrogation and readout of the contents of all the status registers on the chip, and enables the device to be programmed in software. Data registers in the tuning PLLs are loaded to tune in a specific channel.

The sync separator

The purpose of the *synchronising separator* is to slice the sync pulses off the composite video waveform, separate them into line and field, and feed each one to its individual timebase. The process must be immune to changes in the amplitude and picture composition of the video signal.

As explained in Chapter 1, the sync pulses are arranged to fall beyond the black level and to occupy 30% of the total amplitude of the composite video. A clipping network is therefore all that is required in order to separate the sync pulses away from the video information.

Once the sync pulses have been separated from the composite video, the receiver must be able to distinguish between the two types of sync pulses. While the line sync is represented by a straightforward pulse, the field sync is identified by a series of pulse-width modulated (PWM) waveform consisting of five consecutive broad pulses at twice line frequency (Figure 13.14). The field flyback lasts for 25 complete lines, giving a total field blanking time of $25 \times 64 = 1600\,\mu s = 1.6\,ms$. In order to ensure continuous line synchronisation throughout the field blanking period, it is necessary for line triggering edges to occur where a line sync pulse normally appears. The $2 \times$ line frequency broad pulses ensure this takes place. The extra edges during the field flyback are disregarded by the line oscillator. Before and after the five broad field pulses, five equalising pulses are inserted to ensure good interlacing.

Flywheel sync separator

Random pulses due to noise and other interference are sometimes present on the composite video signal. These pulses are similar to the sync pulses themselves and may trigger the timebase at the wrong time. In the case of the field timebase, the integrator which has a slow response removes most of the noise. However, a random pulse occurring near the end of a field causes what is known as frame or *field slip*. To avoid this, a noise gate (also known as a *noise canceller*) may be included to obtain a noise-free output from the sync separator. In the case of the line timebase, noise causes what is known as line tearing as some lines are displaced with respect to the others. On vertical objects, the effect of line tearing is illustrated by ragged edges. Line tearing may be avoided by the use of a *flywheel synchronising* circuit. The principle of flywheel synchronisation is similar to that of the mechanical flywheel which, due to its large momentum, maintains an average speed unaffected by random changes. The flywheel sync circuit maintains an average frequency of the sync pulses by monitoring and taking the average frequency of a number of incoming line pulses so that a random pulse will have very little effect on the frequency. A block diagram for a flywheel synchroniser is shown in Figure 13.15. It consists of a flywheel discriminator followed by a reactance stage which controls the frequency of the line oscillator. The flywheel discriminator itself consists of a phase comparator or discriminator and a smoothing circuit. A control voltage proportional to the timing, i.e. phase, between the line oscillator and the incoming sync pulse is obtained from the phase comparator. The voltage is then smoothed by the use of a low-pass filter. For good noise immunity, the flywheel discriminator should have slow response, which

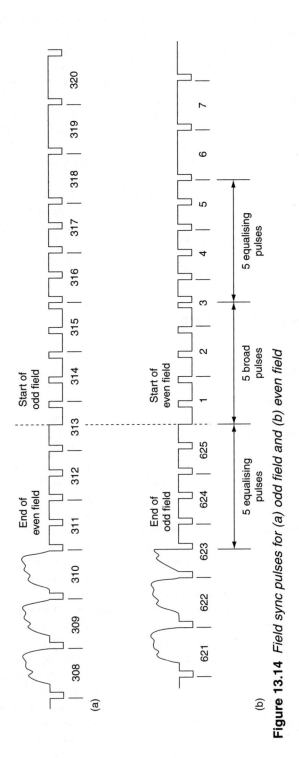

Figure 13.14 *Field sync pulses for (a) odd field and (b) even field*

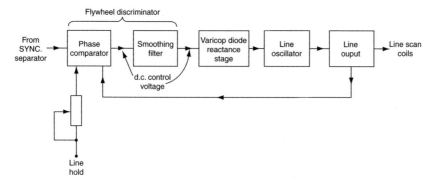

Figure 13.15 *Flywheel sync separator*

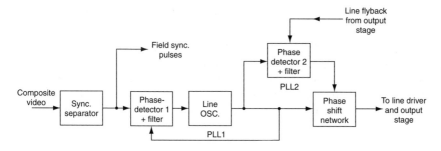

Figure 13.16 *Sync separator chip*

means a long time constant for the smoothing circuit so that the average frequency is taken over a large number of line sync pulses. However, the time constant also determines the pull-in range of the flywheel discriminator. The pull-in range is the range of oscillator frequency drift over which the discriminator will pull the oscillator into lock without having to adjust the manual line hold control. A short time constant improves the sensitivity; hence it widens the *pull-in range* of the discriminator. A compromise has to be struck so the oscillator has stability within the pull-in range of the discriminator.

Figure 13.16 shows the basic arrangement of a *sync processing* chip in which PLLs act as flywheel synchronisers. The stability of the line flyback is secured by the use of two PLLs, PLL1 and PLL2. Phase detector 1 compares the phase of the square wave output of the line oscillator with the line sync from the sync separator, so it ensures the line oscillator is running at the correct frequency and phase. The second phase detector compares the phase of the line oscillator with the line flyback pulse from the line output stage. Any phase error is then corrected by the phase shift network which is in essence a PWM that changes the width and hence the phase of the line oscillator square wave output. An additional third PLL may be added in to control the sensitivity of PLL1. Upon switching on, changing channels or any change in signal strength resulting in a momentary loss of line oscillator lock, it is desirable for PLL1 to have a fast response, so that good pull-in and quick lock are obtained. When the oscillator has been brought into lock, a slow response is desirable. To do this, a third phase detector compares the phase of the line flyback from the line output stage with the sync from the sync separator. An output is produced when the two pulses are in phase in which case the control circuit changes the sensitivity of PLL1 and the time constant of the filter.

Sandcastle pulse

Some sync separator processing systems incorporate the line and field pulses in a single multi-level pulse known as a *sandcastle pulse*. A typical three-level sandcastle pulse is shown in Figure 13.17. The highest level, 7.5 V, provides the narrow burst gating pulse whose average duration is 4 μs. It is generated by level detection of the line sawtooth signal. The intermediate level, 4.5 V, is derived from the line flyback and has duration of 12 μs. At the lowest level, 2.5 V, we have the field blanking pulse with duration of 21 lines. A level detector or slicer may be used to extract the required pulse from the sandcastle combination as and when required.

Field timebase

In the case of a CRT display, the function of the field timebase is to deflect the electron beam relatively slowly (cycle time 20 ms for PAL and 16.7 ms for NTSC) from top to bottom of the viewing screen, then rush them quickly back to the top during the field blanking interval. During the downward stroke, the line timebase draws over 300 scanning lines on the screen, and the timing is such that they are distributed evenly over the viewing area. The lines of one field are traced out in the spaces between

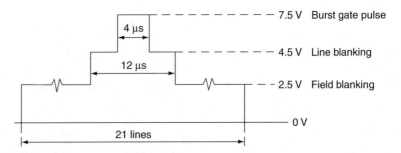

Figure 13.17 *The sandcastle pulse*

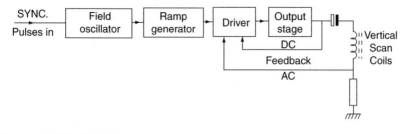

Figure 13.18 *Field timebase*

those of the previous field to satisfy the interlace requirement. The triggering of the field timebase by the broadcast sync pulses is critical for good interlacing performance.

Figure 13.18 is a representative block diagram of a field timebase system. It starts with a free-running oscillator capable of being triggered by the separated broadcast field sync pulses at 50 Hz (or 60 Hz for NTSC) rate. The timing pulse produced by the oscillator triggers a precision ramp generator whose output is a sawtooth waveform with a period of 20 ms (or 17.6 ms for NTSC) and excellent linearity. It is amplified by a driver stage for application to the power output stage and thence to the field coils in the deflection yoke. The a.c. and d.c. feedback are used respectively for linearity correction and stabilisation of the working conditions.

Field output stage

The field output amplifier is made up of a class B complementary-symmetry transistor pair whose mid-point feeds the deflection coils via a large d.c.-blocking electrolytic coupling capacitor.

A typical class B field output stage is shown in Figure 13.19, in which VT8/VT9 is the driver combination and VT10/VT11 is the complementary output pair. At the start of the scan, VT8 base voltage is low, 'turning off' VT8 and VT9. VT9 collector is almost at d.c. supply potential turning VT10 fully on, driving current into the field scan coils. VT11 is off. At the mid-point of the scan, the centre of the picture, VT8/9 begins to conduct, turning VT10 off and VT11 on; this provides the scan current for the second half of the picture. W1/W2 provides a small forward bias for the output transistors to prevent cross-over distortion. Resistor R4 provides d.c. feedback for bias stability.

A more advanced field output amplifier employs a *switched-mode modulator* (Figure 13.20) in which a pulse train is fed into an LC low-pass filter. As can be seen, the pulses are smoothed with the capacitor charging up to the mean value of the input pulse. Since the mean value is determined by the mark-to-space ratio of the input pulse, a varying mark-to-space ratio produces a varying charge across the capacitor. A linearly

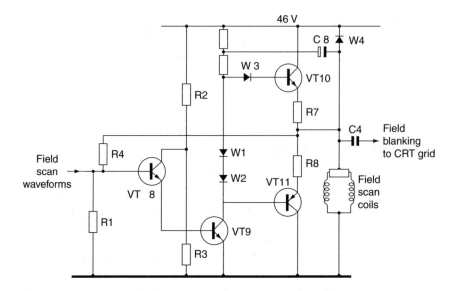

Figure 13.19 *Field output stage using discrete components*

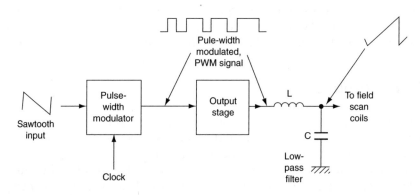

Figure 13.20 *Pulse-width modulated (PWM) field output stage*

increasing capacitor voltage, i.e. a timebase ramp scan, may thus be produced if the mark-to-space ratio of the input pulse is gradually increased. The PWM waveform is first fed into the output stage to turn the active device on and off before going into an LC low-pass filter, which produces the sawtooth waveform. The output device, which may be a transistor or a thyristor, thus acts as a switch. The clock pulse which drives the PWM may be a separately generated waveform or it may be derived from the line scanning pulse. Since the active element is a switch, its power dissipation is extremely low. When the switch is on, its resistance is very small and therefore its power dissipation is very low. When the switch is open, current ceases and power dissipation is nil.

Line timebase and drive

The purpose of the line timebase is to provide the appropriate deflection current through the line scan coils. As in the case of the field scan, the current waveform required to produce linear deflection is a sawtooth. However, at the relatively high line frequency, the reactance of the coil XL is very high compared with its d.c. resistance, so the resistance is insignificant. The line scan coils may then be treated as purely inductive. Given that V_L is proportional to the rate of change of current ($V_L = L \times (di/dt)$) it follows that a linear current waveform in a pure inductor is obtained when a constant or d.c. voltage is applied across it. To obtain a sawtooth current waveform, the voltage waveform must be the pulse shown in Figure 13.21. For the scan period AB, the current is increasing at a small and constant rate. Consequently, voltage V_L is a small positive value which remains constant for that duration. For the flyback period BC, the current is decreasing at a high and constant rate; V_L is again constant but this time large and negative.

Consider the tuned circuit in Figure 13.22. When the switch is closed, a step waveform is applied across the tuned circuit and energy is fed into it. Oscillation, known as ringing, takes place at a resonant frequency $f = 1/2\pi\sqrt{LC}$ These oscillations take place because electromagnetic energy in the scan coils is continuously transferred from electrostatic energy in the capacitor to electromagnetic energy in the inductor and vice versa. Current therefore flows from the coil to the capacitor and back again. Theoretically, this ringing would continue indefinitely since there is no power or energy loss in either a pure inductor or a pure capacitor. However, due to losses caused mainly by the very small resistance of the inductor, ringing gradually dies out; producing what is known as *damped oscillation* (Figure 13.23).

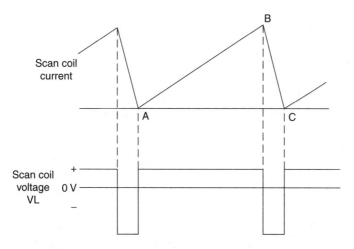

Figure 13.21 *Line scan current and voltage waveforms*

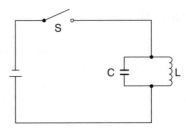

Figure 13.22 *Applying a step voltage to an LC tuned circuit*

Figure 13.23 *Damped oscillation*

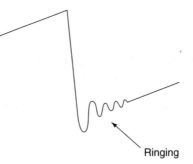

Ringing

Figure 13.24 *Ringing*

Similar oscillation or ringing occurs when a sharp change in voltage is applied across an inductor. The tuning frequency in this case is the self-capacitance of the coil as well as any stray capacitance due to other components. In the case of the line scan, ringing occurs at the beginning of each flyback (Figure 13.24) with a consequent distortion on the left-hand side of the picture on the tube face.

It is possible to remove the effect of ringing by shunting the scan coils with a damping resistor. This, however, will result in a large waste of energy, reducing the power available for beam deflection and reducing the angle of deflection of the tube.

To avoid ringing but without the loss of energy, an *efficiency diode* may used. This technique is based on making use of the energy stored in the scan coils due to flyback to provide the first half of the scan. It involves a

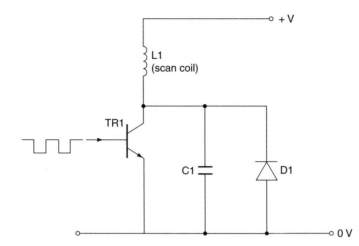

Figure 13.25 *Use of efficiency diode in line timebase circuit*

switching network which directs the transfer of energy to and from the scan coils to obtain the required waveform. A circuit using a parallel transistor-diode switch is shown in Figure 13.25, in which L1 is the scan coil, D1 is the efficiency diode and TR1 is the line output transistor. When TR1 is switched on by a positive edge to its base at time t1 (Figure 13.26), a constant h.t. voltage is applied across L1. A linearly increasing current is therefore obtained, forming part AB of the scan. The current continues to rise until point B, when at time t2 a negative step to the base switches off TR1. At this point, TR1 collector suffers a sudden jump from almost chassis potential to +h.t. This positive voltage ensures that D1 remains non-conducting. C1 is now effectively connected across L1. The large change in current in L1 produces ringing at a frequency determined by C1 and other stray capacitors. Energy due to the sudden change of current through L1 is transferred to C1 to commence ringing oscillation at point B. When C1 is fully charged (point C on the flyback), the charging current drops to zero as the ringing comes to the end of the first quarter-cycle of oscillation. The second quarter-cycle begins as energy from C1 is transferred to L1. The current reverses as the capacitor begins to discharge. When the cycle reaches its negative peak at the end of the first half-cycle (point D), the current begins to decrease, attempting to go to zero again. The rate of change of current di/dt suffers a change of direction. Before the negative peak (point D), the rate of change of current is positive. This induces an e.m.f. across L1, which makes TR1 collector (and D1 cathode) positive, ensuring the diode is off. At the negative peak itself, $di/dt = 0$ and the induced e.m.f. is also zero. After the negative peak, the current begins to decrease. The rate of change of current is therefore negative, reversing the induced e.m.f. and making TR1 collector (and D1 cathode) negative. The diode conducts. Its effect is similar to TR1 conducting, placing the h.t. across L1

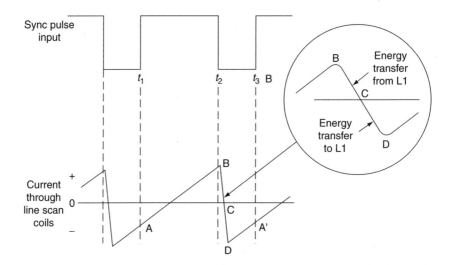

Figure 13.26 *Current waveform using efficiency diode circuit in Figure 13.25*

to start the scan. Current through L1 rises linearly to form about 30% of the scan up to A', when TR1 is switched on at time t3 by a positive edge to the base, and so on.

S-correction

Recall that in order to compensate for the flat surface of the tube face, correction of the line scan waveform is necessary. This correction, known as symmetrical or *S-correction*, becomes increasingly important with wide-angle deflection tubes.

As can be seen from Figure 13.27, an equal angular deflection of the beam scans a smaller distance at the centre compared with the distance it scans at the two ends of the line. Thus, to obtain a linear picture scan using a flat tube face, a non-linear angular deflection is necessary. The purpose of the non-linearity is to slow the rate of change of the angular deflection at both ends of the scan (Figure 13.28). Since the current through the line scan coils is responsible for the angular deflection, the corrected waveform must be of the same shape.

The scan part of the S-correction waveform approximates a half-cycle of the half-line frequency sine wave in Figure 13.28. The scan part may be simply obtained by connecting a tuning capacitor in series with the scan coils. The value of the capacitor is chosen so that it resonates with the scan coils at a frequency slightly higher than half the line frequency, approximately 7.8 kHz. This provides time duration from A to B of

$$\frac{1}{2} \text{ period} = \frac{1}{2} \times \frac{1}{7.8} = 64 \,\mu s \,(\text{approximately})$$

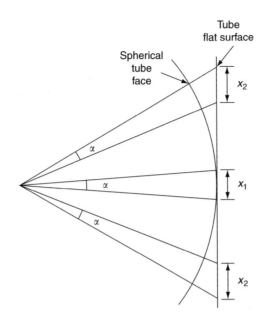

Figure 13.27 *The effect of a flat tube on linearity*

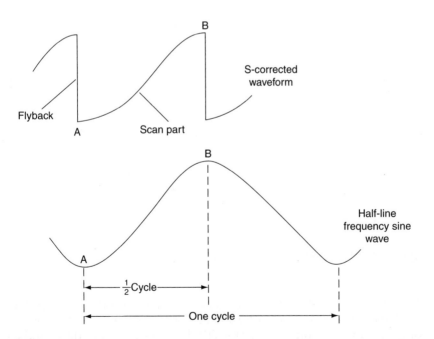

Figure 13.28 *The relationship between S-correction and half-line frequency waveforms*

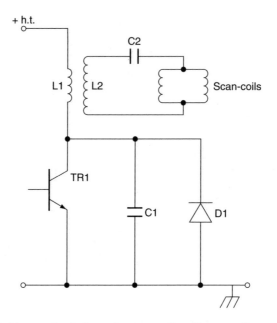

Figure 13.29 *Line output stage incorporating S-correction capacitor C2*

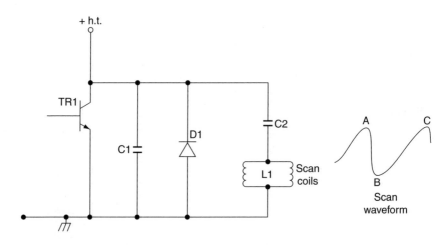

Figure 13.30 *Directly coupled line scan coils*

In the line output stage of Figure 13.29, C2 is the S-correction capacitor which has a value of between 1.5 and 3 μF. The scan coils are transformer-coupled to the line output transistor TR1. D1 is the efficiency diode and C1 is the flyback tuning capacitor.

Modern TV receivers employ direct coupling (Figure 13.30). At the end of the scan, TR1 is switched off, tuned circuit L1/C1/C2 is pulsed into oscillation to provide the flyback AB. At the negative peak (point B), diode D1 conducts placing C2 across L1 to commence an oscillation at a frequency of

$$\frac{1}{2\pi\sqrt{LC}} = \text{half-line frequency}$$

The first half-cycle of this oscillation, BC, provides the scan. At approximately one third of the scan, TR1 is switched on and takes over from the diode, and so on.

Further simplification of the line timebase circuit may be obtained by using the b–c junction of the output transistor TR1 as the efficiency diode (Figure 13.31). The polarity of the b–c junction is the same as the polarity of an efficiency diode, had it been connected; the *n*-region collector (cathode) is connected to h.t. whereas the p-region base (anode) is connected to chassis via the secondary winding of T1. At the end of the scan, the b–c junction is forward biased in the same manner as an efficiency diode.

Line output transformer

Although the scan coils are not normally transformer-coupled, a transformer known as the *LOPT* (line output transformer), pronounced 'loptie' is employed to provide a number of functions including the e.h.t., the

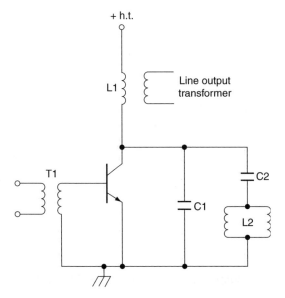

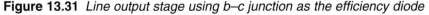

Figure 13.31 *Line output stage using b–c junction as the efficiency diode*

auxiliary d.c. supplies, the boost voltage, the gating pulses for a.g.c. and the reference pulse for the flywheel discriminator (Figure 13.31). In choosing the value of the tuning capacitor for the line output stage, the inductance introduced by the LOPT must be taken into account.

One important function of the LOPT is to provide the high d.c. supply voltage of 30–90 V required by the video, line and field output amplifiers. For receivers operating from the mains supply, this d.c. voltage may be obtained by rectifying the mains voltage. However, this is not possible for battery-operated receivers, so a boost voltage from the LOPT is used.

The boost voltage is obtained by the use of an efficiency diode and a storage capacitor. Consider the circuit in Figure 13.32, in which L1 is the primary winding of the LOPT, D1 is the boost diode, C1 is the storage or boost capacitor and TR1 is the line output transistor. D1 is connected in such a way that it only allows charging current to flow through C1 and prevents the discharging current from flowing through L1, thus maintaining the charge across C1. When TR1 is switched on, D1 conducts; this places C1 across L1. Ringing occurs and electromagnetic energy in L1 is transferred to C1, which charges up to h.t. When C1 attempts to discharge through L1, the current reverses and D1 stops conducting. Provided C1 is large, in the region of 200 μF, it will retain the charge across it. When TR1 is switched off to start the flyback, D1 remains off. At the end of the flyback, TR1 collector goes negative due to the reversal of the rate of change of current in the scan coils. D1 and D2 conduct. Energy in L1 is transferred

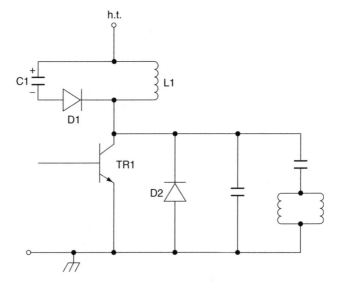

Figure 13.32 *Line output with boost capacitor C1 and boost diode D1*

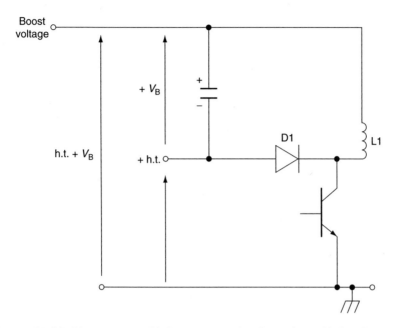

Figure 13.33 *Line output with boost capacitor in series with h.t. line*

to C1 to replace any loss in its charge. Excess energy is fedback into the h.t. supply line. C1 thus remains charged up to h.t.

It is possible to connect the boost capacitor in series with the h.t. line to produce a boosted voltage (Figure 13.33). The voltage V_B across the capacitor, which could exceed the h.t. potential, is added to the h.t. line.

14 Television receivers: colour processing

You will recall from Chapter 2 that the chrominance information is contained in a 4.43-MHz modulated sub-carrier which forms part of the incoming composite video signal. Colour difference signals $B' - Y'$ and $R' - Y'$ are used to modulate the sub-carrier, which is then suppressed to leave two quadrature components, U and V. At the receiver, the chrominance information is separated from the luminance signal by a *comb filter*, decoded and applied to a matrix network, which performs the operation necessary to reproduce the original RGB colour signals that can be applied to a colour tube. Figure 14.1 shows a block diagram of the major functional units required for TV video processing. It consists of two major processing sections: luminance and chrominance. *Chrominance processing* itself consist of four distinct parts: a colour burst section, a colour decoding section, a matrix and a colour drive amplifier.

Colour burst processing

Colour burst processing separates the sub-carrier burst signal from the chrominance information so it may be used to recreate the sub-carrier which has been suppressed at the transmitter. The sub-carrier has to be restored in both its frequency and phase to ensure correct colour reproduction. Two sub-carriers at 90° to each other have to be produced, one for each colour difference, $B' - Y'$ and $R' - Y'$. Furthermore, in the PAL system, the sub-carrier for the $R' - Y'$ demodulator has to be phase reversed on alternate lines. The burst processing section is also used to provide *automatic chrominance control (a.c.c.)* as well as the *colour killer* signal for monochrome-only transmissions.

The colour burst, which consists of about 10 cycles of the original sub-carrier, is mounted on the back porch of the line sync pulse. A burst gate amplifier triggered by the line sync is used to separate the burst from the composite video. The burst gate amplifier is turned on by a delayed line flyback pulse. The delay in the flyback pulse is necessary because, having been placed on the back porch arrives immediately after the line sync. The delay ensures that the amplifier begins to conduct on the arrival of the burst. The burst gate amplifier allows the burst to go through a phase-locked loop (PLL) network known as the automatic phase control (a.p.c.) consisting of the phase discriminator, filter and the voltage-controlled oscillator (VCO).

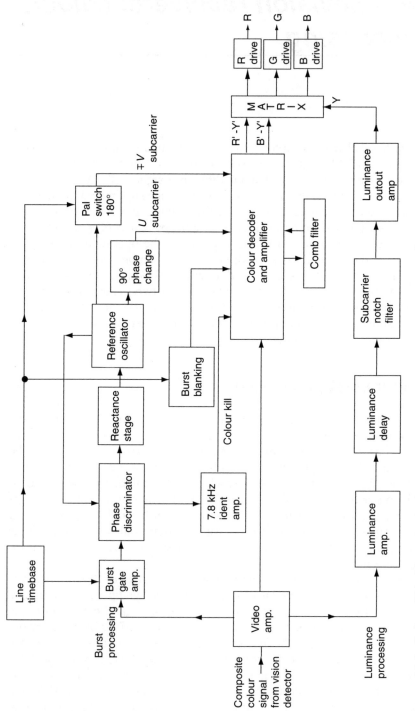

Figure 14.1 *Colour TV video processing*

The phase discriminator compares the phase of the burst with the 4.43 MHz output of the voltage-controlled crystal oscillator. If there is an error, a correction voltage is produced, which after going through a low-pass filter is fed to the VCO to bring the frequency and phase of the oscillator into step with the burst. For a more precise and stable sub-carrier, the VCO natural resonant frequency is doubled to $2 \times 4.43 = 8.86$ MHz. The 4.43 MHz sub-carrier is obtained by dividing the output of the oscillator by 2. For PAL, the burst signal is not of constant phase but swings to convey information of the phase reversal of the V component at the transmitter. For this reason, the phase discriminator must compare the phase of the oscillator with the average phase of the burst signal.

Finally, the sub-carrier for the $B' - Y'$ demodulator is obtained by the insertion of a simple $90°$ phase shift at the output of the oscillator. The sub-carrier for the $R' - Y'$ demodulator must suffer a phase reversal on alternate lines. To achieve this, a square wave at half-line frequency is needed. A component of this frequency is present at the incoming signal because of the $90°$ phase change of the colour burst on alternate lines. Since one complete swing of the burst phase takes place every two lines, the frequency of the 'swing' is half the line frequency, i.e. $15.625/2 = 7.8125$ kHz, which is normally quoted as 7.8 kHz and referred to as the identification or ident signal. After amplification, the ident signal is fed into a PAL switch that reverses the phase of the 4.43 MHz oscillator signal on alternate lines.

Two other functions are derived from the burst processing section: colour killing and a.c.c. The purpose of the colour killer circuit is to close down the chrominance amplifier path on monochrome-only transmissions to prevent random colour noise appearing on the screen. The ident signal is therefore used to provide a normal bias for the chrominance amplifier, which will be turned off if ident is absent.

The a.c.c. prevents varying propagation conditions from changing the amplitude of the chrominance signal in relation to the luminance. To realise this, the gain of the chrominance amplifier is made variable by a control voltage in a similar way to the control of the gain of the i.f. stage by an a.g.c. signal. The a.c.c. control voltage must be proportional to the amplitude of the chrominance signal. This voltage cannot be derived from the actual chrominance signal during the active picture scan as this varies in amplitude as the colour information changes. It is derived instead by monitoring the amplitude of the colour burst. A fall in the amplitude of the burst signifies an attenuated chrominance. This is corrected when the a.c.c. control voltage into the chrominance amplifier increases its gain, and vice versa.

Colour decoding

As stated earlier, the chrominance information forms part of the composite video, from which it has to be separated before demodulation. The chrominance information is centred on a 4.43-MHz sub-carrier with a

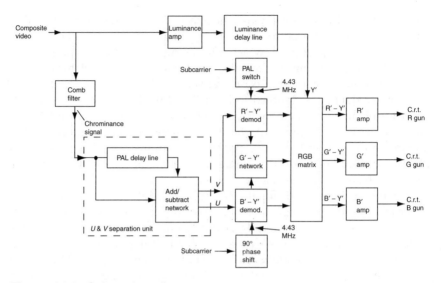

Figure 14.2 *Colour decoding*

bandwidth limited to +1 MHz. The first task of the chrominance decoder is therefore to separate the chrominance from the composite video which is carried out by a comb filter (Figure 14.2). The next task is to break up the composite chrominance signal into its two component parts, U and V, each of which must be demodulated separately to obtain the original colour difference signals. The weighted colour difference signals, U and V, are separated from each other by a unit consisting of a delay line (usually known as PAL delay line) and an add/subtract network. Two separate signals, U and V, are produced, which are fed to their respective demodulators, $B' - Y'$ and $R' - Y'$. Each demodulator is fed with a 4.43-MHz signal at the correct phase from the reference oscillator and burst channel. Two gamma-corrected colour difference signals, $B' - Y'$ and $R' - Y'$, are thus obtained. The third colour difference signal, $G' - Y'$, is obtained from the first two by the $G' - Y'$ network as shown. The three colour difference signals, together with luminance signal Y', are then fed into the RGB matrix. By adding Y' to the three colour difference signals, the original gamma-corrected R', G' and B' colours are reproduced, and after amplification they are fed directly into the appropriate CRT gun. As a consequence of the colour decoding process, the chrominance component suffers a delay with respect to the luminance part. A luminance *delay line* is therefore inserted into the luminance signal path to ensure that both signals arrive at the matrix at the same time.

The chrominance channel must also provide a facility for some or all of the following functions:

- Colour kill to turn the chrominance amplifier off during monochrome-only transmission.

- Manual saturation (or colour) control to allow the user to change the colour intensity of the display by varying the gain of the chrominance amplifier.
- Automatic colour control, a.c.c. to vary the gain of the chrominance amplifier.
- Burst blanking to turn the amplifier off during the sub-carrier burst. Failure to do this will result in a greenish striation appearing on the left side of the screen.
- Inter-carrier sound rejection. The 6 MHz sound inter-carrier must be removed by one or more 6 MHz traps in the amplifying stage.
- d.c. clamping to reintroduce the d.c. level lost during the processing channel.

It is necessary to d.c. clamp all three colour signals to ensure a common black level for the red, green and blue guns. Any drift in the d.c. level of one amplifier with respect to any of the other two would lead to overemphasis producing an unwanted colour tint. Where a.c. coupling is used throughout the channel, d.c. clamping is carried out at the RGB drive amplifier stage. However, in modern receivers, d.c. coupling is employed for the early part of the processing system.

The comb filter

Recall that in TV broadcasting, the colour content of a picture is used to modulate a 4.34-MHz (3.795 MHz for NTSC) sub-carrier. The resulting side frequencies appear in clusters which fit neatly within the clusters produced by the luminance side frequencies. Some kind of filtering is therefore necessary to separate the two components. If no filtering is used, the colour information is interpreted as spurious fine details and the result is a grainy appearance over the entire picture. The question is what type of filter.

The cheapest solution is to use simple filters such as a bandpass or a notch filter that pass only the coarse and medium horizontal detail (lower about 3.5 or 3 MHz for PAL and NTSC, respectively) to the luminance circuits and pass the bulk of the higher frequencies as colour information. For low-frequency luminance, the picture would suffer very little colour contamination or cross talk but at high frequency video where there are fine details such as striped clothing, noticeable artefacts in the form of 'rainbow swirls' appear in the picture at that location. Conversely, medium detail colour information cannot be used because there is too much luminance contamination. Cross-talk from chroma C into luminance Y produces waves of dots in the picture detail while cross-talk from Y into C results in spurious tints appearing as 'rainbow swirls' (Figure 14.3).

To avoid Y/C cross-talk, the two components must be separated as cleanly as possible. Hence, the use of comb filters.

You would recall from Chapter 2 that in order to avoid a dot pattern of alternate black and white dots caused by the insertion of the chrominance

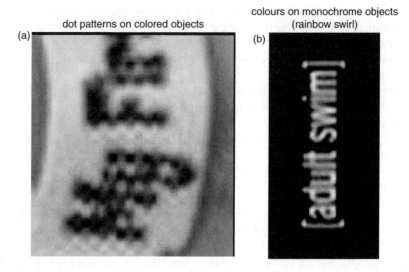

Figure 14.3 *Composite video artefacts without a comb filter*

information within the luminance bandwidth, the sub-carrier value was selected very carefully to ensure that the number of sub-carrier cycles per line ends with a half-cycle. This ensures that the sub-carrier clock is always 180° out-of-phase with the sub-carrier of the previous line in the same field. This fact is used by the comb filter to separate the two components Y and C. The technique involves the simple process of adding and subtracting the video contents of two successive lines. If line 1 in the odd field contains Y + C and line 3 contains Y − C, then by adding and subtracting we get

$$(Y + C) + (Y − C) = 2Y$$
$$(Y + C) − (Y − C) = 2C$$

Similarly for the even field.

However, life is not as easy as that. The above assumes the Y and C components of two lines are the same. In practice, both Y and C change line-by-line as the picture content varies and along each line resulting in severe artefacts. For the effect on colour, consider a simple still picture, with a red square on top of a green square. In the middle of either square, consecutive horizontal scan lines are the same and the addition and subtraction work adequately. However, where the squares meet, we pick up a pair of scan lines that differ profoundly, red versus green. What comes out in the picture is a yellow line. This is the result of two differing scan lines mixed together and yielding a third unrelated colour. More correctly, we see a yellow boundary two lines thick because the commingling of a red line and a green line happens twice per frame, first during the odd interlaced field and again during the even interlaced field. The yellow lines have a fine undulating dark–light pattern due to imperfect removal of the colour information from the luminance information. This is known as *dot crawl*. The undulations usually shift during the 25 or 30 frames per second refresh cycle. This artefact is entirely due to actions of the comb filter. Without a comb filter, there would be a perfectly sharp red to green horizontal boundary. To overcome these drawbacks, several techniques are available for TV manufacturers to use. There are four types of comb filters: *2L, 3L, 2D-3L and 3D*.

The 2-line comb filter

This is the simplest and cheapest comb filter. It uses the simple add and subtract technique mentioned above. The Y and C components of consecutive lines in a frame are added and subtracted to produce luminance chrominance components, respectively. Compared with the notch filter, the 2-line (2L) comb filter improves Y/C cross-talk reducing the 'rainbow swirls' effect and improves horizontal resolution.

The 3-line comb filter

This technique uses three lines giving the first 50% weighting, the second 100% weighting and the third 50% weighting. If all three lines are identical, the result is perfect since the two half strength lines are out-of-phase with the full strength middle line. Where the lines are not the same, the result is a vast improvement on the 2-line filtering. Let us go back to the picture of a red square above a green square. Starting from the top, we get perfect red mixtures until we get down to the boundary between the two squares. When we first encounter a green line we take it at half strength. The two preceding lines are a red line at full strength and another red line at half strength. In this 75% red–25% green mixture, the red overwhelms the green. As we move one line further down, we pick up the next green line at half strength, use the previous green line at full strength, and the line before that is a red line at half strength. This time the green is at 75% thus overwhelming the red at 25%. Again this process happens twice, first in the odd field and then in the even field. So, the total number of finished scan lines derived from less than 100% red or less than 100% green in this example is four compared with the two in the 2L filter.

The 3-line (3L) comb filter provides better resolution than the 2L type, sharper horizontal boundaries and reduced 'rainbow swirl' effect.

The two-dimensional 3-line comb filter

This uses three lines in the same way as the 3L type. However, the weighting is not fixed. It is changed from left to right along the scan lines depending on the similarity of the three lines. For example, where just the last two of the three lines are the same, the comb filter mixes them 0, 50 and 50% to get near perfect filtering. Moving further along the line, the first two lines might be the same and the mixture is switched to 50, 50 and 0%. There may be intermediate mixtures too, used if the filter logic 'was not sure' whether the line above or the line below was a better match. Dot crawl can be eliminated over most of the picture and be made almost unnoticeable in the most difficult places. Crisp colour boundaries are achieved for most of the content of the picture. However, if neither the line above nor the line below matches, these comb filters cannot improve that part on the picture. An artefact will appear just as with the 2L comb filter. The 2D-3L type of comb filters are called two-dimensional (2D) because the mixture is varied vertically by using or not using the predecessor and/or successor lines and changed numerous times horizontally along the line.

The 3D (motion adaptive) comb filter

The 3D comb filter technique is qualitatively different. It uses the fact that the sub-carrier clock is always 180° out-of-phase with not just the sub-carrier of the previous line but of the previous frame as well. Thus, if line L_x in frame n has chroma C, then on the same line L_x on the previous frame $n-1$ would have chroma –C; the chrominance component is phase reversed on adjacent frames.

The 3D comb filter uses adjacent frames (next and/or previous frames) to get a scan line that has the same content as the current line for correct Y/C separation. If there was no motion of the subject matter, the corresponding line in the next frame (625 PAL and 525 NTSC lines away) will have the same content as the line being processed. Its colour content, however, is phase reversed. So, the sum of these lines is pure luminance and the difference is pure colour. Of course, if there is movement at that spot, the lines will differ and should not be commingled. The filter logic senses that, foregoes the 'third dimension' and go back to a method that 2D filters use. A good 3D comb filter contains within it a 2D comb filter. While there are elaborate processes for detecting motion, possibly involving three frames instead of two, all that has to be done is sense whether portions of the lines are the same or different. This process is not perfect and is made more difficult if the picture is noisy (snowy). The circuits would also have to store an entire frame in a rolling data buffer since a line could not be drawn on the picture tube until the corresponding line in the next frame has arrived.

A typical 3D comb filter arrangement employed by Sony is illustrated in Figure 14.4. The composite video CBVS is fed into the colour decoder (IC1) from the video selector at pin 44. It comes out at pin 3 to go to IC2 (3D) and IC3 (3L) for comb filter processing. Two separate combinations of Y and C (Y_D and C_D from IC2 and Y_L and C_L from IC3) are made available for the 2-channel demux to select from. The selection is made by a signal from the

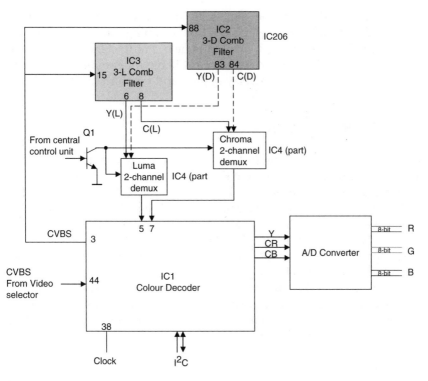

Figure 14.4 *3D comb filter arrangement*

central control unit (CCU) to transistor Q1. A logic high signal will turn the transistor on and the demux will select the outputs from the 3D comb filter and vice versa. The CCU will assess the movement between successive frames and if there is movement, the outputs from the 3L filter will be selected, otherwise, that from the 3D filter will be chosen. The output from the colour decoder is in the form of Y, C_R and C_B (or YP_rP_b) which in this case is digitised by a analogue-to-digital converter (ADC) for further processing.

Comb filter delays

Mixing two adjacent scan lines can be achieved by the introducing time delay. One kind of time delay is the use of tiny loudspeaker that plays the video signal into a tiny glass chamber where a tiny microphone picks it up at the other end. Another is similar except that a thin glass fibre transmits the signal from the speaker to the microphone like a toy tin can and string telephone. Still others (CCDs, or charged coupled devices) use silicon chips with a chain of 'charged cells' so as to make the signal travel slower as is passes through. The circuits are arranged so that the video signal goes through both the apparatus described above and through a straight wire.

At the other end where the signals are mixed, when the current line is coming down the wire, the previous line is coming out of the delay apparatus. This happens for each line in the field. For the 3L comb filter, there are two delay apparatus, so we have available for commingling one line delayed twice, the next line delayed once and the third line not delayed.

U and V separation

Figure 14.5 shows how the PAL-D system separates the U and V components; the chrominance signal of the previous line is stored or delayed for one line duration (64 μs for PAL and 63.59 μs for NTSC). The delayed signal is then added to and subtracted from the signal of the current line to produce separate U and V signals. The precise value of the required delay is very slightly less than one line duration; this is because the number of cycles per line always ends with a half cycle and the delay must correspond to a whole number of sub-carriers.

Let us assume that the chrominance signal going into the delay line driver in Figure 14.5 is unswitched, i.e. $U + V$. The chrominance signal at the output of the delayed line is that of the preceding line, i.e. switched line $U - V$. Addition of the two signals gives

$$(U + V) + (U - V) = 2U$$
$$(U + V) - (U - V) = 2V$$

Alternatively when the input to the line driver is a switched line $(U - V)$, the stored signal is unswitched $(U + V)$. The result of the adder remains as

$$(U - V) + (U + V) = 2U$$

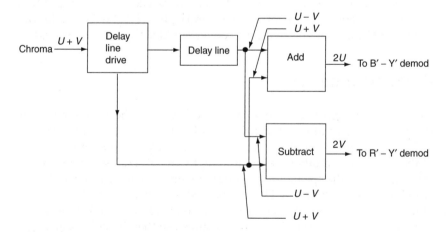

Figure 14.5 *U and V separation*

But the subtraction is reversed

$$(U - V) - (U + V) = -2V$$

The U and V separator thus retains the phase reversal of the V component and also removes the phase error.

Colour difference demodulation

The weighted chrominance components U and V from the separator unit are applied to their individual colour difference demodulators to obtain B' − Y' and R' − Y'. The third colour difference signal, G' − Y', is obtained by a G' − Y' matrix (Figure 14.6). For the B' − Y' demodulator, the sub-carrier is shifted by 90° for the R' − Y' demodulator, the sub-carrier is phase reversed every line through the PAL switch. Recall that the U and V components of the chrominance signal are the weighted colour difference signals, whereby

$$U = 0.493(R' - Y')$$

and

$$V = 0.877(R' - Y')$$

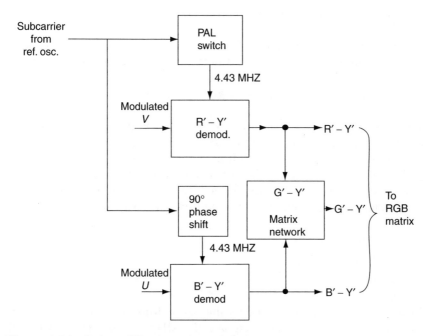

Figure 14.6 *Colour difference demodulation*

It follows that, before $B' - Y'$ and $R' - Y'$ are recovered, the weighted components must be deweighted. This is usually carried out at the demodulation stage by the inclusion of colour difference amplifiers which provide the $B' - Y'$ channel with more gain than the $R' - Y'$ channel.

In amplitude modulation, the information is contained in the change of the amplitude of the peak of the carrier. When the carrier is suppressed, the modulating information continues to reside in the changing amplitude of the modulated signal. To recover the original information, the amplitude of the modulated signal has to be detected when the carrier is at its peak. In order to do this, a synchronous detector is used. In essence, the colour difference synchronous demodulator is a switching device which detects the level of the incoming modulated U (or V) signal every time the regenerated sub-carrier is at its positive peak.

Matrix network

The purpose of the *matrix* is to reproduce the original RGB signals to be fed to the tube electrodes. The first stage is to obtain the green colour difference $(Y' - G')$ which will be used to produce the RGB signals.

The proportions of $R' - Y'$ and $B' - Y'$ components which are necessary to obtain the third colour difference component $G' - Y'$ may be mathematically determined as

$$G' - Y' = 0.51(R' - Y') - 0.186(B' - Y')$$

However, if $G' - Y'$ matrixing takes place before deweighting of the U and V components is carried out, different proportions are necessary

$$G' - Y' = -0.29(R' - Y') - 0.186(B' - Y')$$

Since the required $R' - Y'$ and $B' - Y'$ levels are both less than unity, they may be derived by using a simple resistor network.

Now to the RGB network itself at the c.r.t., a picture is produced by modulating the three c.r.t. electron beams using the gamma-corrected R', G' and B' signals. These are derived by the addition of the colour difference signals to the gamma-corrected luminance signals Y'

$$Y' + (R' - Y') = R'$$
$$Y' + (G' - Y') = G'$$
$$Y' + (B' - Y') = B'$$

During addition, the difference in the bandwidth of the luminance signal and the colour difference signals (5.5 MHz for the luminance and 1 MHz for the colour difference) introduces *glitches* or *notches* at points of fast colour transition. The narrow bandwidth of the colour difference signals restricts the rise time of rapidly changing signals. When they are added to the rapidly changing luminance signal, a 'glitch' is introduced. On a colour-bar display, the effect of the glitch appears as a narrow dark band between the colour bars.

The addition may be carried out by a dedicated RGB matrix before the colour signals are applied to the cathode of the c.r.t., a technique known as direct drive. Alternatively, matrixing may be carried out by the tube itself, a technique known a *colour difference drive*. Colour difference drive involves feeding the colour difference signals to the control grids of the c.r.t. while applying a negative luminance of $-Y'$ to the cathodes. Applying a negative luminance to the cathode is equivalent to applying a positive luminance to the grid; it results in the mathematical addition of the two signals.

In the colour difference drive, separate colour difference and luminance signals are fed into different electrodes of the c.r.t. This creates problems with timing, accentuated because the two signals have different bandwidths. Further problems arise from the relative sensitivities of the electrodes, which have to be the same for all levels of the input signals. Direct RGB drive overcomes these problems by performing the necessary matrixing close to the point of demodulation. For these reasons, direct RGB drive is used in modern TV receivers.

Beam limiting

CRT-type TV receivers employ some form of beam current limiting to ensure the electron emission (beam current), hence the brightness does not exceed a pre-determined limit. Beam limiting is optional for monochrome TVs but essential for colour TVs. A very high beam current overloads the line output amplifier and the extra high tension (e.h.t.) tripper, causing deterioration in focus; it overdrives the c.r.t., causing limited highlights; and it leads to excessive power dissipation in the mask within the tube, which may cause misconvergence of the three primary colours on the screen. Beam limiting involves sampling or monitoring the strength of the beam current directly at the cathode, or indirectly by monitoring the d.c. current taken at the line output stage or the e.h.t. winding. The beam current itself is then controlled by reducing the black level (i.e. the brightness) or the amplitude of the luminance signal (i.e. the contrast) or both.

Figure 14.7 shows a circuit of a typical beam limiter arrangement used in colour TV receivers; the e.h.t. current is monitored by a diode D1 placed between the earthy end of the e.h.t. overwind and chassis. Two types of current flow through D1: forward current ID of about 600 μA to chassis, caused by voltage V_{cc}, and beam current IB flowing in the opposite direction. When the beam current goes above 600 μA, D1 switches off and its

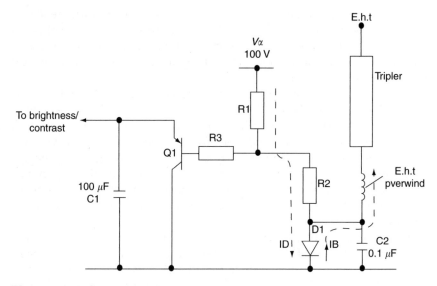

Figure 14.7 *Current limiting circuit*

anode goes negative; this reduces the base voltage of emitter follower TR1, hence it varies the brightness/contrast controls. For larger tubes, the pre-set beam current limiting is raised to 1 mA. Decoupling is carried out by C1 and C2.

Signal-processing IC package

All the low-level signal-processing stages in a colour TV are amenable for integration on a single chip, an example of which is given in Figure 14.8. At the top left-hand corner, the IF signal from the SAW filter enters the vision IF amplifier and PLL demodulator. From here is produced an automatic frequency control (a.f.c.) potential for the tuner – it counteracts drift of the input carrier or the tuner itself. Baseband video signal, complete with inter-carrier sound, passes down to the video amplifier block. Taking first the *'auxiliary'* outputs from here, a sample goes up to the a.g.c. section, whose link to the tuner feeds back information on signal strength at high input levels. A video mute block prevents noise in the form of 'snowy screen' being displayed in the absence of a transmission, and from this the video signal comes out of the IC on pin 6. The sound carrier is selected by a ceramic filter (*'sound bypass'*) to re-enter the IC on pin 1, where it encounters a limiter stage to clip off any amplitude modulation. A PLL FM demodulator follows from which the baseband (mono) audio signal passes through a mute/pre-amp/de-emphasis stage on its way out of the IC on pin 15. It is intercepted by a switch – under I²C bus control – to

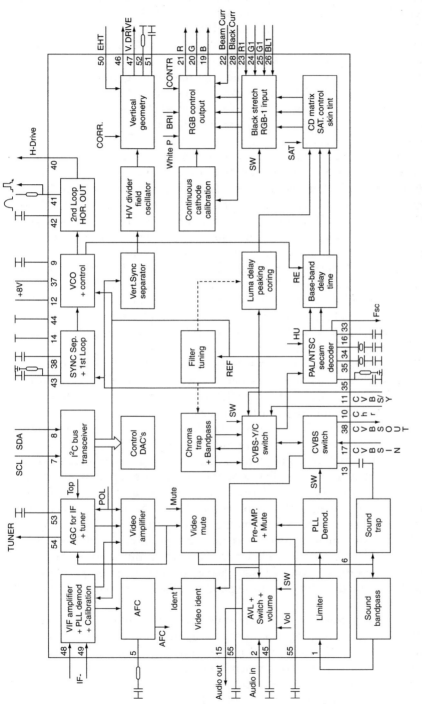

Figure 14.8 *Signal processing IC package (internal architecture of 'jungle' IC for a TV receiver. Operation is explained in the text)*

enable other sources of audio, entering on pin 2, to be selected (along with external video on pin 17) when the set is switched to AV mode.

The main path for the video signal is out of IC pin 6 and through a sound trap, sharply tuned to 6 MHz, into chip pin 13. Here it encounters two routing switches, both working under I²C-bus instructions. The first of them selects between internal and external (coming in on pin 17) composite video; the second selects composite or Y/C signals, the latter entering on IC pins 11 (Y) and 10 (C). We shall meet Y/C signals later. Pin 38 gives a route out for the composite signal, typically via an AV socket. The chroma trap and bandpass block is used to separate the video and colour sub-carrier signals contained in a composite video signal. The entire PAL decoder, also capable of NTSC and SECAM operation, is contained in the bottom centre block in the diagram. The delay line is catered for by on-chip components. The components at IC pin 36 provide time-constants for the sub-carrier flywheel filter, while 4.43 and 3.58 MHz crystals at pins 35 and 34 respectively regenerate colour sub-carriers for PAL and NTSC signals. In the latter case, hue control is affected via the I²C control bus. R − Y and B − Y are separated in the internal baseband delay line on their way to a colour difference/RGB matrix in the bottom right-hand side of the diagram. This also incorporates the saturation (colour) control. Luminance signal on its way to the RGB matrix is optimised in the peaking/'coring' block, and delayed to coincide with the lower-bandwidth, slower-rising chrominance signals.

RGB signals from the chip internal decoder pass into the primary-colour switch, where selection is made between them and signals entering on pins 23/24/25, usually from a teletext decoder. '*Black stretch*' provides level-dependent amplification to low-level (near black) video signals if required. The final picture-processing block in the chip is centre-right, RGB output. As well as providing buffered feeds out to pins 21/20/19, this provides brightness and contrast control for each of the three channels: separately and in parallel, so that external RGB input signals, when selected, can also be controlled. Brightness and contrast settings are pulled back when the voltage entering on IC pin 22 indicates that the beam current in the picture tube is near the point at which its shadow mask will hot-bulge. Also present here is an auto greyscale correction system in which the black current sample enters the chip on pin 18 for entry to the cathode calibration block. RGB signals pass through a two-stage amplifier, usually mounted on the tube base panel, on their way to the cathodes of the picture tube.

It remains to trace the path of the synchronisation pulses. At top centre of the diagram of Figure 14.8, the sync separator stage is fed with composite video signal, from which it strips out line sync pulses and feeds them to a comparison stage which keeps the TV line frequency in step with the broadcast picture. Line drive pulses from IC pin 40 switch the line output stage at 15.625 kHz for PAL operation. A second feed from the composite video signal operates the on-chip vertical sync separator. The result is a single 'clean' pulse once per field to reset an H/V divider, a counter circuit which triggers field flyback every scanning line. Thus synchronised to the broadcast signal, the field oscillator's drive waveform passes through a shaping stage (vertical geometry) on its way out of the IC on pins 46 and 47. There are two influences on the vertical geometry corrector: a feed coming into IC pin 50 contains a sample of EHT voltage to prevent the picture size changing with beam current; and from within the chip, a correction factor by which – under I²C bus control – the vertical scanning current can be adjusted in amplitude and shape to render good picture geometry and linearity. Control of all the functions of the IC depends on the I²C bus transceiver near the top centre of the diagram; it governs all the switches in the chip, plus the D-A converters which set parameters like colour, brightness, contrast and sound level as well as factory/technician settings such as picture-white points and vertical scan correction. For information on the SCL and SDA control I²C bus which enter this chip on pins 7 and 8, refer to Appendix 2. To round up some of the pins not so far mentioned, pins 12 and 37 provide a +8 V operating voltage; pin 33 provides a sample of locked colour sub-carrier; pin 26 takes a picture-blanking pulse from the text decoder for use when text is superimposed on a picture; and pin 41 has the dual purpose of dispensing a sandcastle pulse output and receiving a flyback input from the line scan stage. There are other pins and functions of this 56-pin package which are not absolutely essential to the main operation.

RGB output stage

With RGB drive, the output stage must be able to deliver a peak-to-peak signal of 80–150 V to the c.r.t. cathode. The signal drive to each gun is different because of the different efficiencies of the electrodes; the red gun requires the largest drive. The large signal drive requires a high d.c. supply voltage and often two transistors are connected in series to share this high voltage. To ensure adequate bandwidth, series peaking coils may be employed. Power transistors in class A configuration are used with the necessary heat sinks to deliver the relatively large power necessary to drive the red, green and blue guns of the c.r.t. A small load resistance or an emitter follower buffer is used to ensure low output impedance. This low impedance allows for fast charge and discharge of the c.r.t.'s cathode input impedance, giving good frequency response at the upper end of the bandwidth.

Output stages are normally mounted on the c.r.t. base panel to remove the band-width limitations associated with long leads. Early output stages were also used for matrixing by feeding the luminance to the base and the colour difference signal to the emitter of the transistor. *Black level clamping* and brightness control of the three guns were also incorporated at the video output stage as well as greyscale adjustment. The purpose of

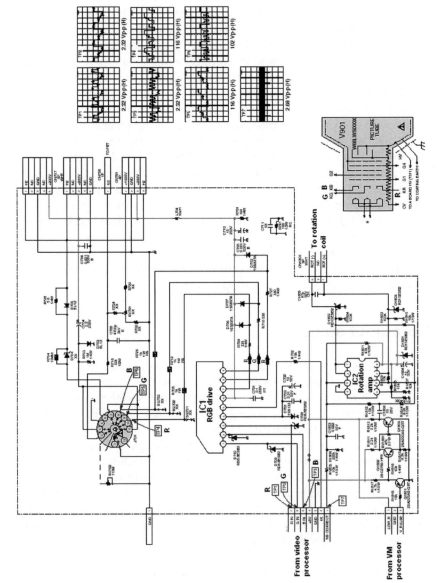

Figure 14.9 *RGB output stage*

greyscale adjustment, or tracking, is to ensure that a purely monochrome picture has no traces of colour tint at any level of brightness from low-lights to highlights across the greyscale display. With the introduction of more advanced integrated circuits, matrixing is now carried out separately and clamping is introduced at the luminance signal channel before matrixing.

A typical RGB output schematic diagram for a 28-in. flat screen cathode ray tube (CRT) is shown in Figure 14.9 together with relevant waveforms. IC1 is the RGB drive chip and IC2 is the rotation amplifier providing the necessary S-correction as well as velocity modulation. RGB inputs arrive at pins 1, 2 and 3 of IC1 and following amplification they are fed to the tube neck panel pins 10, 9 and 8, respectively. Diode 707, 706 and 702 are reversed biased by 200 V d.c. to act as limiting diodes for the R, G and B signals going into the tube electrodes.

15 DC power generation

The various sections of the TV receiver have different power requirements depending on their function and the level of the signal that is being handled. Large-amplitude signals from the video and the line output stages require a high or boost voltage of 150–250 V known as high tension (HT). On the other hand, a low voltage supply of 3–40 V, sometimes known as low tension (LT), is required for those sections such as the front end, video or sound-processing chips and drives and field and line timebase that handle small signal levels. Furthermore, extra high tension (e.h.t.) in the region of 30 kV is necessary for the final anode of the cathode-ray tube (CRT). In this chapter we will cover the principles of power generation and its applications to CRT television receivers. Power requirements for other display systems follow in later chapters.

CRT television power requirements

Apart from the a.c. supply to the tube's heater, regulated power supplies are necessary. Protection measures have to be taken to ensure safe operation of the receiver under normal conditions as well as under faulty conditions such as excessive load current. A great diversity of approaches and techniques are employed in the design of power supplies for TV receivers. The aim is to improve regulation and efficiency and to minimise power dissipation, thus reducing the cost and weight of the receiver.

The simplest power supply is the *unregulated* rectifier circuit shown in Figure 15.1, in which D1/D2 forms a full-wave rectifier, C1 a reservoir capacitor and R1/C2 a smoothing or low-pass filter to remove the 100 Hz ripple appearing at the output. For a more effective smoothing, the series resistor R1 may be replaced by a large inductor. In this simple circuit the d.c. output decreases as the load current increases. For a stable d.c. output, a *regulator* or a *stabiliser* is used. A regulator maintains the d.c. output level as constant against both changes in the mains supply and changes in the load current, known as mains regulation and load regulation respectively.

Linear regulators

A simple linear series regulated power supply is shown in Figure 15.2, in which TR1 is the regulator. Load regulation is obtained by zener diode Z1, which provides the reference voltage for the base of common-emitter

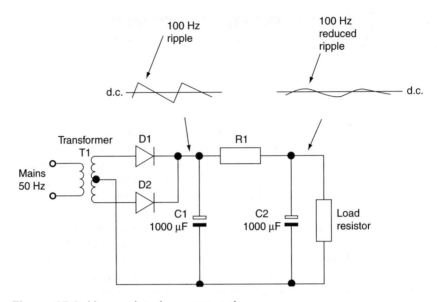

Figure 15.1 *Un-regulated power supply*

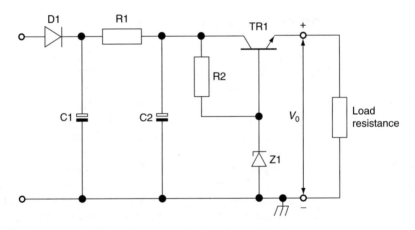

Figure 15.2 *Simple linear regulated power supply*

transistor TR1. Z1 maintains the base of TR1 at a constant potential established by its breakdown voltage. The d.c. output taken at the emitter is thus maintained at 0.6 V below the zener voltage. Changes in the output level produce changes in the b–e bias of the transistor in such a way as to keep the output constant. For higher load currents, a Darlington pair may be used, normally as an IC package. Sensitivity and response may be improved by using an error detector as shown in Figure 15.3. TR1 is the normal series regulator. TR2 compares a portion of the output voltage with the reference voltage of e zener. Changes in the output level are

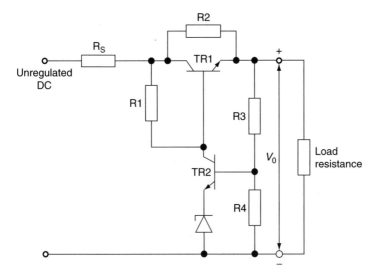

Figure 15.3 *Linear regulator with error detector*

amplified and fed into the base of TR1, which maintains the output constant. A shunt bypass also known as *'bleeding'* resistor, R2, is sometimes connected across the c–e junction of TR1 to reduce the power dissipation of the series regulator by diverting a portion of the load current away from the transistor. It also provides the initial start-up voltage for the regulator by 'bleeding' the current across TR1 to the output; this provides the necessary potential for TR2 and the zener to begin to conduct and start the regulator functioning.

Thyristor-controlled rectifier

An alternative to the transistor, a thyristor (Figure 15.4) may be used as the regulating device for a power supply. A thyristor is triggered into conduction by a positive voltage applied to its gate, provided that its anode is

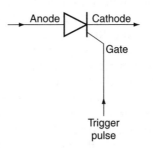

Figure 15.4 *The Thyristor*

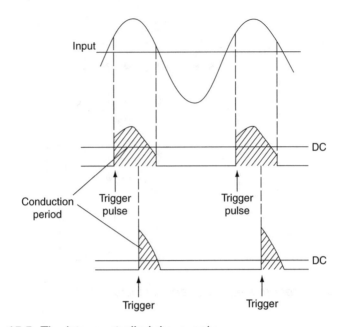

Figure 15.5 *Thyristor controlled d.c. supply*

positive with respect to the cathode. When fed with an a.c. voltage, the thyristor can only conduct during the positive half cycle. The conduction period is determined by the timing of the trigger pulse to the gate. The output level may thus be controlled by switching the thyristor for longer or shorter periods of time as shown in Figure 15.5.

Switched-mode power supplies

In the low and medium power range, 50 W and over, a regulated d.c. source is often required which contains negligible a.c. ripple. For these application, the *switched-mode power supply* (*SMPS*) is often used. Apart from reducing a.c. ripple, switched mode power supplies are more efficient than the linear types, dissipating less heat and are smaller in size. The SMPS is basically a d.c.-to-d.c. converter; it converts d.c. into a switched or pulsating d.c. and back again into a d.c. The switching speed determines the a.c. content or ripple frequency at the output.

The basic principle of a switched mode power supply circuit is shown in Figure 15.6. When the switching element S is closed, current I_1 flows from the positive side of the unregulated input into the load as shown. The magnetic field set up by the current flowing through L1 causes energy to be stored in the inductor. When the switch is open, the current ceases and the magnetic field collapses. A back e.m.f. is induced across the inductor in

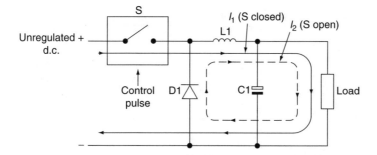

Figure 15.6 *Principle of a switched mode power supper*

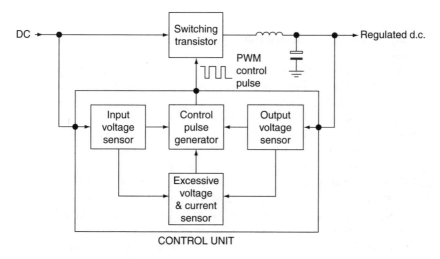

Figure 15.7 *SMPS control unit*

such a way as to forward-bias D1, causing a current I_2 to flow into the load in the same direction as I_1. The energy stored in the inductor when the switch was closed is transferred to the load to keep the load current constant. The ripple at the output has a frequency that is twice the switching speed and is easily removed by the low-pass filter action of L1/C1.

The switching element may be a bipolar power transistor or a power MOSFET; opened and closed at regular intervals by a pulse from a control unit. Regulation is obtained by making the time intervals when the switching element is open and closed (i.e. the mark-to-space ratio of the control pulse) depending on the d.c. output. They may also be made to depend on the input voltage; this guards against changes in the mains (or battery) supply voltage as shown in Figure 15.7. The control unit, which is normally an IC package, may also provide protection against excessive current and voltage. The control pulse is in essence a pulse-width modulated waveform

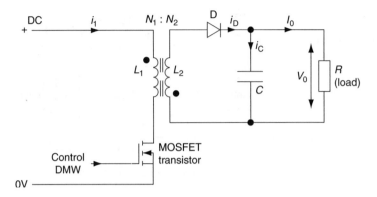

Figure 15.8 *Flyback SMPS*

(PWM) obtained from a free-running oscillator or an oscillator driven by line sync pulses.

There are three main types of switched mode power supplies: *Flyback* or *Boost*, *Forward* or *Buck* and *Cuk* converters. A flyback-type SMPS is shown Figure 15.8 employing an isolating transformer. Apart from the isolation between 'hot' and 'cold' sides, the transformer also provides a simple way of obtaining a range of d.c. outputs by simply employing a number of secondary windings with different turn ratios.

The transformer windings are such that the voltage and current in the secondary L_2 are in anti-phase to their counter part in the primary as denoted by the dots. When the MOSFET transistor is turned on, current flows into the primary of the transformer causing a current to flow in the secondary winding L_2 in the opposite direction. Diode D is reverse biased and energy is stored in inductor L_2. When the MOSFET switches off, inductor current ceases and the back e.m.f. across L_2 forward biases the diode and energy is transferred from L_2 to capacitor C and hence to the load. The process is then repeated when the transistor is switched on again and so on.

A forward-type (Buck) converter is shown in Figure 15.9. When the MOSFET is turned on, the induced secondary voltage turns diode D1 on and energy is stored in inductor L. When the transistor is turned off, inductor current collapses and a back e.m.f. appears across it. D1 is reverse biased and D2 is forward biased. Energy is transferred from the inductor to the capacitor which charges up to keep a stable voltage across the load. L/C also acts as a low-pass filter.

A Cuk converter configuration is shown in Figure 15.10. The operation of the Cuk is slightly more complex than the other two types, but essentially, when the MOSFET is on, current flows in the inductor and energy is stored in L1. When the transistor turns off, energy is transferred from the inductor to capacitor C1 via the transformer windings.

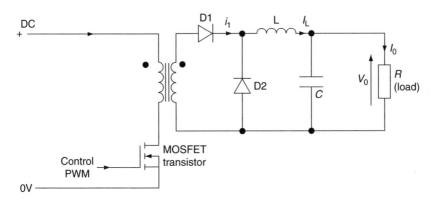

Figure 15.9 *Forward-type (Buck) converter*

A typical flyback converter for a TV receiver, commonly known as a *chopper power supply* is shown in Figure 15.11. Electrical isolation between live mains and the circuits of the TV is provided by the (chopper) transformer whose primary and secondary windings are separated by insulating layers capable of withstanding over 2 kV. To maintain mains isolation, an isolating break in the feedback path is necessary. This may be provided by a transformer, but normally an opto-coupler is used as shown. The opto-coupler consists of two highly insulated halves, linked by an infrared light beam, generated by a light-emitting diode (LED) and monitored by a photo-transistor. An increase in the diode current and hence its emitted light would cause the photo-transistor current to go up as well. The isolating transformer makes it possible to obtain a number of regulated d.c. rails by merely adding secondary windings. The 'chopper' transistor TR1 is normally a bipolar or a MOSFET transistor. Network D3/R1/C1 provides protection for the chopper transistor against excessive rise of collector voltage caused by the turns off. When TR1 is off, its collector voltage rises, turning D3 on. This effectively places *snubbing* capacitor C1 across TR1, diverting the major part of the back e.m.f. energy away from

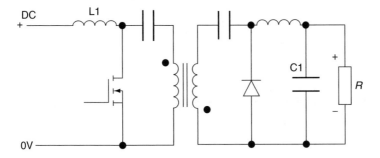

Figure 15.10 *Cuk converter configuration*

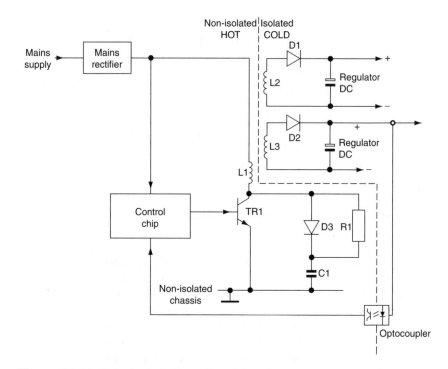

Figure 15.11 *A typical configuration for a chopper power supply*

the transistor. When TR1 is switched on, C1 is discharged through R1. Control PWM pulses arrive at the base of the transistor from the control chip. Both the input and output d.c. levels are monitored by the chip to establish a good regulated output. The output is monitored at one of the d.c. rails and fed back to the control chip via an opto-coulper. A typical PWM control waveform is shown in Figure 15.12.

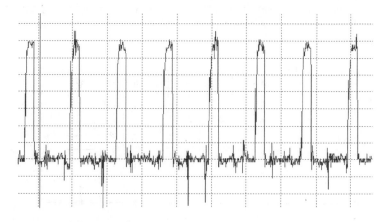

Figure 15.12 *PWM control waveform*

It is normal to derive the control pulse generated by the control chip from the line sync. This makes the functions of the chopper transistor and the line output transformer very similar. This similarity prompted some manufacturers to combine the functions of both into a single transistor, a technique known as *synchronous converter output stage* (*Syclops*). Other techniques involve combining the functions of the chopper transformer and the line output transformer into a single transformer, known as *integrated power supply and line output* (*IPSALO*).

High-frequency chokes

The switching action of the SMPS, which may involve large currents, can introduce interference known as mains pollution in the form of sharp transients, spikes or glitches superimposed upon the mains waveform. This is overcome by the introduction of high-frequency *chokes* or decoupling capacitors at the input terminals; they prevent high-frequency pulses from going back into the mains supply.

Start-up and soft start

The switching element and the control unit require a d.c. supply before they can begin to function. This voltage may be obtained from the nominal 12 V regulated supply for the control unit or other regulated or unregulated d.c. rails. It is normal that, once the control chip is brought into operation, the SMPS itself is used to provide the required regulated voltage to the chip. A slow or *soft start* is desirable for the switching element to prevent it from overworking at switch-on when the output voltage of the SMPS is zero. Soft start also ensures that the auto-degaussing is completed before the tube starts to be scanned.

Standby supply

The standby supply to a TV receiver is a separate power supply. It is separated from the main power supply of the receiver in order to reduce power consumption when the set is in the standby mode as required by legislation. The standby power supply is fed with a.c. from the input mains filter. It is therefore always operational when the mains input is present regardless of the position of the power switch.

The circuit diagram of a standby power unit is shown in Figure 15.13. Rectified mains is applied to the SMPS control TNY256 chip. This IC contains a control circuitry as well as a MOSFET transistor. The power supply provides 9, 5, 3.3 and 25 V (HOT). Feedback is provided by sensing the secondary voltage using a reference zener and an opto-coulper for mains

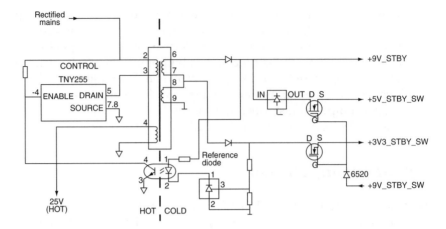

Figure 15.13 *Standby power supply*

isolation. When the output voltage rises above the reference voltage of the diode, its current increases and with it that of the opto-coulper. When this current goes above a threshold of 50 μA, the MOSFET will switch off. When the current falls below 40 μA, the MOSFET will switch on.

Self-oscillating power supply

The functions of regulator and oscillator may be combined in a single transistor to form a *self-oscillating power supply (SOPS)*. The simplest arrangement is that shown in Figure 15.14 in which a secondary winding of the isolating transformer is used as a part of a blocking oscillator. At switch-on the switching transistor TR1 begins to conduct as a result of the forward-bias applied to its base via start-up resistor R1. The collector current increases, which induces a positive voltage across secondary winding S2. This forward biases D1 and further increases TR1 current. When saturation is reached, the increase in current ceases and a negative voltage is induced across S2 to reverse-bias D1, switching off TR1. At this point, the voltage across primary winding P1 reverses and D2 switches on. The tuned circuit formed by primary winding P1 and C1 begins to oscillate, transferring energy from P1 to C1. For the second half of the cycle, when energy begins to transfer back to P1, diode D2 is reverse-biased, oscillations stop and the current in P1 reverses; this causes a positive voltage to be induced across S1, turning TR1 on, and so on. Trigger pulses from a control circuit are used to initiate each cycle to keep the d.c. output constant.

A self-oscillating SMPS circuit based on a Panasonic chassis is shown in Figure 15.15, in which TR3 is the chopper/oscillator transistor and C9 is the blocking capacitor, which discharges through transistor TR2. Pulses derived from the line output stage are fed to the base of TR3, via isolating

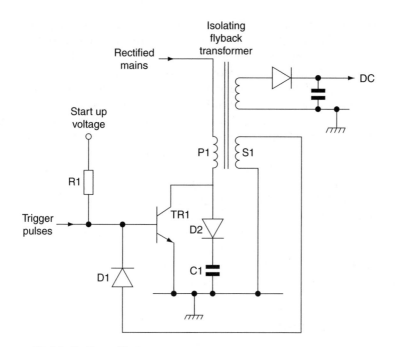

Figure 15.14 *Self-oscillating converter*

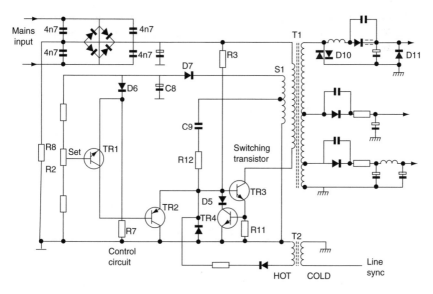

Figure 15.15 *Self-oscillating SMPS*

transformer T2, to turn on TR3 just before its natural turning off point. Regulation is obtained by controlling the conduction of TR2 and thus the discharge of C9. Transistor TR2 is itself controlled by TR1, which senses any change in the voltage developed across C8 caused by changes in the loading of the isolating transformer as well as changes in the mains input. The voltage across feedback winding S1 is rectified by D7. The negative voltage thus produced is used to charge C8 to provide a measure of the loading of the transformer. This voltage is then used to provide TR1 emitter voltage via zener D6. Changes in the loading of the transformer, whether caused by changes in the load current or in the mains voltage, cause the voltage at the emitter of TR1 to change, which determines the base bias of TR2, hence the discharging current of C9. The current through TR1 may also be varied by the 'set' control R2. TR4 provides a degree of protection against excessive current taken by TR3. A large current through R11 will cause TR4 to conduct, taking TR3 base to chassis via diode D5. Further protection is provided by D10 and D11 on the secondary side of the circuit. Transformer T2 provides mains isolation.

Resonant converters

Increased demands for increased power and improve efficiency coupled with stringent electromagnetic interference (EMI) interference levels brought a well-known technology in the field of electric motors and other high-power devices, namely the *resonant converter* to be used in television receivers. Resonant converters will be discussed in detail in the following chapter.

16 Flat panel television receivers

Unlike the cathode ray tube (CRT) display, flat panel displays (FPDs) have a fixed screen format or resolution. If the format of the incoming video is different from the native format of the panel, which it invariably is, image formatting has to take place to re-scale the incoming resolution of the received video to that of the panel. Figure 16.1 shows the main parts of a FPD television.

A FPD TV receiver consists of three main sections:

- Video processing,
- Formatting,
- Display panel.

The video section receives video signals from a variety of sources including

- Tuner (CVBS),
- AV, e.g. SCART (YC, CVBS, RGB, Y, Pr, Pb),
- DVI (*digital video interface*),
- PC (serial, VGA).

One of these inputs is selected by the video selection switch (sometimes called video processor) to be sent to the colour decoder which in conjunction with the comb filter separates the luminance from the chrominance to obtain the three RGB signals. These are then converted into a digital multi-bit stream by the analogue-to-digital converter (ADC). The connection to the next section is normally direct or where a longer connection is needed, as in the case of a PC, it may be attained via a DVI, normally a transition *minimised differential signalling (TMDS)* interface. The formatting section carries out the necessary conversion including *interlace-to-progressive, scan conversion* and *image formatting* to re-scale the image to the native resolution of the panel. The connection to the next stage is made via a DVI normally an LVDS.

The display panel section provides the necessary drive signals for the panel electrodes.

The video processing section is the same as that employed for a CRT television receiver that was explained in detail in the previous two chapters. However, before the video enters the formatting section, it is converted into a 24-bit video with three separate 8-bit groups, one for each primary

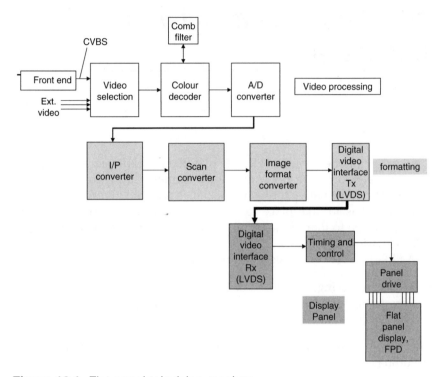

Figure 16.1 *Flat panel television receiver*

colour, R, G and B. Each colour thus has a greyscale of 256 shades resulting in 16.78 million different colours as was explained in Chapter 10.

Video formatting

Figure 16.2 illustrates main functions of a display formatting section of a FPD.

De-interlacing

Unlike the television broadcasts and the CRT in which interlace scanning is used, flat panels are progressively scanned. Hence, the first task of the formatting section is to *de-interlace* the video using an interlace/progressive (I/P) converter.

The process involves the creation of new scan lines and inserts them between the existing lines of an interlaced field. This is not simply inserting even lines of an even field in between the odd lines of the previous odd field. As a result of the time difference between the two fields and the movement that may have occurred during that time, even lines of an even

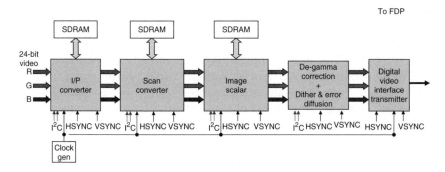

Figure 16.2 *Main functions of a display formatting section*

field are different from the actual missing even lines in a previous odd field.

Several techniques may be used including the very simple line repeat to the very complex motion adaptation and compensation. I/P conversion normally results in artefacts caused by abrupt frame transitions. Some sort of filtering is therefore desirable to smooth out high frequencies.

There are four different techniques available for de-interlacing a video signal: Intra-field, inter-field, motion adaptive and motion compensation.

Intra-field I/P conversion

In this simple technique, data from the same field are used to obtain the missing lines. There are two types of infra-field de-interlacing: *line repeat* and *line average*. In the line repeat also known as line doubling technique, the contents of each line are repeated field by field resulting in the following sequence: 1, 1, 3, 3, 5, 5 and so on for the odd field and 2, 2, 4, 4, 6, 6 and so on for the even field. In the line-averaging technique, also known as '*Bob*' or interpolating technique, a scan line is added to a following line and the average of the two is inserted as the missing line. The sequence for the odd field will thus be

Line 1,
Average of lines 1 and 3,
Line 3,
Average of lines 3 and 5 and so on.

Similarly for the even field.

The major advantage of intra-field I/P conversion is that, unlike other progressive-scan techniques, they do not use multiple fields of data and hence do not require field data storage, translating into lower system manufacturing cost. The penalty paid is in terms of performance, intra-field converters being characterised by horizontal line flicker, still area flicker and a relatively low resolution.

Inter-field I/P conversion

Inter-field I/P conversion, also known as *'Weave'*, combines data from odd and even fields. Two or more fields may be involved. The new line is created by referring to the current as well as to the previous and/or next field. The odd lines of an odd field are incorporated in the even field and vice versa. This technique does not take account of motion between fields resulting; a *'tearing'* artefact may appear in moving areas.

Motion-adaptive I/P conversion

In this technique either intra- or inter-field method is used depending on the motion between successive interlaced fields. A 'motion value' is produced by comparing successive fields and this is then used to determine the *'bias'* towards the intra- or the inter-field method. For large 'motion values' with large variation between the successive video fields, interpolation is biased towards using intra-field components. In this case, use of inter-field data in the presence of motion would result in visible motion artefacts due to the lack of correlation between the current and previous field. However, for small motion values, a high degree of temporal correlation exists between adjacent video fields, and hence a heavy weighting to the inter-field component is given. Motion detection and estimation is therefore critical in this method.

Motion compensation I/P conversion

In this method, the picture content of successive fields are analysed to find out if there is motion from one field to the next. If motion is detected, a full field is then constructed and weaved into the intervening scan lines.

Scan-rate conversion

Once the video has been de-interlaced, a *scan-rate conversion* process may be necessary in order to insure that the input frame rate matches the output display refresh rate. In order to equalise the two, fields may need to be dropped or duplicated. As with de-interlacing, some sort of filtering is desirable to smooth out high-frequency artefacts caused by creating abrupt frame transitions.

Image scaling

Flat panels, plasma, LCD or DLP have a fixed format and a fixed resolution to which all video information must adapt. The purpose of the *image scaling* (or image conversion) chip, normally referred to as the *scalar*, is to convert the incoming signals which may be NTSC, PAL or computer-generated VGA into the native resolution of the flat panel.

Video scaling is very important because it allows the generation of an output stream whose resolution is different from that of the input format. The only way to avoid image scaling is to crop the image to fit within the confines of a smaller panel. This is the cheapest method but not a very satisfactory one.

Depending on the application, scaling can be done either upwards or downwards. It is important for the scalar to distinguish between the video content and the non-video elements such as text. Failure to do so would distort the non-video parts of the picture making text unreadable or cause some horizontal lines to disappear in the scaled image.

The most straightforward methods of scaling involve either *dropping pixels* or *duplicating* existing pixels. That is, when scaling down to a lower resolution, a number of pixels on each line (and/or some number of lines per frame) can be discarded. While this represents a low processing load, the results will yield *aliasing* and visual artefacts.

A small step upward in complexity uses *linear interpolation* to improve the image quality. For example, when scaling down an image, interpolation in either the horizontal or vertical directions provides a new output pixel to replace the pixels used in the interpolation process. As with the previous technique, information is still thrown away, so artefacts and aliasing will again be present.

If the image quality is paramount, there are other ways to perform scaling without reducing quality. These methods strive to maintain the high-frequency content of the image consistent with the horizontal and vertical scaling, while reducing the effects of aliasing. For example, assume an image is to be scaled by a factor of Y in the vertical and X in the horizontal directions. To accomplish this scaling, the image could be up-sampled (interpolated) by a factor, Y, filtered to eliminate aliasing and then down-sampled (decimated) by a factor X. In practice, these two sampling processes can be combined within a single multi-rate filter.

De-gamma correction and error and diffusion

Unlike the CRT, the relationship between the input signal and the brightness obtained from a FPD is linear and since a gamma correction is introduced at the broadcasting stage to compensate for the CRT's non-linear relationship, a *de-gamma correction* is necessary if the signal is to be fed into a panel. The *dither* and *error diffusion* circuits are used to reduce the effect of quantising error which results in a tendency to loose grey scales.

Digital video interface

Direct connection between the processing section and the panel requires a large number of connectors, all carrying high-frequency signals (24 connections for video, three or more connectors for vertical and horizontal syncs and clocks and numerous earth connections). Among other draw-backs, direct connection is susceptible to electromagnetic interference (EMI). For this reason, the preferred method of connecting the processing section to the panel assembly is the use of DVI.

One of the problems of transferring data at a high bit rate is the band-width requirement. For instance, a 24-bit, 640×480 resolution video with a refresh rate of 50 Hz requires $640 \times 480 \times 24 \times 50 = 368.64$ Mbps and for a resolution of 1280×1024 resolution video at a refresh rate of 50 Hz would demand a staggering $1280 \times 1024 \times 24 \times 50 = 1.57$ Gbps.

The normal RS-422, RS-485, SCSI and others have limitations in terms of bandwidth and interference making them unsuitable for use as connec-tors to an LCD, plasma or DLP displays. Hence the need for an interface that is capable of such data rates with low power requirements and low interference. This is what the DVI specifications provide. There are two main types of DVIs: *DVI-D* for digital only and *DVI-I* for both digital and analogue signals.

There are four established techniques: *low-voltage differential signalling* (*LVDS*), *PanelLink* and *TMDS*) and for high-definition video *high-definition multimedia interface* (*HDMI*). In all cases, differential data transmission is used. This is because unlike the single-ended schemes, it is less susceptible to common-mode noise. Noise coupled on the interconnect is seen as a common-mode modulation by the receiver and is rejected. In a single-ended combination Figure 16.3a, noise in the form of a spurious pulse would affect the 'live' wire only, appears at the receiving end and is amplified like the actual wanted signal. However, if a differential signalling is used in which a pair of wires carry signals of opposite polarities, common-mode noise affects both wires equally appearing in the same polarity on both of the wires. When they are subtracted at the receiving end to obtain the signal, they cancel out, a process known as *common-mode rejection*.

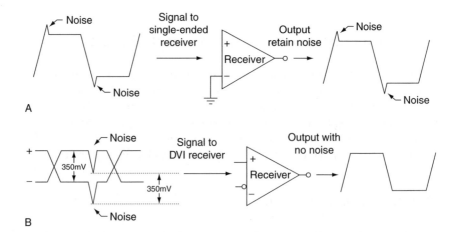

Figure 16.3 *(A) Single-ended communication. (B) Differential signalling*

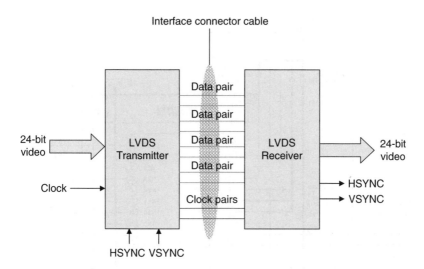

Figure 16.4 *LVDS arrangement*

Low-voltage differential signalling

The LVDS protocol is a unidirectional digital data connection which encodes 24 bits of data using four differential serial data pairs (each pair consists of two twisted wires). The pixel clock is transmitted on a separate differential pair. The differential swing is low (355 mV) and the nominal impedance is 100–120 Ω with a speed of 500 Mbps to 1.5 Gbps (Figure 16.4).

Parallel data are fed into an LVDS transmitter which encodes the incoming data into four serial channels to be sent out on four separate data pairs of wires. The data pairs are not dedicated to any specific part of the video signal with each one transmitting bits from all RGB components. The clock is sent out on a separate pair as shown. At the other end, a LVDS receiver is necessary to decode the data back to its original format or another format as necessary.

The LVDS clock rate is 7× the original *pixel clock*; the pixel clock being the rate at which the pixels are addressed. Given that bits are sent at the LVDS clock rate, then 7 bits are sent every pixel clock pulse. With four LVDS channels, the number of bits per LVDS clock pulse is 4 × 7 = 28 bits. Of these, 24 are used for video. Of the remaining four bits, one bit is allocated to each of the following: frame sync, line sync and display enable (DE). The final bit is a 'custom' bit that may be used by the manufacturer as required for the particular application.

For a basic 24-bit video, a 65 MHz connection can support a standard VGA-resolution panel. Capacity may be increased using an additional four channels, making a total of eight. This is known as *dual-channel LVDS*. Each channel can carry a 24-bit video.

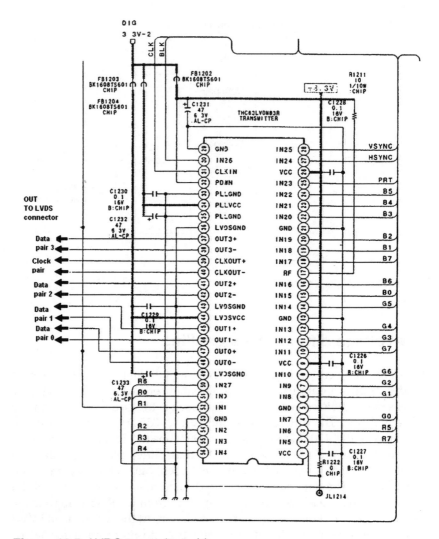

Figure 16.5 *LVDS transmitter chip*

LVDS achieves high aggregated bandwidth, low power consumption and low EMI by virtue of its low switching voltages, soft transition and true differential data transmission.

Figure 16.5 shows a typical LVDS transmitter/encoder chip. The 24-bit video input, H and V syncs and Enable fed into pins marked IN0–IN23. The allocation is as shown with R0 to IN27 (pin 50), R1–R4 to IN1–IN4 (pins 52, 54–56), R7 to IN5 (pin2), R5 to IN6 (pin 3) and so on for the remaining R, G and B bits. Hsync to IN24 (pin 27) and Vsync to IN25 (pin 28) with the bit clock (BLK) to IN26 (pin 30) and finally Enable (PRT) to IN23 (pin 25). The output is a four data differential pairs marked OUT0+ and OUT0− (pins 47 and 48), OUT1+ and OUT1− (pins 45 and 46)

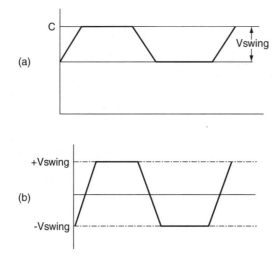

Figure 16.6 *Ideal un-modulated LVDS single-ended and differential signals*

and so on, and a clock pair marked CLKOUT+ and CLKOUT− (pins 39 and 40).

The ideal waveform of a single-ended differential signal, representing either the positive or negative terminal of a differential pair, is in Figure 16.6a. Since the signal is differential, the net signal on the pair has a swing twice that of the single ended as illustrated in Figure 16.6b. The amplitude is in the region of 300–400 mV. The repetition frequency will depend on the LVDS clock which typically is 60 and 80 MHz for standard and high-definition television respectively. This is a good test to see if the LVDS transmitter is functioning normally in no video on the screen.

When video data is transmitted, the LVDS swing is modulated resulting in the waveform shown in Figure 16.7. A typical clock waveform is shown in Figure 16.8.

To observe this differential waveform, an isolated double-beam oscilloscope must be used with its inputs connected to the two data pair wires and their earths shorted.

LDVS point-to-point transmission is suitable for short distances of up to 1 m. For longer cabling, other interface protocols must be used such as PanelLink and TMDS.

PanelLink

This interface is similar to LVDS. The difference is in the number of data pairs; PanelLink has three data pairs, dedicated to red, green and blue respectively compared with the four pairs as is the case with LVDS. Capacity is retained by increasing the data bit per channel to 10 times pixel

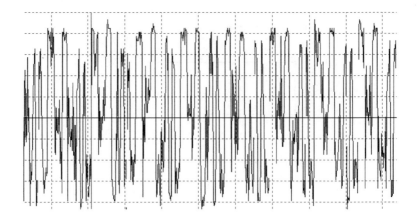

Figure 16.7 *Modulated LVDS data line*

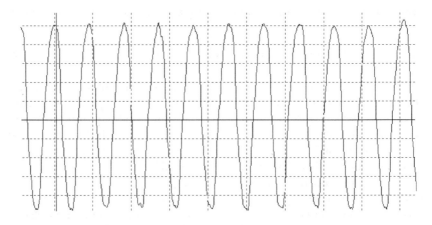

Figure 16.8 *LVDS 60 MHz clock*

clock instead of seven times in the case of LVDS. This is achieved using an encoding technique known as *Silicon Image*. The other essential difference is that PanelLink uses a fixed current instead of a fixed voltage transition. The advantage of this technique is unlike LVDS which is suitable only for short distances; PanelLink may be used for longer distances of up to few meters. The only disadvantage of a fixed current is that a return path has to be provided. The voltage swing for PanelLink is between 500 mV and 1.0 V.

Transition minimised differential signalling

Transition minimised differential signalling, TMDS is a derivative of PanelLink that was developed by VESA for computer, LCD and plasma monitors. It has a speed of 165 MHz going up to 200 MHz. Additional

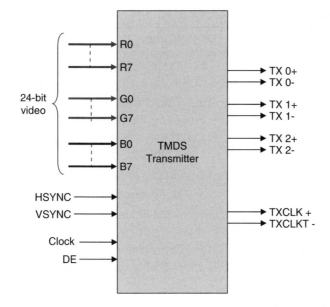

Figure 16.9 *TDMS pin-out*

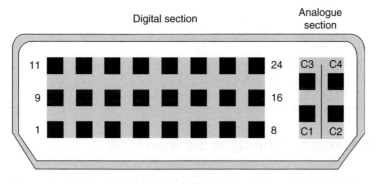

Figure 16.10 *DVI-I (digital and analogue) connector*

pairs may be used for increased capacity. A pin-out of a TMDS transmitter/encoder chip is illustrated in Figure 16.9. The three data pairs are normally marked as TX+ and TX−, one pair for each colour and the clock as TXCLK= and TXCLK−.

DVI connector pin-out

There are two types of DVI connectors: *DVI-I* (digital and analogue) and *DVI-D* (Digital only). *DVI-I* consists of a 24-pin digital section and a 5-pin analogue section as shown in Figure 16.10. The 24-pin digital-only DVI-D connector looks about the same as DVI-I without the 'C' pins. The pin-out assignment is listed in Table 16.1.

Table 16.1 *29-pin DVI connector pin-out*

Pin	Name	Pin	Name	Pin	Name
1	TMDS Data 2−	9	TMDS Data 1−	17	TMDS Data0−
2	TMDS Data 2+	10	TMDS Data 1+	18	TMDS Data0+
3	TMDS Data2 Shield	11	TMDS Data1 Shield	19	TMDS Data0Shield
4	No Connection	12	No Connection	20	No Connection
5	No Connection	13	No Connection	21	No Connection
6	DDC Clock [SCL]	14	+5 V Power	22	TMDS Clock Shield
7	DDC Data [SDA]	15	Ground (for +5 V)	23	TMDS Clock+
8	Analogue vertical sync	16	Hot Plug Detect	24	TMDS Clock−
C1	Analogue Red	−	−	−	−
C2	Analogue Green	−	−	−	−
C3	Analogue Blue	−	−	−	−
C4	Analogue Horizontal Sync	−	−	−	−
C5	Analogue GND Return: (analogue R, G, B)	−	−	−	−

High-definition multimedia interface

The high-definition multimedia interface, HDMI is provided for transmitting digital television audiovisual signals from DVD players, set-top boxes and other audiovisual sources to television sets, projectors and other video displays. It is designed to carry all standard and high-definition consumer video formats as well as high-quality multi-channel audio data. The standard provides content protection to prevent illegal copying. HDMI can also carry control and status information in both directions.

The HDMI protocol supports a number of video formats including:

- $1280 \times 720p$ @ 59.94/60 Hz
- $1920 \times 1080i$ @ 59.94/60 Hz
- $720 \times 480p$ @ 59.94/60 Hz
- $1280 \times 720p$ @ 50 Hz
- $1920 \times 1080i$ @ 50 Hz
- $720 \times 576p$ @ 50 Hz

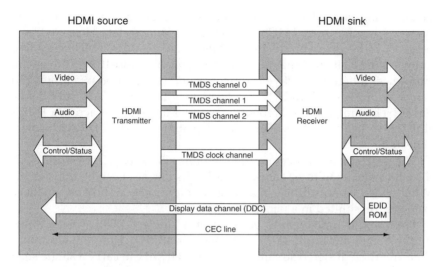

Figure 16.11 *HDMI architecture*

Architecture of HDMI

HDMI system consists of a *source*, also known as encoder or transmitter (TX) and a *Sink*, also known as a decoder or receiver (RX). The data is sent on three pairs of wires carrying a differential TMDS signals, hence the name TMDS channels. A fourth channel carries the clock as shown in Figure 16.11. In addition, a VESA *display data channel* (*DDC*) and a *consumer electronics control* (*CEC*) lines are available as well as a hot plug signal. The DDC is used for configuration and status exchange between a single Source and a single Sink. The DDC is used by the Source to read the Sink's *Enhanced Extended Display Identification Data* (*E-EDID*) stored in a ROM chip in order to discover the Sink's configuration and/or capabilities. DDC specifications are similar to the serial control bus I²C bus specifications outlined in Appendix 2.

The optional CEC is a low-speed bus that works at 400 bps. The protocol provides high-level control functions between all of the various audio-visual products in a user's environment such as one-touch play, one-touch record and menu control.

The hot plug signal detection pin informs the source when it can start to read the E-EDID. A voltage between 2.0 and 5.3 V connected to this pin means that a receiver is connected, and that it could start to read the E-EDID.

HDMI operation

Audio, video and auxiliary data is transmitted across the three TMDS data channels. The video pixel clock is transmitted on the TMDS clock channel and is used by the receiver as a frequency reference for data recovery on the three TMDS data channels. Video data is carried as a series of 24-bit pixels on the three TMDS data channels. TMDS encoding converts the 8 bits per channel into the 10 bit. The 8-to-10 bit conversion introduces a highly effective error-correction code. Video pixel rates can range from 25 to 165 MHz. Video formats with native pixel rates below 25 Mpixels/s require pixel repetition in order to be carried across a TMDS link. $720 \times 480i$ and $720 \times 576i$ video format timings shall always be pixel repeated.

The video pixels can be encoded in either RGB, YC_RC_B 4:4:4 or YC_RC_B 4:2:2 formats. In all three cases, up to 24 bits per pixel can be transferred. While video is transmitted in a serial 10-bit format, audio and auxiliary data are transmitted in a packet structure.

Basic audio functionality consists of a single L-PCM (linear PCM) audio stream at sample rates of 32, 44.1 or 48 kHz. This can accommodate any normal stereo stream. Optionally, HDMI can carry a single such stream at sample rates up to 192 kHz or from two to four such streams (3–8 audio channels) at sample rates up to 96 kHz. HDMI can also carry a compressed (e.g. surround-sound) audio stream at sample rates up to 192 kHz. HDMI can also carry from 2 to 8 channels of one bit audio (1-bit delta-sigma modulated signal stream such as that used by super audio CD).

The HDMI link, which carries both audio and video is driven by a TMDS (video) clock. It does not retain the original audio sample clock. The task of recreating this clock at the receiving end is called audio clock regeneration. This is made simple since the audio and video clocks are generated in the first place from a common clock (known as *coherent clocks*).

For a variety of reasons, an HDMI link may add a delay to the audio and/or video. Due to the uneven transmission of audio data, the delay shall be considered to be the average delay of all of the audio sample packets over the course of three steady-state video frames.

HDMI encoding

Figure 16.12 of an HDMI link shows the three TMDS data channel and the single TMDS clock channel connections between the source and sink. The TMDS clock channel constantly runs at the pixel rate of the transmitted video. During every cycle of the TMDS clock channel, each of the three TMDS data channels transmits a 10-bit character. This 10-bit word is encoded using one of several different coding techniques. The input stream to the Source's encoding logic will contain video pixel data, packet data and control data. The packet data consists of audio and auxiliary data and associated error-correction codes. Each data channel carries one video component (red, green or blue), auxiliary data as follows: audio samples

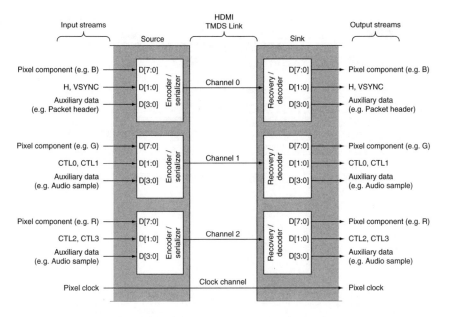

Figure 16.12 *HDMI channels*

and control data in the case of channels 1 and 2 with Channel 0 carrying H and V sync data.

Operating modes

The HDMI link operates in one of three *modes* or *periods*: video data period, data island period and control period. Each period forms a part of a complete TV picture period (Figure 16.13). The video data periods are used to carry the pixels of an active video line. The data island periods are used to carry audio and auxiliary data packets. The control period is used when no video, audio, or auxiliary data needs to be transmitted. A control period is required between any two periods that are not control periods. An example of each period for NTSC television is shown in Figure 16.13. Video is transmitted during the active lines, audio and auxiliary (data island periods) and control data are transmitted during the vertical and horizontal blanking periods as shown.

Content protection, HDCP

Content protection capability is recommended for all HDMI compliant devices. An HDMI compliant source should protect all of the protected

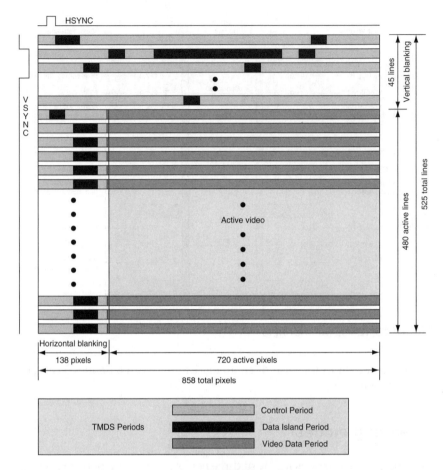

Figure 16.13 *Operating modes of HDMI interface*

audiovisual data. Amongst adequate copy protection technologies that are compatible with HDMI, HDCP is available.

HDMI connectors

There are two types of HDMI connectors:

- Type A connector (Figure 16.14) carries only a single TMDS link and is therefore only permitted to carry signals up to 165 Mpixels/s.
- Type B connector (Figure 16.15) with the dual-link capability which supports signals greater than 165Mpixels/s.

The pin assignment for HDMI connectors types A and B are listed in Tables 16.2 and 16.3. Computability with DVI is essential and this is outlined in Appendix 6.

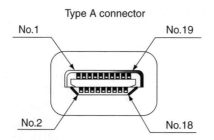

Figure 16.14 *HDMI type A connector*

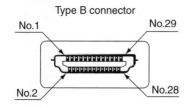

Figure 16.15 *HDMI type B connector*

Table 16.2 *Type A connector pin assignment*

Type A pin	Signal name
1	TMDS Data2+
2	TMDS Data2 Shield
3	TMDS Data2−
4	TMDS Data1+
5	TMDS Data1 Shield
6	TMDS Data1−
7	TMDS Data0+
8	TMDS Data0 Shield
9	TMDS Data0−
10	TMDS Clock+
11	TMDS Clock Shield
12	TMDS Clock−
13	CEC
14	Reserved (in cable but N.C. on device)
15	SCL
16	SDA
17	DDC/CEC Ground
18	+5 V Power
19	Hot Plug Detect

Table 16.3 *Type B connector pin assignment*

Type B pin	Pin assignment
1	TMDS Data2+
2	TMDS Data2 Shield
3	TMDS Data2−
4	TMDS Data1+
5	TMDS Data1 Shield
6	TMDS Data1−
7	TMDS Data0+
8	TMDS Data0 Shield
9	TMDS Data0−
10	TMDS Clock+
11	TMDS Clock Shield
12	TMDS Clock−
13	TMDS Data5+
14	TMDS Data5 Shield
15	TMDS Data5−
16	TMDS Data4+
17	TMDS Data4 Shield
18	TMDS Data4−
19	TMDS Data3+
20	TMDS Data3 Shield
21	TMDS Data3−
22	CEC
25	SCL
26	SDA
27	DDC/CEC Ground
28	+5 V Power
29	Hot Plug Detect
23	No Connect
24	No Connect

Chip count

Modern IC design employs such high degree of integration that most of the functions of the video processing and formatting can be carried out within two or even one single chip. The following are some examples used by some TV manufacturers.

Philips 2-chip solution

This set consists of two chips named by Philips as Hercules (e.g. TDA 120×1) and a Genesis Scalar (e.g. GM1501). The Hercules has its own embedded Flash memory for its start-up and processing routines. CVBS

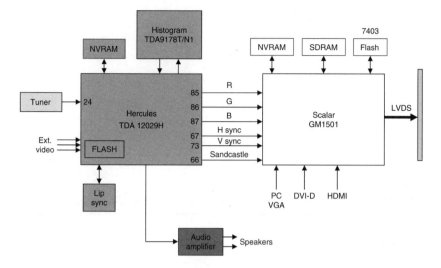

Figure 16.16 *The Philips 2-chip solution: Hercules and GM Scalar*

from the tuner is directly fed into pin 24 and RGB output is obtained at pins 85–87 (Figure 16.16). Non-volatile RAM (NVRAM) stores customer settings such as brightness and colour as well as channel information. The *Histogram* carries out what is known as histogram equalisation of contrast, a process which spreads out the most-frequent intensity values across the picture. This allows areas of lower local contrast to gain a higher contrast without affecting the global contrast. It is most useful in images with backgrounds and foregrounds that are both bright or both dark. The scalar carries out all necessary functions to convert the incoming image format into the panel's native format including interlace-to-progressive and scan conversion. The chip also incorporates an embedded LVDS transmitter/encoder. Flash and NVRAM are external discrete chips. So SDRAM is used for field storage. The full set of functions is listed in Table 16.4.

Panasonic GC3 2-chip solution

This chip set consists of three chips: GC3FM, GC3i and GC3FS (sub-picture) (Figure 16.17).
GC3FM:
      Sync separator
      10-bit ADC
      Colour decoder
      3D comb filter
      NTSC/PAL converter
GC3i:
      Format converter
      Image re-sizing

        Gamma correction
        LCD AI
GC3FS:
        8-bit ADC
        NTSC/PAL converter
        Colour decoder
        Sync separator
        2D comb filter

Table 16.4 *Functions of the 2-chip set Hercules and GM scalar*

Hercules
    Source select
    Video/audio processor
    Sync separator
    Colour decoder
    Gamma correction
    On-chip micro-controller
Embedded flash memory
    GM Scalar
    Video processor
    Format converter
    LVDS transmitter/encoder

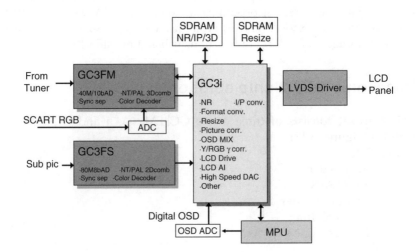

Figure 16.17 *GC3-2-chip solution*

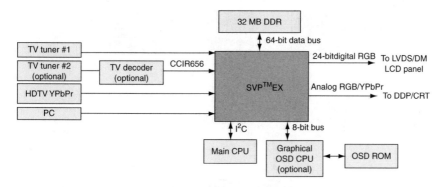

Figure 16.18 *Super video processor*

Samsung single-chip solution

The single chip which caries all processing and formatting functions is a super video processor (SVP). Its functions may be summarised as follows (Figure 16.18):

- 3D colour decoder
- motion adaptive de-interlacing
- digital noise reduction
- image scaling
- APL
- dynamic picture enhancement

A typical configuration is shown in Figure 16.18.

Plasma panel power generation

With the exception of the plasma panel, power generation for FPDs follow very similar lines to those described in Chapter 15. The plasma panel is driven by high voltage, fast switching scan and sustain signals consume a large amount of power generating a large amount of heat. In such conditions it is essential to reduce the amount of heat generated and improve efficiency. In addition to high efficiency, PDP power supplies have to have fast transient response, low noise and low EMI.

A plasma display panel requires the following d.c. power supplies:

- Multi-power for processing chips, audio amplifiers and other semiconductors devices: 3.3 V, 3.6 V, +15 V, −15 V, 5 V.
- Address electrode power supply (70 V).
- Sustain and scan power supply (200 V).

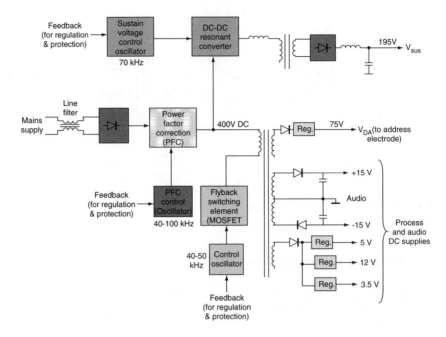

Figure 16.19 *Power generation and distribution for a plasma television receiver*

A PDP power distribution system is shown in Figure 16.19. It consists of the following elements:

- main full-wave bridge rectifier,
- main line filter,
- power factor correction (PFC),
- multi-power supply for ICs, audio, etc.,
- sustain power supply VSUS,
- protection circuits (no shown).

Power factor correction, PFC reduces reactive power and introduces pre-regulation. The sustain power supply provide over 75% of the entire power requirement of the PDP. To ensure high efficiency, low noise and EMI, soft-switching resonant converters are used.

Power factor correction

The purpose of PFC is to reduce the reactive load current which is of no real value to the device. Only resistive power consumption is of any value as far as the device is concerned.

A fully resistive load when fed with a sinusoidal voltage is said to have a power factor of one. Power factor is defined as

$$\frac{\text{Mean power}}{V_{rms} \cdot I_{rms}}$$

If the voltage and current are pure sine waves, the power factor is given as $k = \text{Cos } \theta$, where θ is the phase difference between the current and voltage waveforms. Since for a resistive load, current is in phase with the voltage, it follows that $\theta = 0$ and $\text{Cos } \theta = 1$. Hence the power factor $k = 1$.

For a purely resistive load, power $P = V_{rms} \times I_{rms}$. However, for a reactive load, the power factor has to be taken into account and power is given as

$$P = V \times I \text{ Cos } \theta$$

Thus, with a power factor (Cos θ) of 0.5, only half of the power is useful power and the other half is completely wasted power.

A power factor of one is the highest power factor possible in which all power consumed by the device is useful power. On the other extreme, a purely reactive load, e.g. a pure inductor or a pure capacitor, has a power factor of zero since $\theta = 90°$ and $\text{Cos } 90° = 0$. A power factor less than one can also be obtained if the voltage and/or current waveforms are not pure sinusoidal, if they contain harmonics, even though the load is resistive. This is the case with all types of rectification including controlled rectification. For this reason, the power rectification and supply circuits look reactive to the mains supply with a power factor of between 0.5 and 0.7.

To rectify this situation, PFC is employed to ensure that as little power as possible is wasted. Furthermore, under European law, mains harmonics are restricted for devices of 75 W and over. A PFC circuit is therefore needed to ensure that these harmonics do not exceed the permitted level. A further reason for using PFC is to reduce the amount of imbalance on the three-phase mains power supply caused by low power factors. Such interference becomes noticeable with high power requirements for equipments such as electric motors and PDPs. With PFC, a power factor of 0.9 is possible.

With the simple rectifier circuit illustrated in Figure 16.20, the diodes act as switches which conduct for a short period of time depending on the load current even if the load was fully resistive.

The current flows only when a pair of diodes conducts naturally, i.e. when they are forward biased during the period when the instantaneous value of mains voltages on their anodes exceeds the capacitor voltage on the cathode side. The current waveform is therefore a relatively narrow pulse as illustrated in Figure 16.20. While the fundamental of the current waveform is in phase with the incoming voltage, its harmonics are not resulting in a low power factor. This may be tolerated where power requirements are low (10s of Watts). However, when power consumption

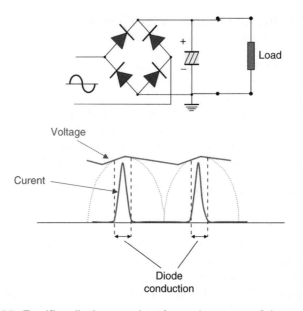

Voltage

Curent

Diode
conduction

Figure 16.20 *Recifier diodes conduct for a short part of the mains input*

is high, in the region of 100s of Watts, then the losses due to low power factors are high and must be avoided. This is carried out by a PFC circuitry.
Instead of the capacitor receiving a charging current once every half cycle,
PFC ensures that the charging current is small but more frequent; 100s of
times every half cycle.

Figure 16.21 shows that when Switch S is closed, I flows through inductor L in a linear manner storing energy in the inductor. When S is open,
current is abruptly cut and back e.m.f. across the inductor provides a forward bias to diode D, inductor L and capacitor C from a resonant circuit
and charging current I_C flows to charge capacitor C. Energy stored in the
inductor is now transferred to the capacitor. When the capacitor is fully
charged and I_C is zero, diode D is reverse biased. At this moment, S is
closed and current begins to flow through L and so on. The average current taken from the mains now follows the shape of the voltage as shown
in Figure 16.21 resulting in a power factor as high as 0.85 or even 0.9. The
switching waveform for S is the pulse modulated waveform shown with
a frequency of 40–100 kHz. Note that the mark-to-space ratio of the PMW
changes with the sine wave input. The off period increases gradually as
the input goes up to peak and decreases as it goes down to zero while the
ON period remains the same throughout.

The main element of a PFC circuit is shown in Figure 16.22. A power
MOSFET, Q_1 is used as the switching element. Q_1 is driven by bipolar tran-
sistor Q_2. The pulse width modulated (PWM) pulses are produced by the
PFC control oscillator. The duty cycle is determined by the shape and

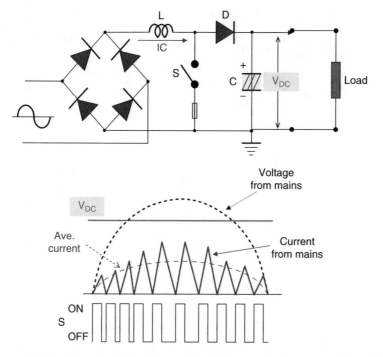

Figure 16.21 *The principle of operation of power factor correction, PFC*

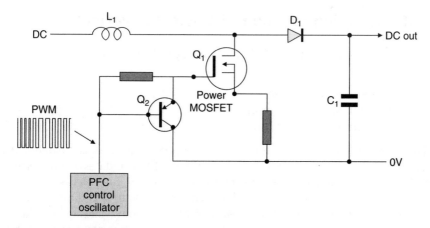

Figure 16.22 *The basic elements of PFC circuit*

amplitude of the mains sine wave as well as the output voltage and current levels. In this respect, PFC acts as a voltage pre-regulator.

Figure 16.23 shows the main elements of a practical PFC circuit used by Panasonic employing two balanced MOSFETs for improved PFC.

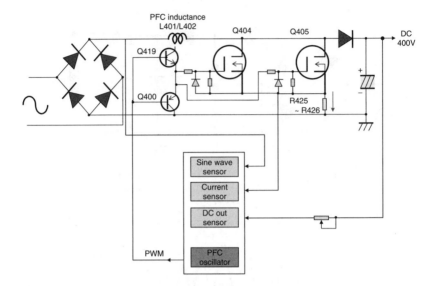

Figure 16.23 *A practical PFC circuit*

Sine wave, current and DC output sensors are used to ensure that the current waveform follows as closely as possible the mains voltage.

Soft-switching resonant d.c.–d.c. converters

The power supply of a PDP TV employs three types of d.c.-to-d.c. converters:

- linear regulated converter,
- switched mode power supply (SMPS),
- soft-switching resonant converter.

The linear regulators and the SMPSs have been discussed in detail in Chapter 15. The type used here is normally the flyback d.c.–d.c. converter. SMPSs suffer from a number of limitations: high EMI, high stress levels on the switching devices and limited switching speeds to less than 100 kHz. These limitations are mainly a result of the switching taking place at a high current and/or voltage levels. Hence these converters came to be known as *hard-switching* converters. When switching at high frequencies, these converters are associated with high power dissipation and high EM interference caused by high-frequency harmonic components associated with their quasi-square switching current and/or voltage waveforms.

To overcome the limitations of the traditional SMPS, a third-generation power converters known as *soft-switching* (*resonant*) converters were introduced in the late eighties.

Like SMPSs, soft-switching power units are d.c.–d.c. converters. D.c. from a rectifier is first converted to a.c. through a switching element. This a.c. is fed into the primary of a transformer with a number of secondary windings for multi-outputs. The secondary outputs are then rectified in the normal way to produce the various d.c. voltage levels. The new element that resonant converters bring is switching takes place when the voltage across the switching element is zero, known as *zero voltage switching* (ZVS) or when the current through the switching element is zero, known as *zero current switching* (ZCS). ZVS and ZCS are produced by using resonant circuits; hence these converters are also known as resonant or quasi-resonant converters. The result is low switching power dissipation and reduced component stress. These in turn result in increased power efficiency, reduced size and weight, faster responses and reduced EMI problems. The reduction in losses due to zero voltage or ZCS makes it possible to utilise much higher switching frequencies in 100 s kHz or even few MHz. And since the size and weight of the magnetic components (inductors and transformers) and capacitors are inversely proportional to the switching frequency, the higher the latter, the smaller the size and weight of the power supply improving its power density. A further advantage of soft-switching resonant converters is, because of their switching frequencies, they can utilise transformer and switching elements' leakage inductance and capacitance respectively as part of the resonant circuit.

Principle of operation of resonant converters

When a resonant circuit is fed with a $+10\,V$ step voltage, it oscillates resulting in what is known as ringing. The capacitor charges up to $10\,V$, at which point current ceases and the capacitor begins to discharge causing current to flow in the opposite direction transferring energy from the capacitor to the inductor. The current continues to flow in that direction until the capacitor is fully charged to $-10\,V$ at which point, current ceases and swings back in the opposite direction and so on as illustrated in Figure 16.24. As can be seen, the resonant waveform has zero current when the voltage is at a peak and zero voltage when it crosses the $0\,V$ line. Resonant converters use this fact to ensure that switching takes place at one of these points.

Steady-state analysis of a basic ZCS resonant converter

Refer to Figures 16.25 and 16.26. In the steady state, load current I_{L2} is constant. The cycle starts when MOSFET Q_1 is turned ON by a control pulse while D_1 is also ON. With capacitor C_1 short circuited by D_1, I_{L1} flows through inductor L_1 and diode D_1. I_{L1} increases in a linear fashion and

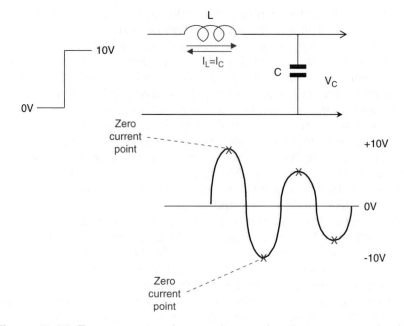

Figure 16.24 *Zero current and zero voltage points in a resonant circuit*

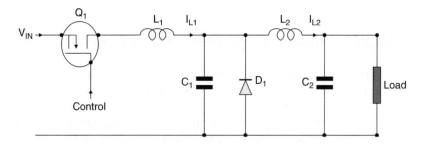

Figure 16.25 *Zero current switching (ZCS) resonant converter*

energy is stored in the inductor. When $I_{L1} = I_{L2}$, diode current drops to zero and the diode turns OFF naturally (ZCS). With D_1 open circuit, L_1 and C_1 form a resonant circuit. I_{L1} increases to a peak in a sinusoidal manner and energy is transferred to the capacitor as I_{C1} charges capacitor C_1. Once I_{L1} reaches its peak (at which point the voltage across C_1 is equal to the input voltage V_{IN}), it begins to drop and when $I_{L1} = I_{L2}$, C_1 is fully charged and I_{C1} drops to zero. When I_{L1} drops further below I_{L2}, I_{C1} is reversed and the capacitor begins to discharge transferring energy to inductor L_2. When I_{L1} drops to zero, MOSFET Q_1 switches OFF naturally (ZCS) keeping I_{L1} at zero. The capacitor continues to discharge and when its voltage falls to zero and $I_{C1} = I_{L2}$, diode D_1 switches ON naturally (ZCS) to short circuit C_1 and break the resonant circuit. This state continues until Q_1 is switched on by a control pulse to commence the next cycle and so on. As can be seen,

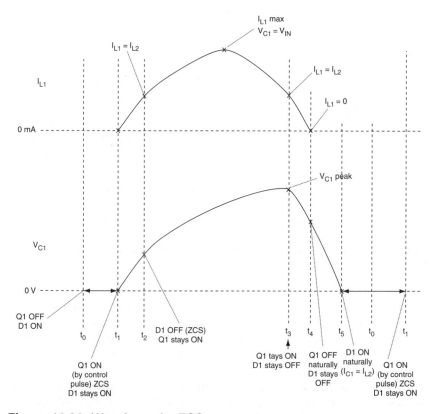

Figure 16.26 *Waveforms for ZCS resonant converter*

the period Q_1 remains ON is fixed by the resonant frequency of L_1 and C_1 while the time it is OFF is determined by the control pulse which is varied as necessary to regulate the voltage. This type of resonant converter is known as '*fixed on-time, varied off-time*'. The ON period of the power MOSFET switch is the resonant period of L_1/C_1, known as the *tank*. For heavy loads, the resonant off-time is made shorter.

Zero voltage switching

A resonant converter which uses ZVS is illustrated in Figure 16.27 with L_2/C_2 forming a low-pass filter. In the steady-state condition, load current I_0 is equal to the current through L_2. Starting with the MOSFET on, its current I_T and I_{L1} are equal to I_{L2} and Diode D_2 is reverse biased. The voltage across the MOSFET is zero. When the transistor is turned off by a control pulse to its gate, current is diverted into C_1. Capacitor C_1 charges up until, when fully charged, diode D_2 becomes forward biased and conducts and C_1 and L_1 begin to oscillate going up to a peak after which, the voltage

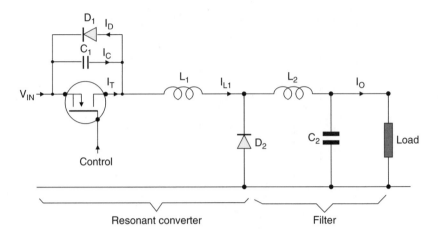

Figure 16.27 *Zero voltage switching (ZVS) resonant converter*

across C_1 attempts to reverse and diode D_1 starts conducting. During this time, the MOSFET is triggered to conduct, but remains off while D_1 is on. As soon as D_1 turns off when I_{L1} drops to zero and begins to reverse, the transistor turns on (ZVS) and remains on until it is turned off by a pulse and so on. The MOSFET turns off naturally and is turned on by a pulse, i.e. *'fixed off-time, varied on-time'*.

A circuit diagram for a resonant converter used to supply the sustain voltage for a 42 in. plasma panel is shown in Figure 16.28. Two pairs of back-to-back Power MOSFETs Q12/Q15 and Q13/Q14 are used as the switching transistors driven by Q10 and Q11. They are fed with 400 V d.c. from a flyback converter. Two anti-phase PMW control pulses arrive from the control panel to Q2 and Q3 to control the switching of each pair. The output for the sustain and the scan electrodes of the plasma panel are provided full-wave rectifiers connected to the secondary of the output transformer. Feedback to the control panel is obtained via Q6 and Q7. Although power MOSFET have been used almost universally, current-driven insulated gate bipolar transistors (IGBT) (Figure 16.29) may also be used.

Energy recovery

Stray capacitances, being of very small value consume negligible power that is normally neglected. However, with few million pixels and each pixel consisting of three cells, the total amount of stray capacitance of a PDP mounts up to a comparatively high level. In a VGA-resolution PDP for instance, the number of stray capacitors is $852 \times 480 \times 3 = 1.23$ million. Coupled with high-amplitude drive pulses, power consumption by these stray capacitors becomes too large to be neglected. Hence the need to recover the energy stored in stray capacitors to be re-used for discharge.

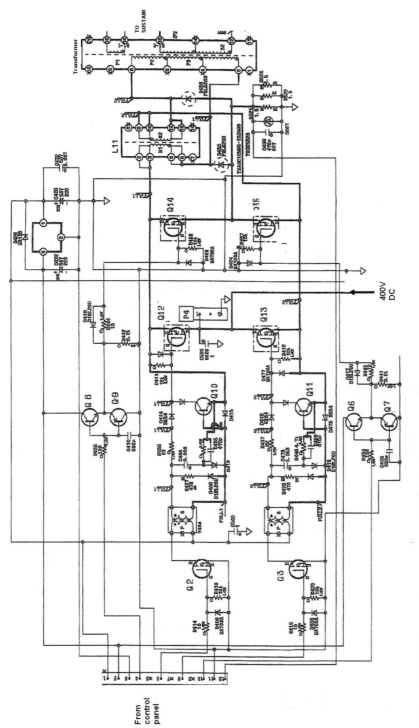

Figure 16.28 *Resonant converter circuit diagram*

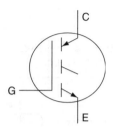

Figure 16.29 *IGBT symbol*

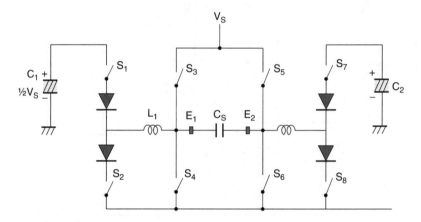

Figure 16.30 *Energy recovery*

The main energy consumption is in the sustain drive and hence the main energy recovery is carried out at these drives. However, energy recovery for the scan drive may also be implemented.

The principle of energy recovery is to use an inductor to resonate with the stray capacitance and at the moment when energy is transferred from the stray capacitance to the inductor, resonance is terminated by a controlled switching element, normally a MOSFET. Subsequently, the energy stored in the inductor is transferred to a PFC storage capacitance.

Figure 16.30 illustrates the principle of energy recovery. C_s is the stray capacitance between electrodes E_1 and E_2. C_1 is the PFC storage capacitor into which recovered energy from the stray capacitor will be deposited. The sequence starts when the rising edge of the sustain pulse is applied between the sustain and scan electrodes (E_1 and E_2) and continues as follows:

Step1   At rising edge of sustain pulse:
              S_1 and S_6 close
              L_1/C_s resonate
              C_s charges up. This is the energy that has to be recovered
Step 2 S_1 and S_6 open
              S_3 close
              C_s is clamped to V_s

Step 3   At the falling edge of the sustain pulse:
S_3 open
S_2 and S_6 close
L_1/C_s resonate
C_s discharges
Energy is transferred from C_s to L_1

Step 4   At peak of L_1/C_s resonance cycle:
S_3 open
S_4 and S_1 close
C_s clamped at 0 V by S_4
L_1/C_1 resonate
Energy is transferred from L_1 to C_1
Energy recovery accomplished

Fault finding on plasma television

Before looking into faults and fault-finding techniques, let us examine the sequence of events following a cold switch of a plasma television receiver. Figure 16.31 shows the component parts of a Plasma TV receiver capable of receiving analogue terrestrial broadcasts. The audio section is not shown as it is relatively independent once stereophonic audio is extracted

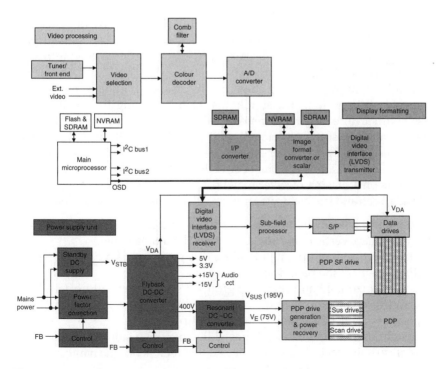

Figure 16.31 *Component parts of a Plasma television*

following demodulation at the tuner/front end. Most modern plasma receivers include a DVB decoder circuit which can then receive and decode digital terrestrial TV broadcasts, known as free view in the UK. Integrated digital television (iDTV) will be considered later in Chapter 18.

When the TV receiver is switched on, d.c. power builds up in three distinct stages: PFC followed by the Flyback d.c.–d.c. converter followed by the resonant converter. First the PFC control oscillator generates the necessary PWM waveform to regulate the PFC section and produce a d.c. output that is fed into the flyback converter. Once this d.c. voltage is up, the flyback converter begins the switching operation to build up the 400 V d.c. output. This d.c. voltage is then used as the d.c. input for the resonant converter which generates the d.c. voltages required to generate the sustain and scan waveforms. Switched d.c. is obtained from the secondary of the flyback transformer to generate the necessary processing and audio d.c. power.

The build-up of the d.c. power is followed by the microprocessor start-up routine. First the microprocessor will download the start-up program from the Flash memory chip into SDRAM to commence the start-up routine. If Flash is faulty or the program is corrupted, the process will be halted and the set will go into standby. The microprocessor will also examine its NVRAM for individual and other specified settings to incorporate them into the start-up routine. Once again, if NVRAM is faulty or its contents are corrupted, start-up will be halted and the set will go into standby. The start-up routine involves testing and initialising all programmable and processing chips to be ready to receive and process the video and audio information. This process is invariably carried out by I^{2}C serial control bus. More than one I^{2}C is used. As a result of increased integration, some of the chips shown as discrete units may be incorporated into a single chip as was indicated earlier. For instance flash and NVRAM may be imbedded within processing chips. In this case, upgrading or re-programming Flash is no longer a simple matter of changing the memory chip itself with an already programmed Flash. Upgrading must now be carried out by running software on a PC and downloading it on the imbedded Flash via a serial port. Other advanced integration includes incorporating the scan, the I/P and the format converters into a single image scalar chip.

Faults on a plasma receiver may be divided into several categories.

Data, scan and sustain drive faults are as follows (refer also to Chapter 10):

- Single black horizontal line suggesting a fault in a scan drive.
- Single black vertical line suggests a fault in a data drive.
- Black vertical band points to failure of a group of data lines corresponding to where the band is on the screen.
- Black horizontal band points to grouped scan drives failure. Check the scan pulse waveforms at the output of the drives corresponding to the position of the band, line by line to confirm.
- Sustain or common drive failures will generally result in a blank screen as all sustain electrodes are connected together, hence the name common.

However, in some cases, the sustain electrodes are divided into separate groups, a sustain failure may result in a horizontal black band. Check both scan and sustain waveforms.

• Vertical black, white or colour lines 1 or 2 cm apart. The cause is a failure in the shift register embedded within one of the data tape connector which forms part of the PDP. The panel has to be replaced.

In all of the above, the fault could be in the display panel itself, in which case, the panel must be replaced.

Pixel defects

This may be a single pixel failure or several pixel failure or cluster failure. They both point to a fault in the panel itself. Examples of these have been illustrated in Chapter 10.

Picture faults

These include *no picture* or *broken picture* and/or *multiple images* and/or *colour distortion*. The classic no picture is when the set is stuck in standby. In this case there will be no sound either which generally but not exclusively points to a power supply fault. This is covered later in this chapter. No picture with normal sound suggests a video, a display formatting, a sustain, a scan or a data drive, a power supply or a display panel fault. Press Menu or another option for an on-screen display. If that appears on the screen, then the fault precedes the image scalar. If OSD is not displayed, the possible fault may lie in the image scalar, the LVDS chip and connector, the scan, sustain or data drives, the power supply or the PDP itself. Check the output of the LVDS transmitter at the LVDS connector. A square wave will be present on the data line if video is missing and the LVDS transmitter is working properly. Also check the LVDS clock. Check the inputs and outputs of the scalar chip, sustain and scan drive waveforms and the power supply.

A fatal fault in one of the processing chips such as colour decoder, image scalar, LVDS encoder or sub-field processor will normally result in no picture with sound OK. A fatal fault in memory chips (flash, NVRAM, SDRAM) will also result in no picture and no sound with the set normally stuck in standby. Similarly, the absence of clocks, control signals such as Enable, Chip Select and signals on I^2C bus lines will result in no picture or stuck at standby. However, a partial fault such as a dry joint in a memory or processing chip, a data or address line stuck-at-zero or one, or a loose connection will result in non-fatal fault with a picture breakup and/or multiple images and/or colour distortion. Typical symptoms of non-fatal faults are illustrated in Figure 16.32.

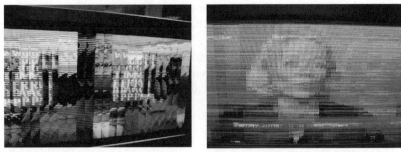

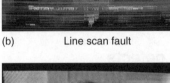

(a)   Image format fault   (b)   Line scan fault

(c) LVDS connector dislodged slightly   (d)   One colour bit (R0) missing
     – one data pair disconnected

(e)   Scan Memory fault (address line shorted to 0V)

Figure 16.32 *Some typical non-fatal faults. Image format fault. Line scan fault. LVDS connector dislodged slightly – one data pair disconnected. One colour bit (R0) missing. Scan memory fault (address line shorted to 0 V)*

DC power failure

These faults normally lead to an overall failure of the d.c. power distribution system apart from Standby and possibly audio power. Failure in one section of power generation would inevitably lead to the other parts turning off as the protection circuit kicks in. The task is to find the part of the power distribution system that is the initial cause of the failure. The following is a suggested series of tests for plasma screens:

- Is the set in Standby?,
- Check d.c. outputs,
- Check the d.c. voltage (around 400 V) from Flyback switching transistor,
- Observe the d.c. build-up of the various d.c. power generators (PFC, Flyback converter and resonant converter) as well as their control signals in turn immediately after switch on from cold. The manner in which the d.c. outputs and the PWM signals are built up would give a good indication as the location of the fault. PWM control signals and d.c. outputs should be present for a very short time before the set trips and goes into standby.

Fault finding in an LCD television receiver

Fault-finding procedure for LCD receivers follows closely that described for plasma panels.

Figure 16.33 shows the components parts of an LCD TV receiver. The start-up sequence is the same as that for plasma screens except for the absence of the resonant converter supplying the sustain voltage.

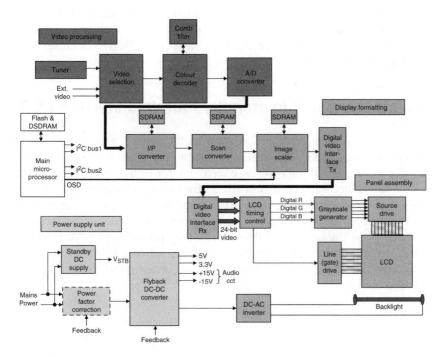

Figure 16.33 *Component parts of a LCD television*

For LCD panels, a similar fault diagnosis procedure may be followed:

- Is the set in Standby?,
- Check DC outputs,
- Check voltage from Flyback switch,
- Check backlight drive voltage by placing a CRO probe in the proximity of the CCF tube step-up transformer. A sine wave signal should be picked with a frequency in the region of 50 kHz,
- Check the PWM tube brightness control pulse using a CRO or a logic probe,
- Check the inverter oscillator DC supply,
- Check inverter oscillator output.

PFC is not shown, although many large screens would incorporate it in their power distribution network.

Faulty CCF tube

For LCD panels, a fault in one CCF tube (if more than one is used), would normally result in the inverter being turned off by the micro-controller either immediately or after a short delay. With one tube malfunctioning, the brightness will still be distributed evenly and the picture will look normal with almost un-noticeable low brightness. However, once one tube goes down, an error signal is sent to the micro-controller and the inverters will be turned off and the set would probably go into standby. Check the over-voltage protection/error signal. Check the PWM tube brightness control waveform; if present, then the fault is in the oscillator.

The classic stuck in standby fault (plasma and LCD)

Invariably, a fault in one section such as the power supply, micro-controller and even video and formatting section leads to the set going into standby as a result of the extensive feedback protection system employed in PDP sets. Once into standby with all other power turned off and all processing and control halted, it is difficult to find the original cause of the fault. However, tests during the start-up sequence in the period between switch on and the set going into standby may provide important clues as to the location of the fault. The precise start-up process differs from one set to another, but they all follow a general sequence which has been summarised above.

17 TV sound, mono and NICAM

Back in Chapter 1, we briefly touched on sound systems for TV. Here the subject will be examined more closely, with particular regard to NICAM for terrestrial analogue TV broadcasts. There remains one other facility, namely teletext provided by analogue terrestrial TV broadcasting. Teletext is a system of broadcasting information in the form of pages of text and graphics. Teletext information is specially coded and transmitted during the unused scanning lines at the start of each field. It is quickly being overtaken by the far superior interactive facilities provided by digital TV broadcasting.

TV FM mono system

As described earlier, the monaural sound is transmitted on its own frequency-modulated RF carrier with ±50 kHz maximum deviation at a level 10 dB below that of the vision carrier. In the USA, the spacing between the sound inter-carrier and the vision carrier is 4.5 MHz above the vision carrier. For the UK, it is 6 MHz above the vision carrier frequency and we will use that for descriptive purposes.

At the receiving end, a sound intermediate frequency (SIF) of 33.5 MHz is obtained. This low-level constant-amplitude signal goes to the vision demodulator where it beats with the vision carrier to reproduce the frequency-modulated 6 MHz signal which is then selected by a filter, amplified and demodulated. The inter-carrier technique has several advantages. One of these is that tuning errors and drift in the local oscillator have no effect on the inter-carrier frequency. This is because a drift in the oscillator frequencies causes the two IFs to change by the same amount, keeping the difference between them constant. Frequency drift does not have the same effect on the vision carrier because the vision carrier is amplitude modulated. Furthermore, the sound carrier benefits from the gain provided by the vision IF amplifier; and the sound processing circuit is a simple one.

The inter-carrier 6 MHz beat frequency is obtained from a device such as a diode which, when used over the non-linear part of its characteristic, produces the sum and difference of two separate input frequencies. The resulting 6 MHz beat frequency retains the FM information which, when demodulated, reproduces the original sound signal. Ideally, the sound inter-carrier should have constant amplitude. However, some amplitude modulation will be present, caused by the amplitude-modulated vision IF.

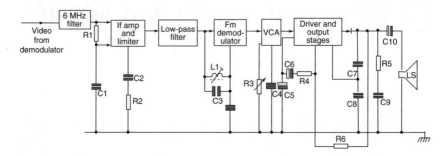

Figure 17.1 *Mono TV sound retrieval system*

This interference will cause what is known as *vision buzz* on the audio output, unless it is removed by a limiting circuit before demodulation.

Receiver circuit

Figure 17.1 shows a typical inter-carrier sound system for a basic TV set. The 6-MHz sound carrier is selected from the demodulated vision carrier by a ceramic filter, resonant at 6 MHz and having a bandwidth of about 200 kHz. The separated sound carrier passes through several stages of amplification followed by a limiter. The limiter clips off any amplitude modulation caused by the vision signal: the AM rejection is about 55 dB. This clipping action produces a square wave output, rich in harmonics of the 6 MHz carrier frequency. They are suppressed by a low-pass filter which restores the carrier wave shape to approximately a sine wave for application to the FM demodulator. The reference carrier is generated by the high-Q tuned circuit L1/C3.

Next comes a voltage-controlled amplifier (VCA) whose gain depends on the d.c. voltage set by R3. The demodulated audio signal is de-emphasised next, by an RC network. The audio signal is now ready for application to the driver and output stages. The output stage generally consists of a push–pull pair of transistors operating in class B (or class D) mode, the d.c. mid-point voltage being isolated from the loudspeaker by coupling capacitor C10. Resistors R4 and R6 form part of a negative feedback loop to control the gain of the output stage. Capacitors C7 and C8 set the amplifier's upper frequency limit. R5 and C9 suppress any tendency for HF oscillation due to the inductive load of the loudspeaker and its wiring.

Frequency demodulation

In frequency modulation, the carrier frequency deviates above and below its centre frequency in accordance with the amplitude of the modulating signal. The FM detector or demodulator thus has to convert the frequency deviations back into the original signal.

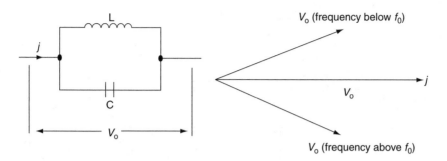

Figure 17.2 *Principle of frequency demodulation*

There are two main types of FM demodulator that are used in TV receivers: the *ratio detector* and the *quadrature (coincidence)* detector. The ratio detector uses discrete components and has the advantage of providing its own rejection of amplitude modulation. Although more complex, the quadrature detector lends itself more easily to IC packaging and hence its extensive used in modern TV receivers.

The operation of the FM detector is based on the fact that the impedance of a tuned circuit is resistive at the resonant frequency but becomes inductive at lower frequencies and capacitive at higher frequencies. Consider the LC circuit in Figure 17.2. At the tuned frequency f_0, the circuit is purely resistive with voltage V_0 in phase with the current i. If the frequency falls below the resonant frequency, voltage V_0 leads the current and if the frequency rises above the resonant frequency, the voltage lags the current. The amount of phase shift is determined by the deviation away from the tuned frequency of the circuit. If the input to the circuit is the FM carrier, then provided the circuit is tuned to 6 MHz, the phase shift represents the original modulating sound signal. In the FM detector, this phase shift is translated into a voltage variation to reproduce the audio signal.

Figure 17.3 shows an IC package (TBA 120) incorporating a quadrature detector together with an amplifier. R302 is the volume control and S300/C304 forms a 6-MHz tuned circuit external to the chip.

De-emphasis

It is common for FM broadcasting to introduce pre-emphasis at the transmitter; high-frequency audio signals are boosted in comparison with middle and low frequencies. The purpose is to swamp most of the noise, the greater part of which tends to be high frequencies. At the receiving end, the signal must be subjected to de-emphasis by attenuating the high-frequency treble, content in a similar but opposite way to the emphasis at the transmitter. Emphasis is carried out by the use of a filter and is expressed in terms of the time constant of the filter. In the UK, a pre-emphasis of 50 μs

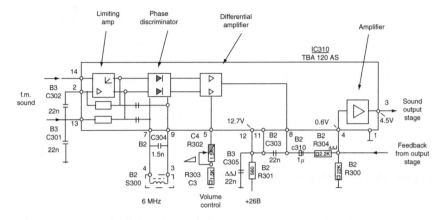

Figure 17.3 *Sound FM detector IC*

is used. A de-emphasis filter with a time constant of 50 µs must therefore be used in the receiver to restore the audio frequencies to their original relative levels.

NICAM

The system of transmitting sound for TV broadcasting employing a sound FM inter-carrier provides a good quality sound given good quality sound amplification at the receiver. However, it is incapable of producing hi-fi quality and unable to carry stereophonic sound. Stereo sound has been successfully transmitted by VHF radio broadcasting using analogue modulation of an FM carrier. Such a system does not readily commend itself for TV broadcasting because of its bandwidth requirements. It is not possible to add a second sound carrier without causing unacceptable interference to either the vision or the primary 6 MHz sound carrier. To avoid this, compromises would have to be reached, and this would defeat the original aim of stereo hi-fi sound transmission.

After years of research and development, BBC engineers came out with a radically new sound system for TV broadcasting using state-of-the-art technology. The system became known as *NICAM 728* or NICAM for short. NICAM stands for near instantaneous companded audio multiplex and 728 refers to the data rate of 728 kbps. It provides two completely independent sound channels so that dual-language sound tracks may be transmitted in place of stereophonic sound. It can carry data in one or both channels and is completely separate and independent of the existing FM monophonic sound channel.

In the NICAM system, A and B channels are digitised at a sampling rate of 32 kHz. The digitised analogue sound signals are grouped into data blocks of 704 bits each and the data blocks are organised in 1 ms frame structure (Figure 17.4). Each data block is preceded by a *frame alignment*

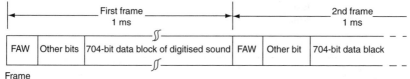

Figure 17.4 *NICAM 1 ms frame*

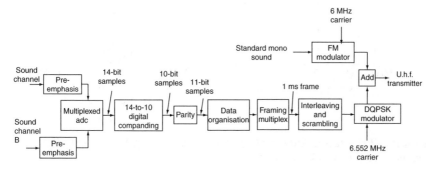

Figure 17.5 *Outline of NICAM encoder*

word (*FAW*) to inform the receiver of the start of each frame. The total frame is then used to modulate a 6.552-MHz carrier, which falls just outside the normal 6 MHz FM sound but remains within the total TV channel bandwidth of 8 MHz. For stereo transmission, the two sound channels are multiplexed, digitised and transmitted in turn.

The basic outline of the NICAM stereo system is shown in Figure 17.5. The two analogue sound channels, A and B, are pre-emphasised before going into an analogue-to-digital converter (ADC) with a sampling frequency of 32 kHz and a Nyquist rate of 30 kHz resulting in a maximum audio frequency of 15 kHz. Groups of 32 samples of each channel are grouped together to form the basic data segment of the system. The samples of each segment are then converted into 14-bit codes. This is followed by a 14-to-10 companding network which compresses the 14-bit codes into 10 bits without any significant loss of quality. The error detection parity bit is then added, resulting in 11-bit samples. Next, the channel data segments are organised into data blocks. Each block consists of two segments, one from each channel, a total of 2 × 32 × 11 bits = 704 bits.

These 704-bit chunks of data form the basic block (sound + parity) of the NICAM broadcast data frame. Frame alignment and control bits (24 bits in total) are then added to 704 + 24 = 728-bit frame.

The time duration of each frame is 1 ms resulting in a bit rate of 728/1 ms = 728 kbps.

Framing is followed by interleaving and scrambling. *Interleaving* is necessary to ensure that error bursts are distributed among several samples

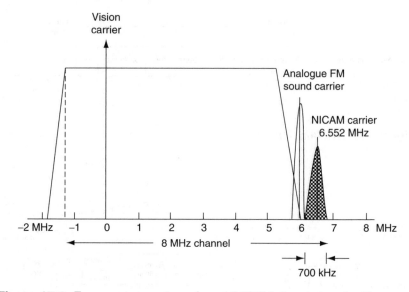

Figure 17.6 *Frequency spectrum for a UHF TV signal showing the position of the NICAM carrier*

which are far apart. *Scrambling* avoids the uneven distribution of energy which follows the process of modulation. The companded, interleaved and scrambled data frame is then used to modulate a sub-carrier that is 6.552 MHz above the vision carrier (Figure 17.6). The modulation technique selected for NICAM is differential quadrature phase shift keying (DQPSK) which was described in Chapter 8. This type of modulation is very economical in bandwidth requirements. It is used here in order to squeeze yet another signal in the tightly packed 8 MHz bandwidth allocated for each TV channel. Before transmission, the modulated NICAM carrier is passed through a sharp cut-off, low-frequency filter to ensure the NICAM frequency spectrum does not overlap with the analogue FM carrier, a process known as spectrum shaping. The NICAM carrier at +6.552 MHz is broadcast at the low level of −20 dB with respect to the vision carrier, representing a power ratio (peak vision carrier to NICAM carrier) of 100:1. The analogue 6 MHz FM sound carrier is retained for compatibility and both sound carriers are added to the video signal for UHF transmission.

We shall now describe the new techniques introduced by the NICAM system in detail.

14-to-10 digital companding

A 14-bit digitiser provides 16,384 quantum levels, which is adequate for high-quality sound reproduction. If fewer than 14 bits are used, the quantising error can become audible in the form of a 'gritty' quality for low-level signals, an effect known as *granular distortion*.

The use of 32 kHz sampling with a coding accuracy of 14 bits per sample would require a data bit rate of approximately 1 Mbps, and consequently a very large bandwidth which could not be accommodated within a single TV channel. For this reason, near instantaneous digital companding is used, which enables the number of bits per sample to be reduced from 14 to 10 with virtually no degradation in the quality of sound reproduction. Consequently, the data bit rate of the system is markedly reduced.

Unlike the analogue companding described earlier, which has the aim of improving the signal-to-noise ratio (SNR), the purpose of *digital companding* is to reduce the number of bits per sample. Furthermore, because all the operations of digital companding are performed in digital form, the compressor at the transmitting coding stage and the expander at the encoder receiving stage can be matched precisely, without the mist racking that is associated with analogue companding.

The companding technique used in NICAM is based on the fact that the significance of each bit of a binary code depends on the sound level which the particular sample code represents. For instance, assuming a peak analogue input of 1 V, then with a 14-bit code the quantum step is given by $\text{quantum step} = 1V / \text{quantum levels} = 1V/16,384 = 61\,\mu V$

This is the value of the least significant bit (LSB). The second LSB has a value of $2 \times 16 = 32\,\mu V$, and so on. It can readily be seen that for a loud sound, i.e. a high-amplitude sample, say 500 mV or over, the effect of the three or four LSBs is imperceptible and may be neglected. However, for delicate or quiet passages with sample amplitudes in the region of a few hundred microvolts, the LSBs are all important. NICAM companding reduces the 14-bit sample codes to 10-bit codes in such a way that for low-level signals the receiver is able to recreate the original 14-bit samples, and for high-level signals it is possible to discard between one and four LSBs as irrelevant.

Each segment of 32 successive audio samples is investigated to find the largest sample in that segment. The amplitude of this sample is then used to indicate the audio strength of the whole segment in five coding ranges (Figure 17.7). Coding range 1 represents a segment where the largest sample falls between the maximum amplitude and one-half the maximum amplitude, range 2 is from one-half to one-quarter maximum amplitude, range 3 is from one-quarter to one-eighth, range 4 is from one-eighth to one-sixteenth and range 5 represents one-sixteenth of maximum amplitude to zero or silence. The shaded bits in Figure 17.7 show the bits actually transmitted for each range. In each case the most significant bit, the 14th, is the sign bit and is therefore retained to indicate a positive or negative value. The 13th bit is discarded if it is the same as the 14th, the 12th is likewise discarded if it is the same as the 13th and 14th, and similarly with the 11th and 10th bits. Where high-order bits are discarded, NICAM provides a labelling technique known as *scale factor coding*, which enables the missing bits to be reconstituted at the receiver. When the

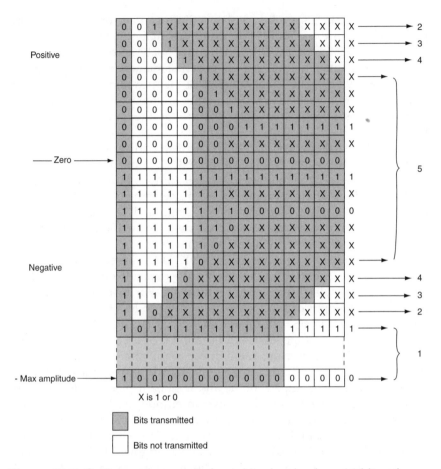

Figure 17.7 *Coding and compression table showing how 14-bit codes are reduced to 10 bits*

discarding of the high-order bits is completed, coded samples of between 10 and 14 bits are left, depending on the sequence of the high-order bits. Where a code has more than 10 bits, a sufficient number of bits are removed to reduce the size of the code to 10 bits; the first bit to be removed is the LSB and successive bits are removed by working upwards.

It follows, therefore, that for a segment of 32 samples falling in the largest amplitude range, range 1, the four LSBs of each sample are discarded and lost for ever. In the case of segments falling in range 2, the bit next to the most significant bit, the 13th bit of each sample, is discarded along with the three LSBs. Although the three LSBs are lost, the 13th bit is reconstituted at the receiver, since it always has the same value as the most significant bit, and so on for ranges 3, 4 and 5.

The next stage is the addition of the parity bit to each sample code, resulting in an 11-bit word. One parity bit is added to the 10-bit sample to check the six most significant bits (MSBs) for the presence of errors. The remaining bits, the five least significant one, are transmitted without a parity check. Even parity is used for the group formed by the six most significant bits. Subsequently, the parity bits are modified to introduce greater error protection and correction as well as coding range information.

The decoder at the receiving end needs to know the number of high-order bits that have been discarded, so they may be reinserted. This is carried out by labelling each coding range with a code known as the scale factor. The scale factor is a 3-bit code which informs the decoder of the number of discarded high-order bits. To save on bandwidth, this information is conveyed without the use of additional bits. Instead, the information is inserted by modifying the parity bits, a technique known as signalling-in-parity.

Interleaving

The parity system gives good protection against corruption of the data words for single-bit errors, but impulsive interference can take out complete data words, making necessary further protection. It is achieved by a simple interleaving system in which the data is written into memory and then read out non-sequentially according to an address sequencer held in ROM. The data readout order ensures that bits which were originally adjacent become separated by at least 15 other bits. An error burst in the received data can corrupt several consecutive bits, but when the words are reassembled at the receiver the errors are distributed among several of them so that the damage to each is usually minor and capable of repair by parity correction and/or error concealment as necessary. The digital signal now has enough protection to enable it to pass unscathed through all but the worst propagation conditions.

Framing

Along with the audio data, it is necessary to send 'housekeeping' data to control the decoding process. Figure 17.8 shows the composition of a

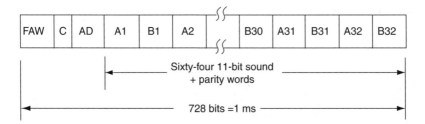

Figure 17.8 *NICAM data frame*

single stereo broadcast data frame, each of which spans 1 ms and contains 728 bits. The frames are sent end-to-end with no gaps between them.

The block starts with a FAW which synchronises the decoder at the receiver. It is the same in every frame, and consists of the 8-bit sequence 01001110. Next comes five control bits, C, to define the type (stereo, bilingual, data) of signal being sent and thus to control the decoder's operating mode and the switching and routing of its output signals. The next data block AD is reserved for 'additional data' and can be used for various purposes.

The initial section of the frame has used 24 bits. The remaining 704 convey the stereo or dual-channel sound data in sixty-four 11-bit words, the A channel (stereo left) and B channel (stereo right) samples being sent alternately throughout the period, 32 of each. The two sound channels (plus the control/data preamble) are thus transmitted as a single serial data stream, and the bit rate of each is approximately doubled as a result. This is a form of *time-division multiplex* (*TDM*) and the required time compression is achieved by writing each set of data, A and B, simultaneously into memory then reading them out alternately at double speed. TDM is used in many data storage and transmission systems to match signal density to channel bandwidth. The order of A and B samples shown in Figure 17.8 represents the sequence before bit interleaving takes place. Only the 704 bits of sound data are interleaved.

NICAM reception

At the receiver, the tuner converts the vision carrier and the FM sound inter-carrier to an IF of 39.5 and 33.5 MHz, respectively. The NICAM carrier is 6.552 MHz away from the vision carrier, so it is converted to an intermediate frequency (IF) of 39.5 − 6.552 = 32.948 MHz (normally referred to as 32.95 MHz.

This is demodulated by a DQPSK detector and applied to the NICAM decoder, which reverses the processes carried out at the transmitter to recreate the 14-bit sample code words for each channel. This is then followed by a digital-to-analogue converter (DAC), which reproduces the original analogue two-channel, left and right, sound waveforms.

The basic elements of NICAM sound reception in a TV receiver are shown in Figure 17.9. Following the tuner, a special surface acoustic wave

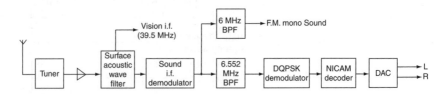

Figure 17.9 *NICAM receiver*

(SAW) filter provides separate vision and sound IF outputs. A sharp cut-off removes the two SIFs, 33.5 MHz for mono and 32.95 MHz for NICAM, from the 39.5 MHz vision carrier. The SAW filter provides a separate path for the FM and NICAM carrier IFs. In the sound IF demodulator, the 39.5 MHz pilot frequency is used to beat with the FM SIF and with the NICAM IF to produce 6 MHz FM and 6.552 MHz DQPSK carriers. Sharply tuned filters are then used to separate the two sound carriers. The FM carrier goes to a conventional FM processing channel for mono sound and the 6.552 MHz NICAM phase-modulated carrier goes to the NICAM processing section. This consists of three basic parts. The DQPSK decoder recovers the 728 kbps serial data stream from the 6.552 MHz carrier. The NICAM decoder descrambles, de-interleaves, corrects and expands the data stream back into 14-bit sample code words. Finally, the DAC reproduces the original analogue signals for each channel.

DQPSK decoder

The phase demodulator or detector works on the same principles as an FM detector in which a variation in phase (or frequency) produces a variation in the d.c. output. In the case of two-phase modulation, the d.c. output of the detector has two distinct values, representing logic 1 and logic 0. However, in the case of quadrature, i.e. four-phase modulation, the output of the detector is ambiguous. The same output is obtained for a 90° and 270° phase shifts. Similarly for phase shifts of 0° and 180°. In order to resolve the ambiguities, a second phase detector operating in quadrature (90°) is used. Figure 17.10 shows the main elements of a DQPSK demodulator using an in-phase phase detector (PDI) and a quadrature phase

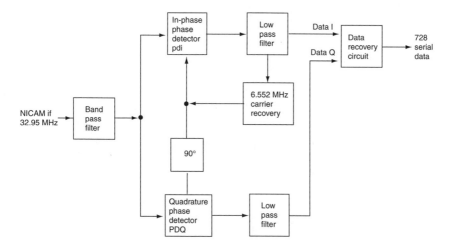

Figure 17.10 *The DQPSK demodulator*

detector (PDQ). The outputs from the two-phase detectors, data I and Q, are fed into a data recovery circuit which reproduces the original 728-bit serial data stream. The 6.552 MHz reference carrier frequency is generated by a carrier recovery circuit which includes a crystal-tuned voltage-controlled oscillator (VCO) and a phase-locked loop (PLL).

The data recovery circuit includes a second PLL locked to the bit rate of 728 kHz. In order to ensure a 'clean' bit rate clock, a master system clock which is a multiple of the bit rate is used. In this case, a clock frequency of $728 \times 8 = 5824$ kHz = 5.824 MHz is used. The bit rate is then retrieved by dividing the master system clock by eight. The I and Q data streams from the DQPSK detector are fed into a differential logic decoder, which produces the corresponding 2-bit parallel data. The pairs of parallel data are then fed into a parallel-to-serial converter, which reconstructs the original 728-bit serial stream before going to the NICAM decoder.

NICAM decoder

The NICAM decoder descrambles, de-interleaves and reconstitutes the original 14-bit words. It provides data, ident and clock signals to a DAC to obtain the original L/R audio signals. A simplified block diagram of the decoding process is shown in Figure 17.11. The encoded data from the DQPSK detector is fed into the FAW detector for frame recognition and resetting of the descrambler and de-interleaver. The descrambled data is then fed into the de-interleaver, which reproduces the original dual-channel (L and R) data together with an L/R ident signal to select the signal paths as appropriate. De-interleaving is carried out by using a ROM look-up memory. The descrambled data is also fed to the operation mode detector which decodes control bits C0–C4 and provides information to the expander and other parts of the system in terms of the type of transmission, e.g. stereo, mono or bilingual.

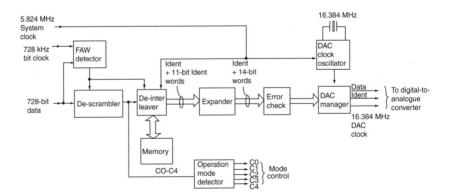

Figure 17.11 *NICAM decoder*

Having restored each 11-bit word (10 + parity) to its correct order, they have to be expanded back to a 14-bit format. This is carried out by an expander, which functions in a complementary manner to the compressor at the transmitter but uses the scale factor embodied in the parity bits. This is followed by an error check circuit, in which the error parity is used to investigate and correct the bit stream. Before leaving the decoder for the DAC, the data is fed to a DAC manager to organise a three-line bus output consisting of a data bit stream, an ident signal and a DAC clock known as DACOSC. At the converter, the DAC clock is subdivided to accurately produce the sampling frequency. There are two formats for the three-line output bus that may be used to feed the DAC: the S-bus for converters using a 16.384-MHz clock and the I²S-bus for converters using an 8.192 MHz clock.

Multi-sound processor

Demodulation and decoding of NICAM sound is incorporated in a single IC package known as a multi-sound processor (MSP; Figure 17.12). SIF goes into the FM demodulator to produce the normal mono sound which is fed into L and R loudspeakers. NICAM inter-carrier IF is extracted and fed into the NICAM decoder which carries out all necessary functions to retrieve the L/R stereo sound, which following DAC is fed into the appropriate L/R amplifiers and subsequently to the speakers.

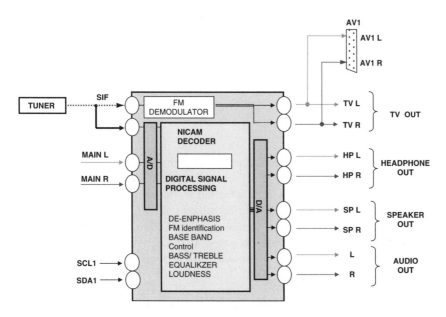

Figure 17.12 *Multi-sound processor (MSP) incorporating FM and NICAM demodulation and decoding*

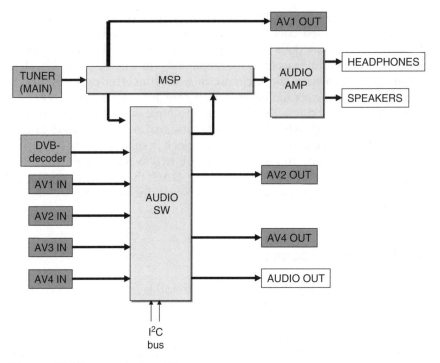

Figure 17.13 *Sound signal flow*

In practice, there are several sources of audio including stereophonic audio from an in-built DVB decoder. A sound selector is therefore necessary as shown in Figure 17.13. The audio switch is controlled by an I²C bus to select a particular sound output. The MSP carries out all necessary demodulation of inter-carriers and decoding of NICAM sound. The output of the MSP is fed into a dual audio amplifier on its way to the loudspeakers.

Multi-standard MSP

Modern MSPs are multi-standard processing chips covering sound processing of all analogue TV standards (PAL, NTSC, SEACAM) as well as NICAM (Figure 17.14). Automatic sound selection is incorporated within the package together with *automatic volume correction (AVC)* to avoid annoying volume changes when the sound source changes. Different sound sources (e.g. terrestrial channels, SAT channels or SCART) fairly often do not have the same volume level. Advertisements during movies usually have a higher volume level than the movie itself. The automatic volume correction solves this problem by equalising the volume level. Processing is fully digital and analogue inputs including SIF and SCART inputs are fed into an ADC before demodulation and processing and subsequently converted back to analogue audio (mono, stereo or multi-channel).

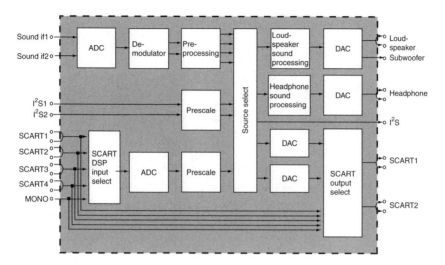

Figure 17.14 *Multi-standard MSP*

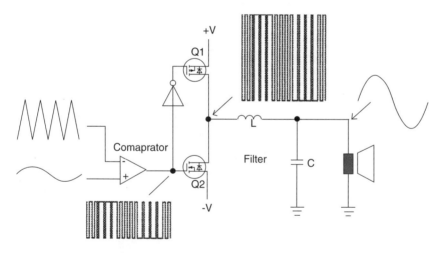

Figure 17.15 *Principles of class D amplification*

Audio amplification

Audio amplifiers invariably use the class D amplification mentioned in Chapter 14. The amplifier consists of a comparator driving two MOSFET transistors which operate as switches as illustrated in Figure 17.15. The comparator has two inputs. One is a triangle wave; the other is the audio signal. The frequency of the triangle wave must be much higher than that of the audio input. If the frequency of the triangular input is 20 or more times higher than the audio input, the output of the comparator is the

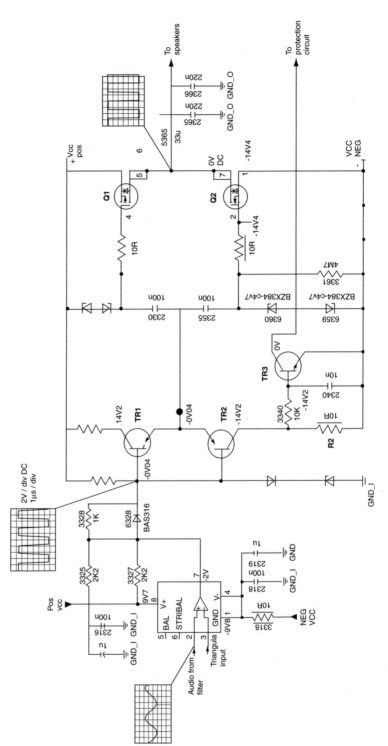

Figure 17.16 *A typical class D audio amplifier*

varying-duty cycle pulse (pulse width modulated, PWM) waveform shown. The output transistors Q1/Q2 switch from 'full off' to 'full on' and then back again, spending very little time in the linear region in between. Therefore, very little power is lost to heat. If the transistors have a low 'on' resistance, little voltage is dropped across them, further reducing losses.

It is necessary, however, to introduce a low-pass filter (L/C in the diagram) to the signal before amplification. The main reason for this filter is that the switching waveform results in maximum current flow. This causes more loss in the load, which causes lower efficiency. The LC filter with a cut-off frequency less than the class D switching frequency (typically 350 kHz), ensures that only the average of the output wave goes through, which is an amplified version of the input signal. In order to keep the distortion low, negative feedback is used. The disadvantage of class D amplifiers is the large output filter that drives up cost and size.

A typical TV audio amplifier circuit diagram is shown in Figure 17.16. Q1/Q2 package forms a class D complementary pair fed with ± 14 V d.c. Audio from the filter enters the voltage comparator from which a PWM waveform is produced. TR1 and TR2 are drive transistors and TR3 is a part of a high-temperature protection circuit. If the current through TR2 increases as a result of say a short-circuited output pair Q1/Q2, the voltage across its collector resistor R2 goes up, and TR3 conducts to activate a d.c. protection circuit (not shown). Because of the symmetrical supply, a d.c. blocking capacitor between the amplifier and the speaker is not necessary. However, it is still necessary for a protection circuit to protect the speaker from increased d.c. voltages.

18 The digital TV reception

Digital TV reception is normally provided by an *integrated receiver decoder* (*IRD*) commonly known as a *set-top box* (*STB*). The STB may be either a standard (SD) or high definition (HD) as well as satellite or terrestrial. *Digital terrestrial television* (*DTTV*) may also be incorporated within the TV receiver in what is known as *integrated digital television* (*iDTV*). In this chapter, we will look at the component parts of a SD and HD STBs for both the terrestrial and satellite transmission media.

The digital receiver/decoder of set-top-box

The set-top-box, STB is a self-contained tuner/decoder device in that it tunes to the required channel, extracts and decodes the selected pro-gramme data, checks the access rights of the user and produces picture, sound and other services as instructed.

Digital signals are fed to the STB in the same way as for analogue TV broadcasting. Thus, in the case of satellite transmission, an outdoor dish and *low noise block* (*LNB*) are necessary to receive the signals and convert them to a suitable intermediate frequency (IF). For terrestrial television, the input to the STB is obtained directly from a terrestrial aerial. And for cable too, the signal is obtained directly from the network. The output from the SD digital decoder is in the form of analogue video and audio signals, that is fed to an analogue TV receiver either directly via a SCART component video or s-video connections or as a UHF-modulated signal using the aerial socket at the back of the receiver (Figure 18.1). In the case of HD, digitised HD video and sound are available on a single HDMI connection or components video and L/R sound outlets. Video and audio from HD decoders are also available on the normal SCART or s-video ports, but this time in SD video only.

System overview

Figure 18.2 shows the basic components of a digital receiver/decoder STB. The channel decoder, also known as the front end, is specific to the broad-casting media: satellite, terrestrial or cable. It extracts the required MPEG transport stream from the RF or IF (in the case of satellite reception) signal. It consists of a tuner, ADC, an appropriate demodulator and a forward error correction (FEC) processor. The transport stream containing

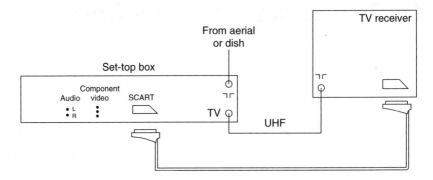

Figure 18.1 *Connection between an STB and a TV receiver*

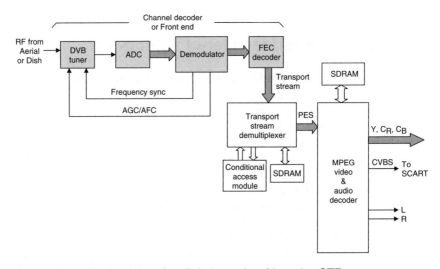

Figure 18.2 *Basic units of a digital receiver/decoder STB*

standard 204-byte packets belonging to one or more television programmes is fed into the transport demultiplexer. Before processing the transport stream, the demultiplexer sends the bitstream to the conditional access module (CAM), which controls the user access rights to the selected programme or service. The CAM interrogates a smart card to find out if the user has an active subscription to the selected programme. If access is granted or the programme is free view, the transport stream is routed back to the transport demultiplexer. The demultiplexer selects the transport packets belonging to the chosen programme and reassembles them to reconstruct the packetised elementary streams (PESs) of the programme. A fast static RAM (SRAM) or synchronous DRAM (SDRAM) memory is used to store selected audio, video and service packets on their way to the MPEG decoder.

The MPEG decoder carries out the video and audio decoding of the audio and video PESs. It converts the video PES data stream back into its original components: Y (luminance) and C_R and C_B (chrominance). The picture is reconstructed from the I, P and B frames. This reconstruction requires simultaneous storage of these frames, hence the need for a large video memory in the form of a DRAM or a faster SDRAM buffer. The three components are made available for direct connection to a TV receiver or they may be sent to an embedded PAL encoder to convert the digital video into an analogue PAL composite video, which is fed directly to a television set via a SCART connector or into a UHF modulator for a UHF output.

The audio PES packets are decoded using the same rules as adopted in the encoding stage to produce left and right analogue audio signals. The same DRAM chip is used for audio buffering which also provides an approximately 1-s delay to ensure audio and video synchronisation. This delay is necessary given that the processing of the video signals takes a longer time to complete than the processing of audio signals. The left and right analogue audio signals from the audio decoder may be fed to a summing amplifier to produce mono sound for the UHF modulator. A separate stereo output (L and R) is also provided, which may be sent to the SCART connector.

The system requires a high level of programming and control. This is performed by a 16-bit or a 32-bit microprocessor (not shown). Control is exercised using normal control signals such as read/write; reset and *interrupt request (IRQ)* as well as the I²C serial control bus. A non-volatile flash memory chip is used to store start-up and other software programs. Upgrading of the software may be carried out off-air by first loading the new software into the DRAM chips and then transferring it to the non-volatile flash memory.

DVB satellite channel decoder

The precise construction of the channel decoder depends on the type of receiver, satellite (DVB-S) or terrestrial (DVB-T). For cable digital video broadcasting (DVB-C), the front end is similar to that used for satellite reception except for the input carrier frequency.

The function of the front end of a DVB STB is to extract the multiplexed transport stream from the modulated carrier received from a satellite dish or a terrestrial aerial. It consists of four stages:

- A tuner
- An analogue-to-digital converter (ADC)
- A demodulator: quadrature phase shift keying (QPSK) for satellite reception; orthogonal frequency division multiplex (OFDM) for terrestrial reception
- An FEC processor error correction

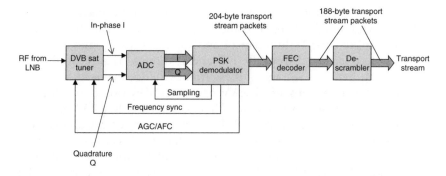

Figure 18.3 *The basic components of a channel decoder of a digital satellite television receiver*

Figure 18.3 shows the basic components of a channel decoder of a digital satellite television receiver. The input to the tuner, a self-contained isolated unit, is in the form of a carrier known as the first IF from the LNB of a satellite dish. It is frequency modulated by a signal, itself a QPSK-modulated carrier. The tuner down-converts the input carrier to a second IF and reproduces the two original QPSK-modulated carriers: in-phase (I) and quadrature (Q). The phase-modulated I and Q carriers have the distinctive waveform as shown in Figure 18.4. The I and Q carriers are

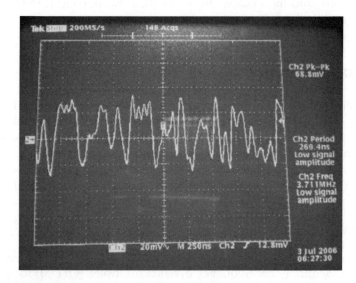

Figure 18.4 *Typical I or Q carrier waveform*

digitised by a dual-ADC (one for each carrier) into two 6-bit parallel streams. The sampling rate is set by a signal from the QPSK demodulator, which controls an in-built voltage-controlled oscillator (VCO). The sampling rate is normally set to twice the *symbol rate*; the symbol rate is the rate at which the phase of the carrier changes. The symbol rate itself is determined by the broadcaster and has to be set at the receiver before the tuner can select a channel. Thus, for a *symbol rate* of 27,500 kilo symbols per second (not to be confused with the 27 MHz video sampling clock), the sampling rate will be set at $2 \times 27.5 = 55$ MHz. Where the symbol rate is low, a sampling rate of three times the symbol rate is usually used. Cable broadcasting uses a sampling rate of four times the symbol rate.

The demodulator carries out the phase detection of the I and Q signals, samples the recovered phases and quantises each phase change into a 3-bit code. The demodulator normally carries out several other functions including, AGC to control the gain of the tuner, control of the VCO and the sampling clock for the dual ADC.

As was explained in Chapter 17, the recovery of the original bitstream from a phase-modulated I and Q carriers involves two distinct stages: *phase recovery* and *data recovery*. Phase recovery identifies the phase change of the carrier and data recovery reproduces the original bitstream, a process known as symbol-to-bit mapping. The FEC decoder recovers the modulating data bits from the QPSK carriers and arranges the information in a way that is suitable for presentation to the error detection/correction stage. The FEC error detection/correction stage consists of three parts. First comes a soft decision-making decoder, the Viterbi, which uses the original convolution coding to determine whether a received bit is logic 0 or logic 1. Next comes the deinterleaver; it rearranges the symbols and bits into the original order before they were interleaved at the transmitting stage. The final part is the *Reed–Solomon (RS) decoder*, which performs the hard decisions. Hard decisions determine whether a packet contains errors. If it does, the FEC unit will attempt to correct them. If it fails to do so, the FEC decoder will flag the packet as erroneous that must be discarded. At the end of the process, a transport stream that consists of a series of 188-byte transport packets is obtained. Before the transport packets are fed to the next demultiplexing stage, the pseudo-random energy dispersal is removed by a descrambler using the reverse of algorithms applied at the transmitting stage.

In practice, the dual-ADC, the demodulator and the FEC processor are integrated into a single package together with an embedded processor. The block diagram of a typical demodulator/FEC processing chip shown in Figure 18.5 is capable of processing BPSK, QPSK, 8-PSK and H-8PSK modulation. It has its own embedded dual DAC, tuner interface and microprocessor. I and Q carriers go in the input and a transport stream is available at the output.

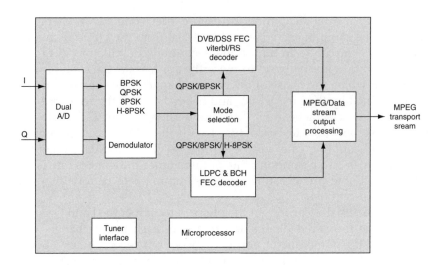

Figure 18.5 *DVB-S demodulator/FEC processor*

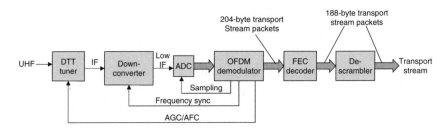

Figure 18.6 *DTTV channel decoder*

DVB terrestrial channel decoder

The main components of a DTTV front end or channel decoder are shown in Figure 18.6. The tuner receives UHF-modulated signals from the terrestrial aerial, selects the appropriate channel and produces a modulated VHF IF. Before the IF can be processed, it is fed into a *down-converter* to produce what is known as a low IF. The down-converter removes the UHF carrier and retains the baseband signal with a centre frequency of 4.75 MHz or thereabouts, depending on the chipset used. The low IF is digitised by an ADC using a sampling frequency derived from the OFDM demodulator. The OFDM demodulator retrieves the original modulating transport stream and sends it to the FEC decoder.

The down-converter consists of a mixer followed by a lowpass filter. The function of a mixer is to change the frequency of an input signal. This is carried out by 'mixing' the original signal (frequency f_1) and a separate signal (frequency f_0) obtained from a local oscillator. The two signals are multiplied by the mixer to produce two beat frequencies: the sum $f_1 + f_0$

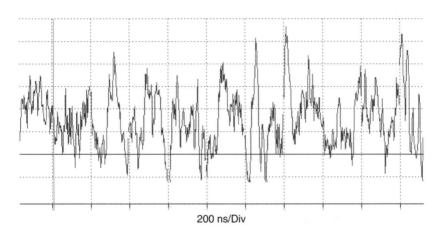

200 ns/Div

Figure 18.7 *Low IF (baseband) OFDM waveform 16-QAM*

and the difference $f_1 - f_0$. If the local oscillator frequency $f_0 = f_1$, then $f_1 - f_0 = 0$ and $f_1 + f_0 = 2f_1$.

Now, if the input signal contains a side frequency $(f_1 + f_m)$ (where f_m is the modulating signal) as well as f_1, then the output will contain the following beat frequencies:

$$(f_1 + f_m) + f_1 = 2f_1 + f_m$$
$$f_1 + f_1 = 2f_1$$
$$(f_1 + f_m) - f_1 = f_m$$
$$f_1 - f_1 = 0$$

And here where the lowpass filter comes in. If the cut-off frequency of the lowpass filter is arranged to remove $2f_1$ and all frequencies above $2f_1$, the output will contain the modulating frequency f_m only. Each one of the OFDM carrier frequencies would thus produce one modulating side frequency, f_{m1}, f_{m2}, f_{m3} and so on. The summation of all these frequencies is the baseband having the distinctive wave shape as illustrated in Figure 18.7.

OFDM demodulation

You will recall that in the OFDM system, a large number of carriers (1705 and 6817 carriers in the 2K and 8K modes, respectively) are used. Each carrier is phase modulated by a symbol consisting of 4 bits (16-QAM) or 6 bits (64-QAM). The purpose of the demodulator is to retrieve the 4-bit or 6-bit data represented by the phase shift of the carriers. The basic principles of the OFDM demodulator are illustrated in Figure 18.8. Following digitisation of the low IF baseband by the ADC, original OFDM carriers are reconstructed

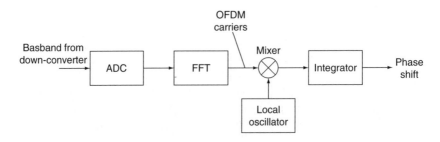

Figure 18.8 *OFDM demodulator*

using a fast Fourier transform (FFT) processor which analyses the incoming signal into its frequency components. The QAM-modulated OFDM carriers are then fed into a mixer.

Let's consider the operation with one OFDM carrier. The mixer is fed with two signals: an OFDM carrier and local oscillator signal of equal frequency, f_c. The output from the mixer is the sum and difference of the two frequencies. If the OFDM carrier contains no phase shift, an output signal with a beat frequency of $2f_c$ will be obtained. However, if the OFDM carrier is phase modulated, the phase angle will be produced at the output together with the $2f_c$ beat signal.

This is explained mathematically as follows:

If input 1 (phase modulated) is $\cos(\omega_c t + \theta)$ and input 2 (oscillator frequency) is $\cos \omega_c t$, where $\omega_c t = 2\pi f_c$, then by mixing the two signals, we get

$$\cos(\omega_c t + \theta) \times \cos \omega_c t = \frac{1}{2}\cos(\omega_c t + \theta = \omega_c t) + \frac{1}{2}\cos(\omega_c t + \theta - \omega_c t)$$

$$= \frac{1}{2}\cos(2\omega_c t + \theta) + \frac{1}{2}\cos\theta$$

The task of the demodulator is to separate the phase angle from the twice-carrier frequency signal. The obvious way of doing this is to use a filter. While this is possible for one carrier, it is not practical for several thousand, as is the case with the OFDM technique. The process is further complicated by the fact that each local oscillator frequency will beat with all the other OFDM carrier frequencies, producing millions of beat frequencies of differences and sums. A technique has to be devised to remove all beat signals and leave only the phase shift angles. This is where the integrator comes in using the fact that the integral of a sine wave is the area under the sine wave curve and thus the integral of a sine wave over one complete cycle is zero; there is as much area above the zero line as below it. The same applies over a 2-cycle period and so on. Provided we integrate a sine wave over a complete number of cycles, the result will always be zero. Going back to the OFDM system, the spacing between the carriers is a precise figure, namely 4464 Hz for the 2K mode and 1116 Hz for the 8K mode. Every carrier is therefore a multiple of 4464 or 1116 as appropriate. Thus, if the baseband signal is integrated over

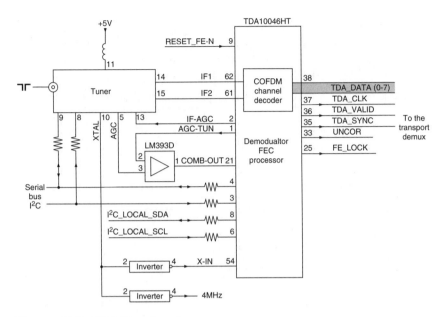

Figure 18.9 *DTTV front end*

one cycle of the carrier spacing ($1/4464 = 224\,\mu s$ or $1/1116 = 896\,\mu s$ for the 2K and the 8K modes, respectively), all the carriers will be reduced to zero. The only element left would be the phase shift of each carrier which represents the transmitted data. The process is repeated for each set of modulated carriers.

A typical DTTV front end is shown in Figure 18.9. The tuner receives a UHF signal from the aerial and produces two IF signals: IF1 and IF2 which are fed to the COFDM demodulator/FEC processor TDA10046HT. The demod/FEC chip is a multi-standard package incorporating its own microprocessor and carry out all necessary functions including demodulation, deinterleaving and error correction to extract the original bitstream. The output is an 8-bit transport stream (TDA DATA 0–7) together with a number of control signals: transport clock (TDA_CLK), data valid (TDA_VALID), sync (TDA_SYC), drop uncorrected data (UNCOR) and bit clock (FE_LOCK). Both the tuner and the demod/FEC are controlled by an I²C bus. The functional block diagram of the COFDM demodulator/FEC TDA10046HT chip is shown in Figure 18.10.

Transport demultiplexing and MPEG decoding

The channel decoder is followed by the transport stream demultiplexing/video/audio MPEG-2 decoding IC package (Figure 18.11).

The input to the chip is an 8-bit wide transport stream obtained from the front end consisting of video, audio or other service information belonging to four, five or more different programmes organised into 188-byte transport

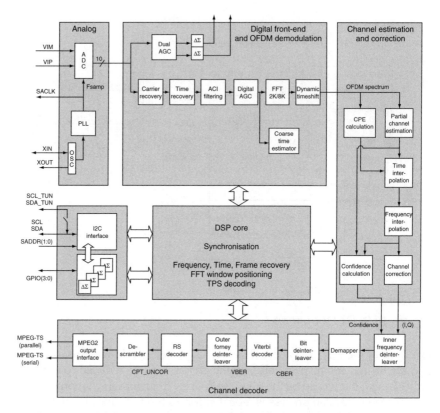

Figure 18.10 *COFDM demodulator/FEC TDA10046HT chip*

packets. The transport stream is examined to ascertain if conditional access scrambling is employed. If it is not, the CAM is bypassed and the transport stream is processed in the normal way. If the programme is encrypted, the CAM ascertains whether the user has access rights to the programme. If so, the transport stream is de-encrypted and returned to the demux for further processing. If the user has no entitlement to view the programme, the transport stream is blocked, giving a blank display on the screen with a message to inform the user of the problem.

The central task of the demux is to extract the transport packets that belong to the selected programme. To do this, the demux first extracts the packet with a packet identification (PID) of zero. This contains the programme association table (PAT) in which the demux will find the PID of the packet that contains the programme map table (PMT). Once the PMT is opened, the demux is able to identify the PID for all the 188-byte PESs of the particular programme (video, audio and service) and proceed to extract them from the transport stream. The 188-byte transport packets are then reconstructed to obtain their original elementary stream packets: video, audio and service PESs.

The function of MPEG video decoding is to reconstruct the picture to its original form. This involves data decompression, inverse DCT and

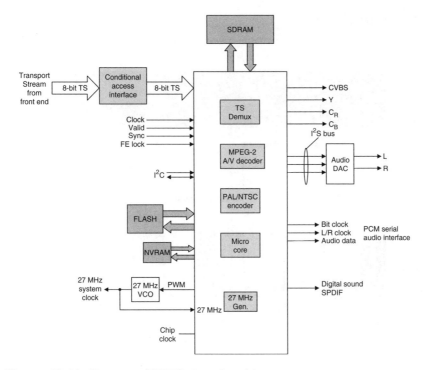

Figure 18.11 *Transport/MPEG decoder chip*

reconstructing the picture from the I, P and B frames to obtain the original Y, C_R and C_B. This process involves the simultaneous storage of a number of frames hence the need for a large memory store, normally one or more SDRAM chips. In order to provide an PAL/NTSC composite video signal, an on-board PAL/NTSC encoder is used. The PAL/NTSC encoder produces a standard 625/525-line, 25/30 pictures per second television signal.

Audio from the demultiplexer in the form of serial pulse code-modulated (PCM) data together with audio control signals are decoded by the MPEG decoder section. The process involves decompression and reconstruction of the original audio L/R audio signals. Information on the actual sampling rate used by the transmitter is provided by the transport demux, which extracts the information from the incoming transport stream. Audio buffering is provided by the SDRAM memory store, which also provides the necessary delay. The audio output may be in the form of a basic serial PCM, *inter-IC sound serial bus* (I^2S) or *SPDIF*.

The serial PCM audio interface comprises three lines: the PCM data line and two control clocks, the left/right and serial clocks. The two audio channels are time-multiplexed with alternate samples of each channel transmitted in series along the data line. The data line is a 16–32 bit coded in two's compliment. The left/right (L/RCK) clock which is known by several names, including word clock, frame clock, frame sync and probably several others identifies which sample is a left and which is a right channel.

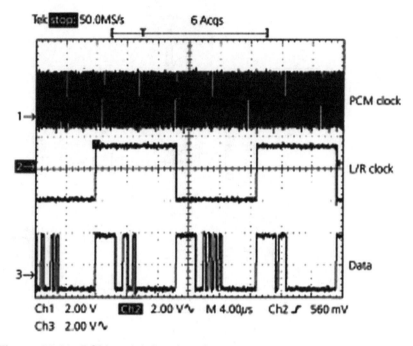

Figure 18.12 *PCM serial data interface*

Since the samples are alternate L and R, it follows that the L/R clock frequency is half that of the sampling rate. The sampling rate may thus be obtained from this clock. In some applications, a fourth control line is added to carry a separate sampling clock known as the sys or master clock. The purpose of the serial clock, also known as the *bit* or *PCM clock*, is to set the speed by which the audio data bits are shifted in or out of the port. The frequency for the *serial clock* is directly proportional to the system audio sample rate and the audio word length with the minimum frequency = sampling rate $\times n$ where n = audio word bit length or $2 \times$ L/R clock $\times n$. A typical PCM serial data and control lines are illustrated in Figure 18.12.

The I²S serial sound bus

The I²S is a particular application of the PCM serial interface described above. It is designed to handle audio data between different devices such as compact disc players, digital recorders and digital television. The bus (Figure 18.13) has three lines:

- *serial data* (SD)
- *word select* (WS)
- continuous *serial clock* (SCK)

The device generating SCK and WS is known as the master.

Figure 18.13 *Unidirectional I²S serial bus*

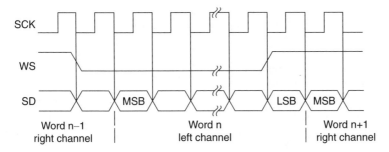

Figure 18.14 *The three lines of I²S serial sound bus*

Serial data is transmitted in two's compliment with the MSB first to avoid incompatibility of word lengths between the transmitter and the receiver. Time-division multiplexing is used to send alternate channels, L, R, L as shown in Figure 18.14. The WS line indicates the channel being transmitted as follows:

- WS low (logic 0) channel 1 (left)
- WS high (logic 1) channel 2 (right).

WS may change either on the trailing or leading edge of the serial clock, but it does not need to be symmetrical.

SPDIF digital interface

SPDIF stands for *Sony/Philips digital interface format*, also known as IEC 958 type II, part of IEC-60958. It is a collection of hardware and low-level protocol specifications for carrying digital audio signals between devices and stereo components. It is a consumer version of the standard known as AES/EBU. SPDIF, as the name suggest, is a purely digital system used to transmit audio signals in a number of formats, the most common being the 48 kHz sample rate format used in DAT and the 44.1 kHz format used in CD audio. Each data bit is transformed into a 2-bit code (10 or 01 representing bit at logic 1 and 00 or 11 representing a bit at logic 0) using biphase mark coding technique in which clock speed is twice the bit rate of the original data. SPDIF which has no defined data rate is designed for transmitting 20-bit audio data streams plus other related information.

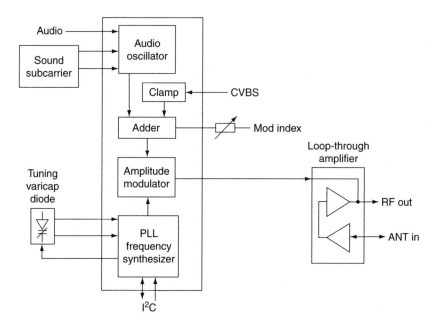

Figure 18.15 *UHF modulator*

Where the source data is less than 20 bits per sample, the superfluous bits are set to zero.

UHF modulator

The *UHF modulator* receives CVBS and audio signals from the demux/ MPEG decoder chip and uses it to modulate a UHF carrier of an unused channel to provide a facility for UHF output to a TV receiver. The UHF modulator (Figure 18.15) has five components:

- An I²C-controlled phase-locked loop (PLL) frequency synthesiser
- An amplitude modulator
- An audio oscillator for the sound sub-carrier
- A video clamp to ensure correct modulation index
- A loop-through amplifier

Tuning of the modulator is carried out by a d.c. voltage (0–30 V) derived from the PLL frequency synthesiser. The audio signal is used to frequency modulate the sound sub-carrier. The modulated sound carrier is added to the clamped video signal and used to amplitude modulate a UHF carrier. The modulated UHF is then fed into the loop-through amplifier, which mixes it with the UHF signal from a normal external antenna. When the digital decoder is on standby, the UHF synthesised oscillator is switched off by a command from the microcontroller via the I²C bus. But the loop-through amplifier operates normally so that the RF signal from the external antenna may loop through to the RF output socket.

The integrated digital television

The iDTV receivers are capable of receiving analogue as well as DTTV broadcasts. They employ two separate tuners: the traditional analogue tuner and a digital DVB tuner for receiving DTTV broadcasts. Two separate aerial input sockets may be available or alternatively a single input socket is used with a loop-through to feed the second tuner.

Two basic configurations are employed by manufacturers depending on the type of display used: CRT or flat panel.

Figure 18.16 shows the basic components of a CRT-display iDTV. The digital tuner contains all the necessary circuitry for the down-conversion of the incoming UHF signal producing an IF of the digital video broadcasts. This IF is an analogue signal carrying the selected digital video broadcast and as such is known as *'digital IF'* but more appropriately named *DVB IF*. The tuner is followed by the OFDM demodulator which extracts the transport stream and feeds it to the highly integrated transport multiplexer/MPEG decoder chip. Analogue CVBS representing the digital broadcast, invariably referred to as *digital CVBS* (a more appropriate name is *DVB CVBS*) is produced as well as analogue DVB L and DVB R audio. The DVB CVBS signal is sent to the analogue/digital (A/D) switch. The other input to the switch is analogue CVBS from the analogue tuner. The switch selects the required signal according to the status of its control lines. The audio signal from the analogue tuner is extracted via a sound trap circuit and fed into the audio section, normally an audio processor, and forwarded to the audio switch.

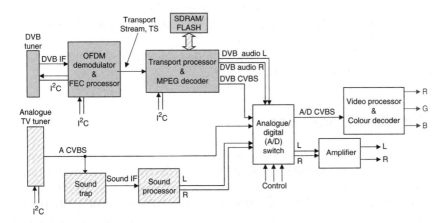

Figure 18.16 *Basic components of a CRT iDTV*

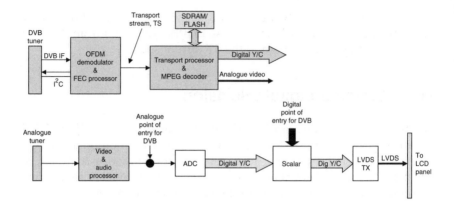

Figure 18.17 *Two possible configuration for flat screen iDTV receivers*

Flat panel iDTV

There are two possible circuit configurations for iDTV flat screen displays (Figure 18.17). As was explained in earlier chapters, for flat display panels, LCD or plasma, the incoming video signal must be converted to the native resolution of the panel before it can be displayed. Formatting involves I/P conversion, scan conversion and image conversion and it may be carried out by a number of different chips or more likely by a single chip usually referred to as a scalar chip. Formatting being a processing activity requires the video input to be in a digitised format. Since the video signal extracted from the DTTV decoder is in digital form, it is possible to feed this DVB video directly into the formatting section. However, a second configuration is possible. This configuration involves converting the DVB video signal into an analogue format resulting in two analogue video signals, DVB video and analogue video. An analogue/DVB video switch then selects between two similar signals to feed into the scalar.

An example of the second configuration employed by a Philips LC4.3E chassis is illustrated in Figure 18.18.

An example of the other configuration employed by a Panasonic LCD television is illustrated in Figure 18.19.

The HD decoder box

The front end of HD satellite decoder (Figure 18.20) is the same as that used for SD broadcasting. The same goes for the transport demultiplexer. The difference arises in the audio and decoding section. With HD reception, a multi-standard decoder chip is used capable of decoding both HD advanced video coding (AVC) and stereo/surround sound advanced audio coding (AAC) as well as MPEG-2 video and audio packets.

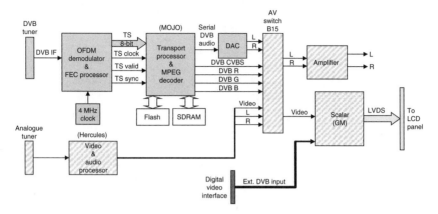

Figure 18.18 *Philips LC4.3E LCD iDTV configuration*

A large SDRAM memory store is necessary. The picture is reconstructed into Y, C_R and C_B, and the sound is reconstructed into a serial PCM (digital) with associated control signals (L/R clock, sampling rate and bit or sys clock). The video and audio signals are fed into a HDMI encoder (or transmitter) which converts them into a serial bitstream suitable for connection to an HD ready television receiver. Also available is SPDIF sound on a coaxial or optical port as well as component video, SCART and so on.

The HD ready television receiver

Invariably, the HD receiver is a flat panel television (Figure 18.21). Video and audio data arriving at the HDMI port go to the HDMI decoder (receiver). The HDMI decoder converts the video back to a digital 8-bit RGB together with the clock, H/V sync and enable. The audio PCM (stereo or surround sound) from the HDMI decoder is fed directly to the multi-sound processor (MSP). Sound from other sources such as optical/coaxial SPDIF or a SCART is also fed into the MSP chip. The centre of the receiver is the image scalar which receives HD video from the HDMI decoder and SD video from the SDTV processing section. The purpose of the image-scaling chip is to convert the incoming signals which may have a variety of resolutions (HD or SD video as well as PC-generated video) into a form that is compatible with the native resolution of the plasma or LCD panel. The output of the scalar is fed into a DVI encoder (normally LVDS) to feed the display panel driver boards. The SD video may be in analogue format from an analogue tuner/video processor or a SCART connector or digitised video from a terrestrial DVB SD tuner/demodulator/decoder.

The pin assignment for an HDMI decoder (receiver) chip is outlined in Figure 18.22.

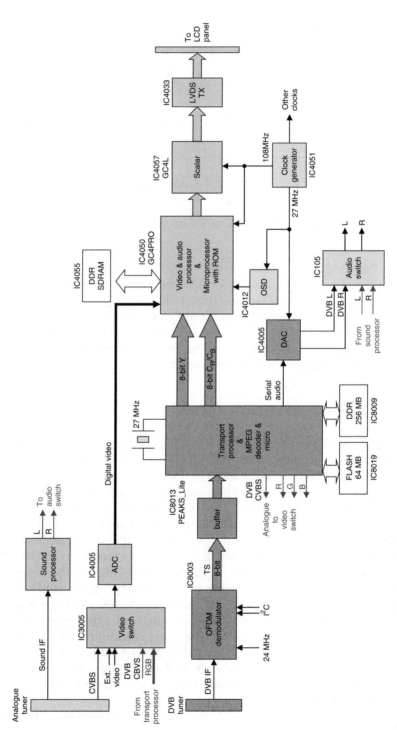

Figure 18.19 *Panasonic LCD iDTV configuration*

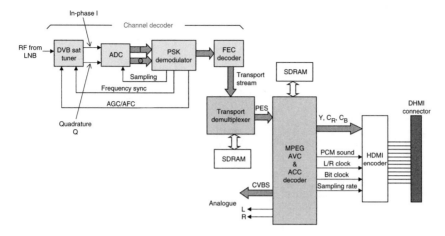

Figure 18.20 *Basic units of an HD satellite decoder*

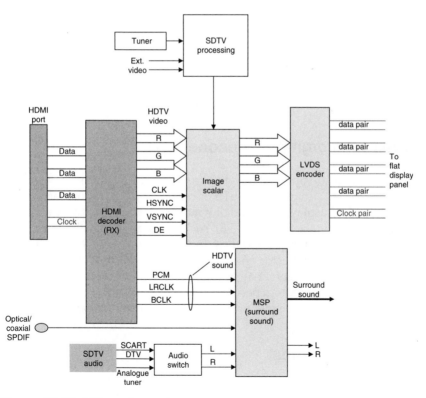

Figure 18.21 *HD ready receiver*

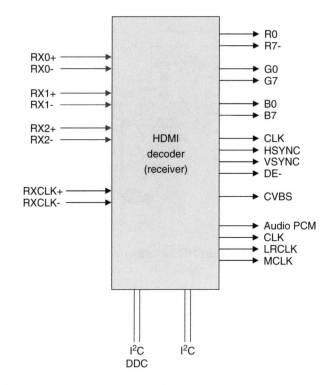

Figure 18.22 *HDMI decoder (receiver) pinout*

Testing the digital TV decoder box

The first step in attempting to service a suspect DTV system is to determine whether the fault lies within the STB or outside it. The external factors that affect the video display and its quality may be divided into two categories: those that precede the STB (*pre-STB*), such as aerial or dish alignment, aerial lead and signal strength, and those that succeed it (*post-STB*), such as the HDMI or SCART connector and lead or the television set itself. Failures caused by the first category may be total video/audio failure, video and/or sound break-up, a symptom peculiar to digital TV reception. Failures in the second category do not include this type of video/audio break-up.

Aerials and dishes

In the conventional analogue system, low signal strength or low carrier-to-noise ratio (C/N) results in adverse picture quality, with ghosting in the case of terrestrial TV or noisy picture sparklies in the case of satellite TV. In digital TV reception, such conditions would cause the picture and sound to fail altogether. Low signal strength would result in failure of the channel decoder to lock to a channel, with a message 'no signal' being displayed. In DTV, the picture is either perfect or non-existent. This is known as the digital cliff.

Unlike conventional analogue reception, where low signal strength or low C/N would result in low signal amplitude, in DTV applications it would result in increased bit error rate (BER). If the BER is too high, the FEC unit cannot correct the errors and the data would be marked accordingly. Normally, the video decoder decodes the video PES packets and stores them in the video memory for display. The video memory would thus be updated as new PES packets are decoded. However, when a PES is marked as erroneous, it is neglected and the relevant part of the video memory fails to update; this produces a freeze in that part of the picture and pixelisation results as illustrated in Figure 18.23. If a whole series of PES packets are marked erroneous, none of the video memory is updated and a picture freeze occurs. This latter effect may be observed immediately after removing the aerial lead. In this case, the freeze continues for a few moments before the picture disappears completely. Where audio PES packets are marked erroneous, they are neglected by the audio decoder and sound break-up results.

A spectrum analyser may measure the strength of the signal from an antenna. The software resident within the STB provides a facility for measuring the signal strength which may be used as an indication of the signal strength but not as an accurate measurement. Low signal strength may be

Figure 18.23 *Pixelisation due to high BER count*

caused by aerial misalignment, low antenna gain or adverse reception conditions.

In terrestrial DTV reception, picture and/or sound break-up may also result from strong reflected waves with long delays. COFDM is designed to avoid the effects of reflected waves provided they arrive at the aerial before the end of the guard period. The actual time delay that may be accommodated will depend on the COFDM mode as well as the selected guard period. For a 2K mode and a guard period of one-quarter, the symbol duration—the maximum delay that could be accommodated is 56 µs. Reflected waves with longer delays are normally too weak to have any effect on the decoding process. However, where high-gain aerials or RF amplifiers are used, these reflected waves with long time delays may be strong enough to introduce uncertainty in the FEC processor, resulting in intermittent video and/or sound break-up. To avoid this, plug-in attenuators may be fitted to the aerial input socket of the STB.

The aerial cable carries signals with frequencies in the UHF or higher bands. Cable attenuation and the physical condition of the lead thus become important factors in signal integrity. This explains why low impedance double-screened coaxial cables are normally used. Bends and kinks in the cable may disturb the standing wave along the cable, causing intermittent video and/or sound break-up. Similar effects may be produced if the cable is squeezed by a very tight clip, for instance.

For satellite reception, the size of the dish, the condition of the F-connectors and the condition of the LNB will have an effect on the signal strength and hence the quality of the video reproduction.

The boot-up sequence

Normally, a STB is never switched off. When not in use, it remains in the standby mode. Its microprocessor, microcontroller and all other processing chips remain set and ready to receive and process data.

However, when an STB is switched on from cold, the reset pin of the microprocessor goes high and the microprocessor searches for the start-up program by placing the start-up address on the address bus. This address is the start of the start-up routine stored in flash memory. The processor then goes through a comparatively lengthy process of setting, initialising, configuring and programming the processor and decoder chips as described in Chapter 16.

When the initialisation process is completed successfully, the channel decoder begins to search for the default channel, known as the *home channel*. If the signal is detected, the channel decoder locks to it; data is received, decoded and processed. Picture and sound are produced. If the home channel cannot be detected, the channel decoder searches for other channels. Failure to lock to any incoming signal is indicated by a 'no signal' message on the screen.

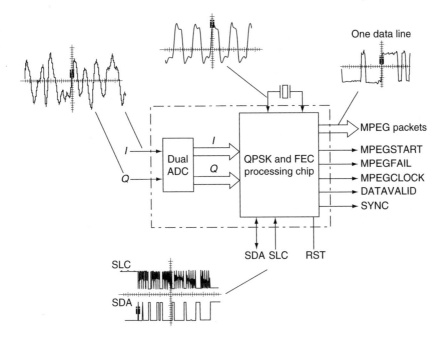

Figure 18.24 *Testing the satellite channel decoder*

Testing channel decoder

The modulated carrier entering the channel decoder is in the form of two quadrature carriers: I and Q in the case of satellite transmission and COFDM carriers in the case of terrestrial transmission. The function of the channel decoder is to digitise the incoming carriers into a multi-bit parallel stream, demodulate them and reproduce error-free MPEG packets.

Testing a satellite channel decoder (Figure 18.24) should start with checking the dual-ADC sampling clock (if a viable test point is available), which would be in the region of 55 MHz (twice a symbol rate of 27.5 Mega symbols per second). The sampling clock should be present regardless of whether the tuner is locked to a channel. A wrong or missing sampling clock would cause a complete malfunction of the channel decoder.

Testing the terrestrial channel decoder should follow a similar pattern to testing a satellite channel decoder. The sampling clock of the ADC if available should be checked first. It is normal to have an 18 MHz clock pulse with amplitude of around 4 V. Further checks involve testing the MPEG packets at the output of the FEC processor chip, at the output of the OFDM demodulator and at the output of the ADC. All data lines should indicate digital activities on a logic probe or an oscilloscope. Typical waveforms are illustrated in Figure 18.25.

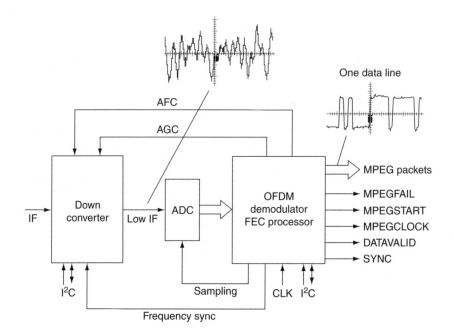

Figure 18.25 *Testing a terrestrial channel decoder*

Testing the transport/MPEG decoder

Figure 18.26 shows a chip set for an HD decoder box comprising of a transport processor and a multi-standard MPEG decoder.

The input to the transport demultiplexer is MPEG-2 or MPEG-4 data packets together with control signals MPEGSTART, MPEGFAIL and MPEG clock from the channel decoder. The function of the demux is to reassemble the video and audio packets of the selected TV programme. Testing the demux involves checking the control signals from the channel decoder followed by the clocks. There are two types of clocks to be tested: the 27 MHz system clock and the chip clock. Each processing chip has its own chip clock (in the region of 50–90 MHz) which is invariably internally generated and must be accurate within 10% or less. The presence of a clock pulse may be tested by a *logic probe*. However, if the frequency has to be measured, an oscilloscope or a frequency meter must be used. An inaccurate clock frequency would cause total failure of the processor chip.

Testing the multi-standard MPEG decoder involves examining the input and output data streams using a logic probe to check for digital activity, or an oscilloscope to display real-time waveforms. The audio output is in the form of a serial L/R multiplexed PCM digital stream. The multiplexing order is determined by an L/R control signal having a frequency equal to half the sampling rate. The 27 MHz system clock and the decoder processing clock should be checked next, followed by the control

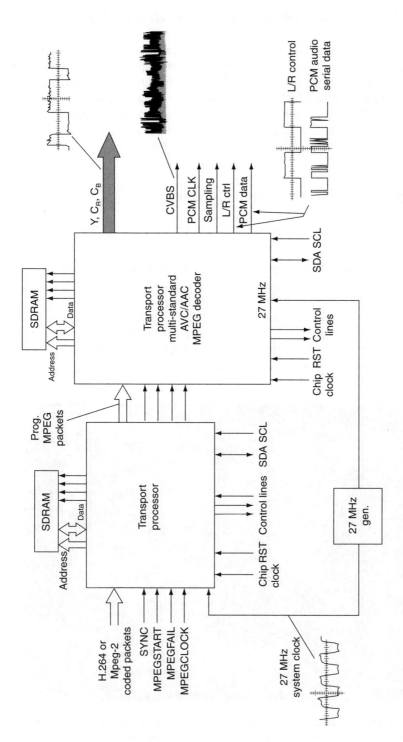

Figure 18.26 *Testing the transport and MPEG decoding*

signals which regulate the transfer of data between the demux and the decoder.

Both the demux and the MPEG decoder have their own dedicated SDRAM memory chips for transport packet storage and video/audio storage, respectively. In many applications, Flash (for start-up and programming routines) and NVRAM (for individual settings) memory chips are also used. Memory chips may be tested by checking for digital activity along their data and address lines as well as the read or *output enable* (OE), *write enable* (WE) and *chip enable* (CE) control lines.

Memory faults and their symptoms

Memory chips perform a number of vital functions in a DTV decoder box including

- Flash: start-up and processing routines for central microprocessing/microcontroller, transport processor and MPEG decoder.
- DRAM/SDARM/DDRAM: micro memory
- SDARM/SRAM: transport packet storage at the demux stage
- SDRAM: video I, P and B frames storage at the MPEG decoding stage
- SDRAM: audio segment storage at the MPEG decoding stage
- SDRAM/DRAM: video buffering

Figure 18.27 *Symptoms of a partial fault of video memory (address line stuck at zero showing image blocks in wrong places)*

Figure 18.28 *A different symptom of the same fault as Figure 18.27 with repeated image blocks*

Memory chips may fail either totally or partially. Partial failure may be caused by a corruption of one or more cells; an address or data bus line stuck at 1 or 0, or shorted pins. A fault, fatal or partial, in Flash or NVRAM would cause the DTV box to go into standby. A fatal fault in the transport processor or MPEG decoder memory chips would result in total picture and sound failure. However, a partial fault would display a constantly changing pattern of picture and sound break-up. The picture break-up is different from the pixelisation caused by a high BER count. In the latter case, all the picture parts are in the correct place, block by block, but some blocks are frozen or obliterated while in the former case, all the picture blocks are clearly present, but some blocks are in the wrong position (Figure 18.27) and/or repeated as illustrated in Figure 18.28.

19 Projection systems

Projection displays utilise an optical system to magnify a small picture created either by a conventional direct-view technology, such as a CRT, or by the use of light *modulating system* (*LMS*). An LMS employs a 3–5 cm microdisplay panel as a *light valve* (*LV*) to modulate the light. A projection display can be operated either in *front-projection* (*FP*) mode, where the viewer and projector are on the same side of the screen, or in *rear-projection* (*RP*) mode, where the viewer and projector are on the opposite side of the screen.

There are three major LV technologies, namely the *digital micromirror device* (*DMD*), the *high-temperature polycrystalline silicon* (*HTPS*) and the *liquid crystal on silicon* (*LCoS*). Each technology has unique properties that determine the image quality.

A microdisplay-based projection system (Figure 19.1) consists of the following major parts:

- Video and audio processing
- Microdisplay drive
- Colour management and image projection

The colour management and image projection system contains the microdisplay, the light source assembly, the light engine and the telecentric optical lens. The input which could be in the form of composite, component or digital video is processed and the image resolution scaled to the native resolution of the microdisplay. RGB signals are then fed to the microdisplay drive on its way to the light engine. Stereophonic audio input is also processed and amplified as shown. Light from a light source assembly is directed to the microdisplay to be projected onto a screen via a telecentric projection lens. User adjustments are provided by position-controlled motors for focus and zoom.

First, we will consider the CRT-based projector.

CRT-based projection systems

The CRT-based projector was the first commonly used system and came to dominate the middle- and low-end RP system market for a number of years. Three single-colour (R, G or B) tubes, each optimised for luminance and beam spectrum of the appropriate primary colour, are used to project three converging images onto the screen. Since the path of the electron beam in the small size CRT is relatively short, the beam spot size can be better controlled, minimising any smearing effects. These features are

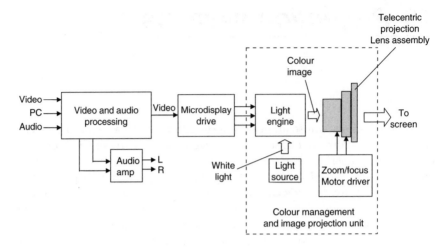

Figure 19.1 *Microdisplay-based projection system*

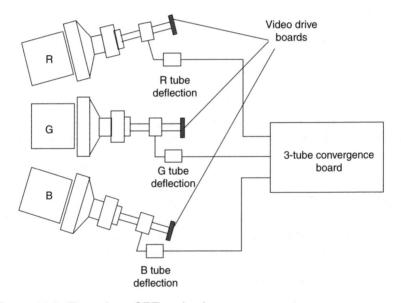

Figure 19.2 *Three-lens CRT projection arrangement*

required to produce high resolution and good colour reproduction with high brightness. There are two configurations for CRT projectors, using either three lenses or a single lens.

In the *three-lens configuration*, each tube is dedicated to a primary colour projecting its own individual image on the screen. The three images are converged by deflection of the off-axis R and B tubes as shown in Figure 19.2 to produce a colour image. Convergence is the most challenging problem in this type of CRT-based projector. For good quality image, it is desirable to

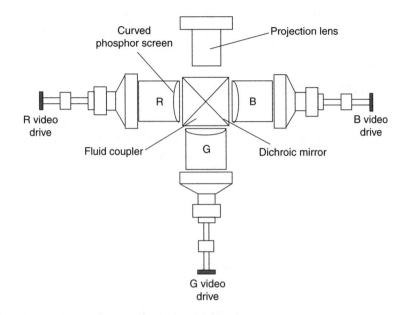

Figure 19.3 *Single-lens CRT arrangement*

converge the images from the three tubes to within one-half pixel. The three-lens arrangement suffers from two main drawbacks: keystone distortion caused by the trapezoidal images produced by the off-axis R and B tubes and colour non-uniformity due to the angular dependence of the reflection coefficient of the optical system. They are obviated by suitable convergence circuitry and by tilting the blue and red lenses, respectively.

In the single-lens CRT configuration, a dichroic mirror is used to merge and converge the three images as shown in Figure 19.3. Even though single-lens CRT projectors are free from convergence errors, there arise other convergence problems caused by optical, electrical and magnetic issues such as the long back focal length.

High performance CRT projection systems have a long and proven track record for professional high-resolution graphics and premium quality video presentations. Their wide range of signal input compatibility, edge and colour matching, excellent black level and geometry correction are key reasons for the continuing success of this technology. In addition, the mature CRT technology provides highly reliable and maintainable projectors at affordable pricing.

However, these projectors are not as bright on images with mostly white content. Furthermore, the Gaussian beam-shape of the scanning electron beam reduces image sharpness when compared with a fixed pixel display operating at its native resolution. Convergence drift and CRT phosphor burn require regular servicing of the projector.

Another disadvantage of CRT projectors is its large size. There is a trade-off between the cabinet size and image quality. A shorter focal length lens decreases the *optical throw distance* and allows a thin cabinet. However, it increases offset angles between tubes, which results in poor image quality due to increased electron beam deflection.

Microdisplay-based systems

Microdisplay-based projectors may function in two ways: *reflective* (such as DMD-based projectors) and *transmissive* (such as HTPS-based systems). A third hybrid *transmissive/reflective* technique is provided by LCoS in which a mirror is placed behind the transmissive LC pixel to reflect the light back through the pixel.

A microdisplay-based system consists of two parts: a *light source* or illumination assembly and a *light engine*. The illumination assembly contains a lamp and appropriate filters, mirrors and lenses to provide light waves to the light engine. The light engine contains the microdisplay and all other necessary splitting, combining and focusing optical arrangements to project a full colour image on a screen.

There are two ways of producing the primary colours from pure white light emerging from the illumination assembly: the first is to use R, G and B filters, and the second is to split the light wave into its three primary components, red, green and blue using *dichroic mirrors*. The use of filters provide a simple method of obtaining colour, but it does reduce the light intensity and hence brightness of the image. The filterless technique normally requires three microdisplays, making the system bulkier and more expensive.

The illumination assembly

A simplified light source assembly is shown in Figure 19.4. The light source could be an incandescent *filament*, a *tungsten-halogen, fluorescent, mercury, metal halide* or *arc-lamp* devices. Each type offers different white tints, luminance per wattage and lifetime.

The light from the lamp is focused by an elliptical reflector onto a set of 'relay' lenses forming a telecentric optical system. A 'cold' mirror, one that does not reflect UV and IR waves, oriented at 45° and a UV/IR filter is placed between the first two relay lenses to remove IR and UV radiation from the light beam. A prism plate component may be used between the second and third relay lenses to recover some of the light that would otherwise be rejected by the polariser in the light engine. This component converts about 55% of the 'wasted' light into useable light.

DMD-based projectors

The working principles of the micro-electromechanical system (MEMS)-based DMD chip also known as DLP has been fully explained in Chapter 12.

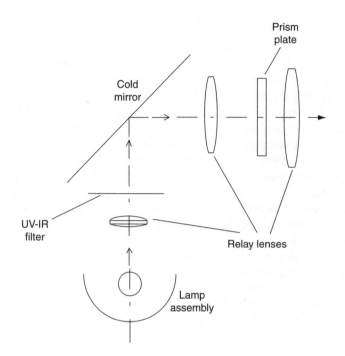

Figure 19.4 *The illumination assembly*

The DMD-based projector is a reflective type in which light is reflected on and off by millions of deflecting micromirror. The deflection of the micromirrors is a result of an electrostatic force between the mirrors and address electrodes connected to a substrate of SRAM nodes. A DMD-based projector system using one chip and three chips were also explained in Chapter 13.

There are several unique advantages of DMD-based (DLP) projection systems. These include:

• Small compact size: This is especially the case with the single-chip arrangement.
• High contrast ratio: This is especially the case with the new generation of DMDs that increase the mirror tilt angle from 10 to 12° and features an absorbing coating to the substrate under the mirrors.
• High aperture ratio: As a result of the small gap between adjacent mirrors (>1 μm) and the ability to mount the driving circuits underneath the deflecting mirrors, the aperture ratios can often exceed 90%. Visible pixel boundaries leading to the *'screen door effect'* are therefore not a problem in DLP systems.
• Good reliability.
• Polarisation independence: With DLP systems, there is no loss associated with any polarisation of the light source.

As for the weakness of the DLP systems, they include the following:

- Manufacturing complexity and cost: The CMOS electronics in the underlying silicon substrate consists of a six-transistor SRAM circuit per pixel, and additional auxiliary addressing electronics. This drives the price of high-resolution DMD chips.
- Colour break-up or rainbow effect (single-chip only): DLP systems based on a single-panel DMD require spinning colour wheel to achieve full colour, resulting in a visible artefact known as colour break-up, or the 'rainbow effect' as was described in Chapter 13.
- Temporal artefacts: The binary nature of the DMD addressing process creates the sensation of temporal modulation similar to that produced by the sub-field addressing technique used in plasma panels. In fast-moving video images, the edges of moving objects can become temporally unstable and appear fuzzy. Improvements in the addressing algorithms have reduced this effect, but it can still be perceived under certain viewing conditions.
- Poor colour saturation (single-chip only): In most single-panel DLP projectors, colour wheels often contain a clear (white) segment to boost brightness. Though the image appears brighter, it reduces colour saturation.

HTPS-based systems

The structure of a polycrystalline silicon LC panel is very similar to that of the active matrix LCD (AMLCD) commonly used for television screen and monitor screens described in Chapter 12. The electron/hole mobility of poly-crystalline silicon is, however, much higher than that of amorphous silicon used in AMLCD, allowing the size of a thin-film transistor (TFT) to be made much smaller. It is also possible to have driver and shift register IC packages embedded on the panel itself. Due to the good aperture ratio, high-brightness polysilicon LC projection systems are feasible. There are two methods of fabricating polycrystalline silicon: low (<600°C) and high (>1000°C) temperature. Because the low-temperature type suffers from higher leakage current, the HTPS is normally used offering good performance in terms of degree of miniaturisation, high definition (HD), response speed and reliability.

Colour filter HTPS

Figure 19.5 shows a colour *combining x-cube assembly* used in a transmissive HTPS light engine using a colour filter. White light from the light source assembly is directed to three HTPS LV panels, one for each primary colour. The LV is sandwiched between two sheet polarisers in the same way as the traditional AMLCD display with the exception that the LV pixels do not sub-divide into three cells. Pixels of a LV contain one cell

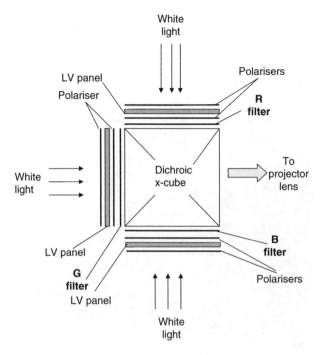

Figure 19.5 *The 3-panel transmissive HTPS x-cube assembly*

each. This reduces the number of data (column) connections by a third. The output light from each LV panel is spatially modulated by their individual red, green and blue data coming from the video processing and formatting board. At this point, all three images are monochrome. They are turned into RGB images by the following red, green and blue filters. The three images converge into a full colour picture by a solid glass '*x-cube*' *dichroic beam combiner*. The colour picture is then projected on a screen via a *telecentric projection lens*. A typical 3-panel light engine assembly is shown in Figure 19.6. Three flexible tape carriers are used to feed video data as well as line select signals from the video and control boards to the LV panel. Shift registers are embedded with the LV driving circuitry to demultiplex the video data using a shift register and drive the panel.

Colour filterless 3-panel HTPS projector

In the 3-panel HTPS projection system that does not use a colour filter, incident white light is split into three primary colours by dichroic mirrors/filters before going to the x-cube core of the light engine. Each primary colour passes through an HTPS LV sandwiched between two-sheet polarisers in the same way as the traditional AMLCD. A dichroic x-cube is used to combine the three colours immediately before the projection lens as illustrated in Figure 19.7.

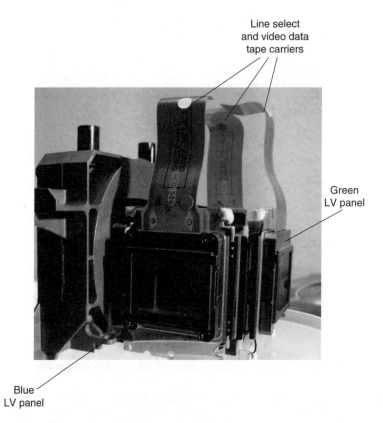

Figure 19.6 *The 3-panel x-cube assembly*

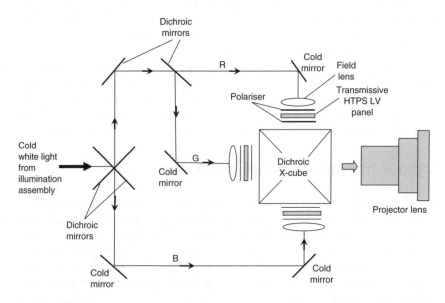

Figure 19.7 *The 3-panel filterless HTPS light engine*

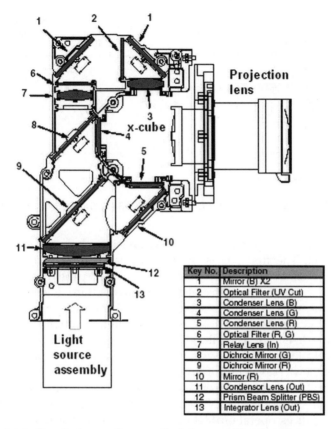

Key No.	Description
1	Mirror (B) X2
2	Optical Filter (UV Cut)
3	Condenser Lens (B)
4	Condenser Lens (G)
5	Condenser Lens (R)
6	Optical Filter (R, G)
7	Relay Lens (In)
8	Dichroic Mirror (G)
9	Dichroic Mirror (R)
10	Mirror (R)
11	Condensor Lens (Out)
12	Prism Beam Splitter (PBS)
13	Integrator Lens (Out)

Figure 19.8 *Break down of a practical 3-panel filterless light engine*

A complete breakdown of a practical 3-panel filterless light engine used by Sanyo is shown in Figure 19.8.

Pros and cons of HTPS projectors

HTPS projectors present several advantages and disadvantages when compared with the DMD-based projector. The advantages include:

- Better colour saturation: Since HTPS projectors use three RGB panels, colours tend to be rich and vibrant.
- Sharper image than equivalent resolution DLP systems, whose tilted mirror pixels can appear blurred at the corners.
- The 3-panel HTPS provide a higher lumen output than single-panel DLP projectors with the same wattage lamp.
- Mature technology.

The disadvantages of HTPS projectors include:

- Screen door effect: As was explained earlier, the black matrix TFT element that is necessary in HTPS panels creates visible pixilation known

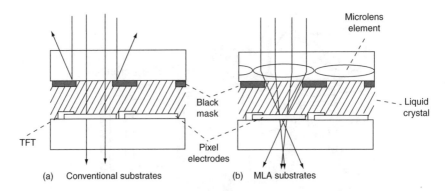

Figure 19.9 *Light throughput is increased using microlens array (MLA)*

as the screen door effect. Several measures have been taken to alleviate this effect. The first one is through increased pixel count. A WXGA projector has over 1 million pixels while a VGA panel has only 300,000 pixels. Second, reducing the inter-pixel gaps. A third development is the use of *microlens arrays* (*MLAs*; Figure 19.9). Though developed primarily to boost efficiency, the MLA concentrates light through the pixel aperture thus de-emphasising the sharp pixel edges.

- Relatively low contrast: System contrast is difficult as a result of the field of view (FOV) of the TN LCD mode.
- Lifetime: LC alignment layers used in HTPS LVs are organic polyimides which are susceptible to UV and deep blue light photochemical damage, which reduces operation lifetimes. UV filters with a long-wavelength cut-off are helpful but tend to reduce the blue content of the final colour images.

Liquid crystal on silicon systems

The liquid crystal on silicon system, LCoS is a hybrid technology in that it is transmissive using LC light modulators and reflective by virtue of a back passive mirror (Figure 19.10). Incident light goes through the LC pixel which controls its brightness by its polarisation state. It is then reflected back by the mirror to pass through the same LC pixel again, a double pass. Being a reflective system, the transistor, shift register and driver circuitry can be constructed beneath the LV giving the LCoS LV a very high aperture ratio approaching 90%. The silicon substrate contains all necessary transistors, data and address drives. Although the traditional analogue addressing is common for LCoS panels, digital addressing has been developed and applied successfully providing stable and uniform grey levels, low fabrication costs and higher reliability.

The fact that the reflected light shares the same path as the incident light necessitates the use of an optical device to separate them.

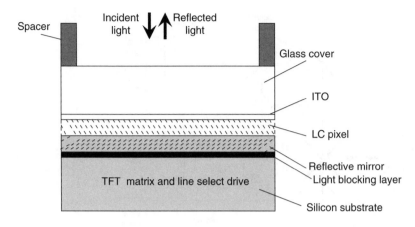

Figure 19.10 *LCoS light valve (LV) pixel*

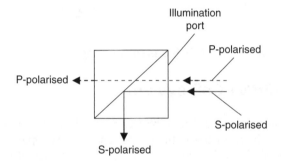

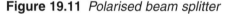

Figure 19.11 *Polarised beam splitter*

This is carried out by a *polarising beam splitter* (*PBS*). The PBS has four sides or ports with the input side known as the *illumination port*. Polarised light entering the illumination port will be reflected or deflected into one of the other three ports depending on its polarisation state. For instance, a P-polarised light entering the illumination port would pass right through to the opposite port, while an S-polarised light would be deflected by 90° as shown in Figure 19.11. Since the incident and reflected light in an LCoS LV are differently polarised, the use of a PBS may be used to separate them.

There are two types of LCoS-based projector architecture. The first uses three PBSs with a combining x-cube (known as $3 \times PBS/x$) in a similar configuration to the 3-panel HTPS-based system. The second uses a single unit composed of PBSs and retarders (known by their trade name of *ColorSelect*™) that carries out both the polarising/splitting functions of the PBS and the splitting/combining functions of the dichroic mirrors and x-cube. The PBS/retarder combination is capable of very compact colour management systems.

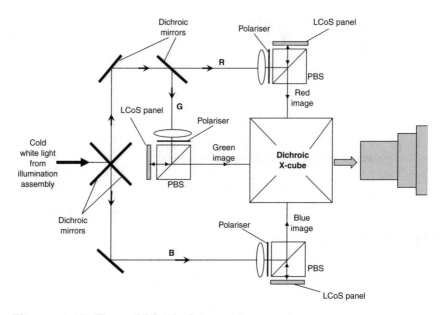

Figure 19.12 *Three-PBS LCoS-based light engine*

The three-PBS LCoS system

The main components of a three-PBS LCoS-based system are illustrated in Figure 19.12. Light from the light source assembly is separated into RGB using dichroic filters and mirrors in the normal way. The three beams are directed towards the three PBSs by several fold mirrors. Before hitting the PBSs, the light beams are polarised by sheet polarisers in order to improve contrast. Each PBS reflects S-polarised light towards its respective panel. The light enters the liquid crystal pixel through the glass cover and is reflected back to the liquid crystal by the mirror surface. If an electricfield is applied to the liquid crystal by a video signal (the 'on' state), the double pass through the liquid crystal rotates the S-polarised light intoP-polarised light which can efficiently pass through the PBS and proceed towards the colour combiner. With no voltage applied ('off' state), the light polarisation is not rotated and the S-polarised light is deflected back on towards the lamp. Light from the red, green and blue channels is recombined in the x-cube dichroic beam combiner to be projected on a screen via a telecentric projection lens.

ColorSelect™ LCoS-based systems

Polarised beam splitters may be used as an alternative to dichroic mirrors to split light into its three primary colours. This is accomplished by polarisation coding. For instance, take the case of S-polarised red and P-polarised green/blue colours. When fed into a PBS illumination port, the PBS will split them with red appearing on one port and green/blue on a separate port as

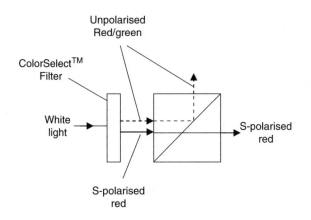

Figure 19.13 *The use of retarder ColorSelect™ filter in light splitting*

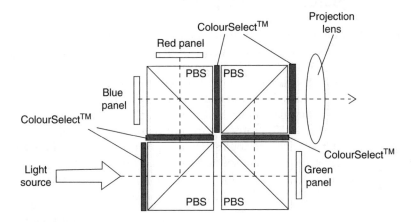

Figure 19.14 *Outline of a 3-panel LCoS using ColorSelect™ filter/PBS optics*

shown in Figure 19.13. Colour polarisation is carried out by a ColorSelect™ retarder filter. The red and green may be further split using a second ColorSelect™ filter/PBS. The ColorSelect™ filter is a multi-layer laminate of retardation films that converts one spectral band, e.g. a primary colour to an orthogonal polarisation while leaving the remaining spectrum (green/blue) unaffected as illustrated in Figure 19.13. The outline of a complete system is illustrated in Figure 19.14. Retarders are used at PBS ports other than the illumination port in order to re-polarise or de-polarise the exiting light.

Merits and demerits of LCoS projection systems

The merits include:

- IC compatibility: The LCoS electronic substrate is compatible with the standard silicon technology, allowing additional driving circuitry to be integrated into the back-plane design.

- Cost effectiveness for high resolution: LCoS can achieve HD resolution (1920 × 1080) in a 0.7″ panel and 1280 × 720 in a 0.5″ panel.
- No screen-door effect.
- Smooth picture: The pixel edges in LCoS tend to be smoother and more natural looking compared with the sharp edges of the micromirrors with DLP.
- High contrast.
- High response speed.

Demerits include:

- Long-term reliability.
- Colour break-up is observed in single-panel system as in single-panel DLP systems.
- Complexity.

Other projector technologies

In addition to the projection systems mentioned above, there are several other technologies under development. These include the colour filterless single lens projector illustrated in Figure 19.15.

In this configuration, three dichroic mirrors are stacked in a non-parallel arrangement to split the white light beam into three primary colours beams. Alternatively, RGB angular separation may be achieved by *holographic* or *blazed diffraction grating*. The three beams are then projected onto a single LC LV from different incident angles. The LV pixels are divided into three RGB cells. A microlens array is used to direct the beams to their corresponding cells. The advantage of eliminating the colour filter is the recovery of

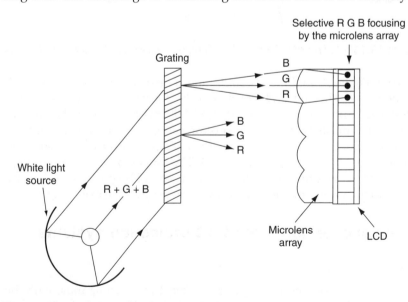

Figure 19.15 *Colour filterless single lens projector*

light absorbed by the filter thus making the single-panel projector as bright as a 3-plane projector without the extra expense and size.

Other technologies include the *polymer-dispersed LC* (*PDLC*) devices which scatter light but appear transparent when a high voltage is applied, *surface-stabilised ferroelectric LCs* (*SSFLCs*) which exhibits two states depending upon the polarity of applied voltage and *actuated mirror array* (*AMA*) based on piezoelectric angular deflection of individual mirrors within an array and light amplifiers in which low-intensity image is amplified for high-intensity projection.

The 3-panel HTPS projector block diagram

Figure 19.16 shows a block diagram for a 3-panel HTPS projector. Inputs from PC via a VGA connection, component video or composite video (CVBS) together with stereo sound may be received and processed by the projector A/V section. External digital video signal from a TMDS receiver (or HDMI for HD operations) is fed directly into the image processor. Composite video/chrominance C is selected by the video switch and processed by the video decoder/comb filter and following interlace-to-progressive conversion passed on to the image processor or scalar. RGB input from a personal computer is first clamped and then converted into a digital format before going to the image processor. The purpose of the image processor is to rescale the image to the native resolution of the LCD microdisplay. The output from the image processor is analogue RGB which are individually gamma corrected on their way to RGB digital-to-analogue converter (DAC) and sample-and-hold ICs. The sample-and-hold (S/H) captures the instantaneous analogue value of each RGB signal and feeds it into the appropriate LCD panel to set the pixel value. LCD panel level shifting is carried out under the direction of a timing chip.

The zoom and focus functions are controlled by the central processing unit (CPU) as instructed by the remote IR control or manual operation of the front panel. The CPU also controls the operation of the fan motors.

A typical optical control system is illustrated in Figure 19.17. An I^2C-controlled I/O expander chip is used to control three separate motors: *focus*, *zoom* and *lens*. The expander received instructions from the CPU via the serial bus to send (+) or (−) control signals to the relevant motor drive IC (Figure 19.18).

Three-dimensional video display

Three-dimensional (3D) video display is not new. As far back as 1946, John Logi Baird demonstrated a 3D image. So far, 3D video has been restricted to a one-off entertainment experience to theatre-goers. Today, however, it

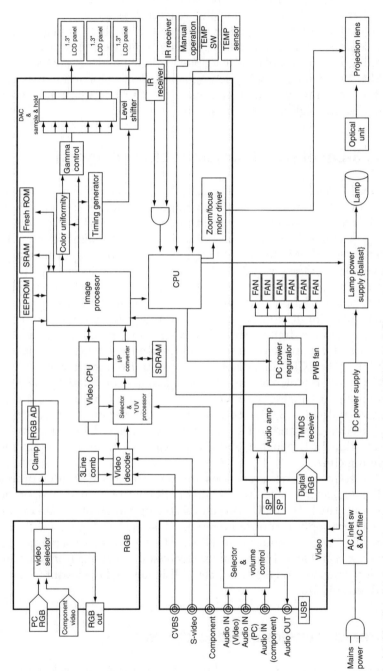

Figure 19.16 *Complete block diagram for a 3-panel HTPS (LCD) projector*

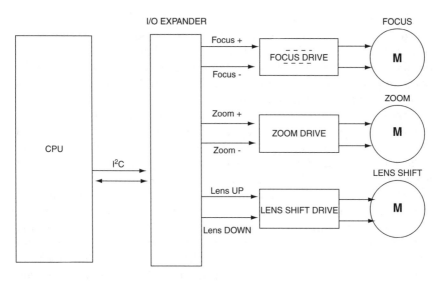

Figure 19.17 *Optical control system*

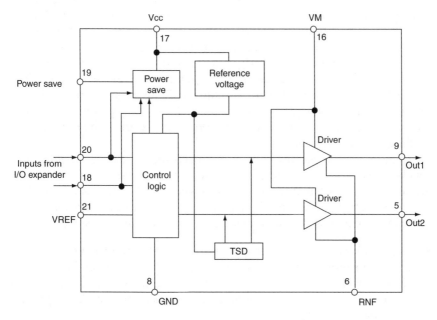

Figure 19.18 *Motor drive IC (BA6920F)*

is seen as the next big thing after flat panels and HDTV to such an extent that more than 100 companies led by Sharp and Sanyo formed the 3D consortium (3DC) and MPEG established a 3D audio/visual coding (3DAV) ad hoc group.

The manner in which we perceive 3D images is the disparity between images seen by each eye. The two images are focussed in the brain giving us a sensation of depth. It is this mechanism that 3D video attempts to emulate. 3D technologies fall into two major categories: *binocular stereoscopic* requiring the viewer to wear spectacles and *autostereoscopic* which do not require the use of spectacles.

There are a number of different types of binocular stereoscopic displays; all use a pair of spectacle which allows one image through to the left eye and another to the right eye. The *polarised light display* (PLD) uses Polaroid glasses to direct the image to the appropriate eye. The *head-mounted display* (HMD) uses two separate monitors with a viewing hood to display left and right pictures, respectively. The two pictures are brought together by optical means. HMDs give an immersive experience and are popular with virtual reality systems. The *split-screen type* is similar to the HMD in that it displays separate left and right pictures. The difference is the split-screen, as the name implies, uses a single monitor to do that with the inevitable reduction in resolution. The *complementary colour type* is the technique used in early 3D cinema. The spectacles have colour filters: red and blue to separate the images going to the left and right eyes. Finally in this category, the *eclipse method* in which the left and right images are presented in alteration instead of simultaneously. The viewer uses special spectacles with occulting aperture that open and close in synchronism with alternating left/right image. Binocular glass-based systems have many disadvantages: users who already wear glasses feel uncomfortable wearing a second pair simultaneously; users are concerned with hygiene issues when glasses are shared. It is also uncomfortable and inconvenient to try for instance, to do other work at the same time such as reading and writing notes on paper.

An autostereoscopic display does not need special glasses. Various optical elements are used to direct each view into space in front of the display. When the user's eyes are in the regions known as *viewing windows*, then each eye receives a different image and the human vision system (HVS) sees 3D. Autostereoscopic displays can be classified into three types:

- *Fixed viewing windows*: The display is designed to produce two windows at fixed positions in front of the display. The user must be located correctly laterally in order for each eye to see the correct image and therefore it is only practical for a single user to use. This may be improved with viewer tracking.
- *Viewer tracking*: The user's head is tracked using the video image so that they always see the correct 3D image. Generally, only one user can view the display at one time.
- *Multiple views*: The display produces a number of views, usually in fixed positions. Several users can then view stereo simultaneously from different pairs of views. If the user moves, then it is quite easy for them to reposition themselves so that each eye is in a different pair on views.

3D LCD panels

Any type of matrix display can be used as the basis of the 3D display. However, the recent fall in cost of LCDs has made some stereoscopic display methods viable for mass production of 3D displays. The display itself is a standard mass produced LCD panel, with some additional optical elements. There are several types of 3D LCDs:

- *Parallax barrier*: This type uses a series of black vertical lines in front of the display. These lines, known as a parallax barriers, are positioned so that each of the eyes only sees half the pixels, every other pixel. By careful selection of the barrier geometry, it is possible to adjust the viewing window position and angles for a 3D viewing. Apart from halving the resolution of the display, the main disadvantage to parallax barriers is that the barrier reduces the brightness of the display. The parallax barriers can be switched off to view a perfect 2D image.
- *Lenticular* or *integral* display uses tiny lenslets (very small lenses) attached to the display with high accuracy to split the image into left and right. Lenticulars are often slanted to improve the transition between viewing zones for a multiview display. The main advantage of lenticulars is that they transmit full brightness. The disadvantages include the fact that it is harder to switch the function off to achieve 2D.
- *Time-sequential displays*: This type employs a directional light source placed behind the display whose direction can be altered. In one frame, the light is directed to the left eye while the left image is displayed. In following frame, the light is directed to the right eye while the right image is displayed. In this way, full resolution is maintained for both views. However, it has the disadvantage that the refresh rate of the display must be doubled.

Power supply requirements

As far as the electronic sections of a projection system are concerned, the power requirements are the normal nominal 3, 5 and 12 V d.c. These voltage rails also serve the motor drivers for the focal unit. The light source, however, needs its own separate power supply incorporating a *'ballast'* as well as power factor correction (PFC) circuits. Essentially, a ballast provides a suitable starting voltage for the lamp and then limit its current flow during steady-state operation. It is normal for the ballast itself to provide the power factor correction. It will also regulate the lamp output against line voltage changes, minimise power losses and obtains a high power factor.

Traditionally, electromagnetic (EM) ballasts were used which are designed to operate at line frequency (50 or 60 Hz) with very low light output per consumed power watt. Current ballasts are electronic switching circuits with a high frequency (20–60 kHz) and increased efficiency.

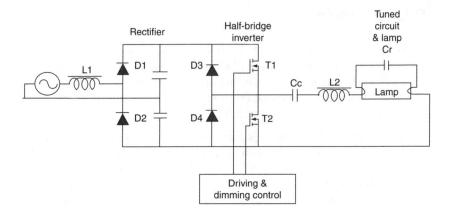

Figure 19.19 *Lamp ballast supply incorporating PFC*

Figure 19.19 shows a typical ballast circuit with half-bridge inverter circuit configuration. The inverter block consists of two controlled switching transistors T1/D3 and T2/D4, a coupling capacitor Ce, a resonant inductor L2, a resonant capacitor C_r and a lamp. The operation of the circuit is similar to that of a resonant converter and PFC described in Chapter 11. The rectifier D1/D2 and filter circuits convert the a.c. power into d.c. The half-bridge inverter converts the d.c. voltage into a high frequency a.c. voltage to start the lamp and stabilise the lamp current. When it is in a steady-state operation, the gating signals from the dimming and control IC determine the conduction of the switching transistors T1 and T2 and diodes D3 and D4. By controlling the duration when T1 is on relative to T2, the power transfer to the lamp can be adjusted. The energy going into the lamp is in the form of chopped d.c. which triggers tuned circuit L2/C_r into oscillation at a frequency of about 60 kHz.

20 DVD

DVD stands for *digital versatile disc* or more commonly known as *digital video disc*. DVD is the most recent generation of compact disc (CD) technology able to store huge amounts of data in GB, comparable to the hard disk commonly used in computers. DVD discs are available in two diameter sizes: the commonly used 12 cm and the 8 cm. The 12-cm DVD disc is the same size of the familiar CD (audio or ROM), but holds up to 25 times more data. This takes the DVD to a qualitatively higher level in terms of its application. For the first time, we a have a CD that can hold over 2 hours of high-quality video with six channels of high-fidelity surround sound. Further development introduced the *blue-laser* based (as opposed to the *red-laser* based *DVD-video*) discs, namely *'HD DVD'* and *'Blu-ray DVD'* with a capacity of up to 50 GB suitable for up to 9 h of HD video. Apart from video/audio applications, DVD discs may be used for a variety of other applications including archiving of books and still pictures as well as a mass storage device for computer applications and educational/training purposes which make use of interactive facilities.

The first DVD format, DVD-video format was introduced in 1996 with its specifications published in Book B. Previous to that, optical disc technology was limited to the CD and the *Laserdisc* using a 780 nm wavelength (infra-red) laser beam. The CD had two main formats: audio CD and CD-ROM. The Laserdisc had a 30-cm diameter and was intended to hold high-quality audio and video of up to 1 h per side. That technology never caught on and was quickly superseded by the far superior DVD.

The DVD-video was joined by the DVD-ROM format (Book A) and these were followed by other standards including DVD-Audio, DVD-R, DVD-RAM and DVD_RW. These were followed by the high Density (HD) DVD and the Blu-ray Disc (BD) formats using the shorter 405 nm blue-laser wavelength.

DVD construction

The basic principle of storing data on a DVD is the same as that of the audio CD and the CD-ROM, namely the creation of tiny *pits* along closely placed spiral track. The pits and the non-pit parts known as lands are translated into ones and zeros by a laser pickup head. Unlike the CD-ROM which has one substrate, the DVD disc consists of two 0.6 mm thick substrates bonded back-to-back (a total of 1.2 mm thickness) giving it the necessary stiffness to avoid disc wobble or tilt.

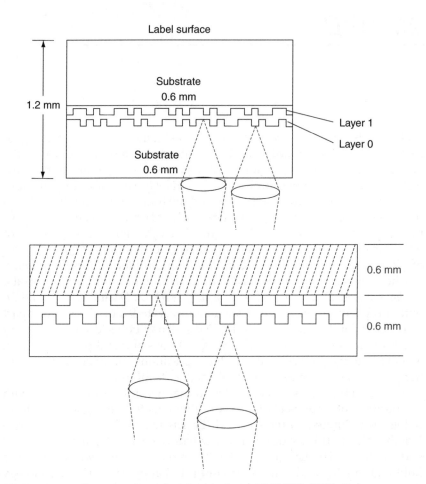

Figure 20.1 *Construction of a double layer DVD/HD DVD disc*

Table 20.1 *DVD construction formats*

Single-sided (SS), single-layer (SL)
Single-sided (SS), dual-layer (DL)
Double-sided (DS), single-layer (SL)
Double-sided (DS), dual-layer (DL) on one side only
Double-sided (DS), dual-layer (DL) on both sides

DVD discs may record data on one side only (*single-sided, SS*) or on both sided (*double-sided, DS*). Each side may have a *single layer (SL)* of recording or two layers (dual-layer, DL) (Figure 20.1). We thus have five different construction formats: SS-SL, SS-DL, DS-SL, DS-DL (dual-sided on one side only) and DS-DL (dual-sided on both sides) as listed in Table 20.1. The most popular are the SS discs. Double-sided DVDs have proved to be unpopular with manufacturers and they are not commercially used.

With DL construction, layer 1 must be read through layer 0. Hence, layer 0 must be semi transparent with a reflection of 20% while that of the second layer should have a reflection of 70%. For double-sided discs, data must be read from both sides of the disc.

DVD capacity

The capacity of a disc is determined by the amount of data that may be stored on it in the form of pits and lands. Increasing the capacity of a DVD thus requires a corresponding increase in the density of the stored data in terms of smaller size pits, which is primarily achieved through the reduction of the dimensions of the pits. The problem however is not in the manufacture of ever smaller pits and ever closer tracks. The problem arises with the pickup head and its ability to read very tiny pits and discriminate between adjacent tracks. While smaller pits permit increased track density and data density along the track, it requires a corresponding smaller, more tightly focused laser read beam in order to resolve the data with the required level of accuracy. In optical storage technology, the diameter of the focused read beam is proportional to the wavelength of the read laser, and inversely proportional to the *numerical aperture* (*NA*) of the focusing lens. The NA of a lens describes it power to converge and focus the light passing through it. Therefore, smaller laser beam spot may be achieved by a combination of shorter laser wavelengths and a larger NA. In the case of DVD video, a wavelength of 650 nm (red laser) and a NA of 0.6 are specified resulting in a spot diameter of 0.54 mm compared with a wavelength of 780 nm and a NA of 0.45 and a diameter of 0.54 mm for CDs.

Further increase in DVD capacity was made possible by the use of *blue-laser diode* (*LD*) with a wavelength of 405 nm. The blue laser is just outside the violet of the light colour spectrum as shown in Figure 20.2 with a predominantly blue colour. The laser beam spot is reduced allowing

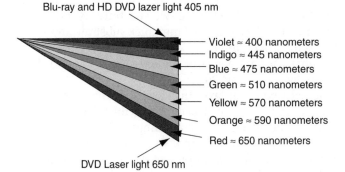

Figure 20.2 *Light spectrum and DVD laser wavelengths*

DVD   Blu-ray/HD DVD

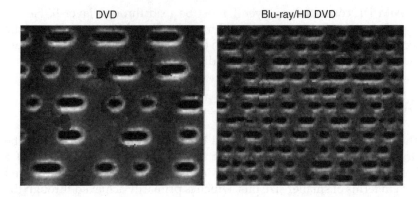

Figure 20.3 *Pits and tracks of traditional 650 nm DVD disc compared with the 405 nm blu-ray/HD DVD disc*

for even smaller pit size and closer tracks as illustrated in Figure 20.3. Based solely on the reduction of the wavelength from that of the red to the blue LD, an increase in disc data density by a factor of 2.6 is possible. Coupled with the use of focusing lenses with increased NA compared to that in DVD video even greater gains in disc capacity have become possible.

In the blue-laser-based disc applications, two competing DVD disc formats are on offer: the HD DVD (supported by Toshiba, NEC, Sanyo and others) and the Blu-ray disc, BD (supported by Hitachi, Panasonic, Sony, Samsung and others). Both formats provide increased data-storage capacity (nominally 25 and 15 GB respectively) compared with the traditional DVD's 4.7 GB nominal capacity.

The higher capacity of the BD compared with the HD DVD is fundamentally due to the higher NA of the focal lens. However, as the NA is increased, the quality of the optical read spot becomes more sensitive to disc tilt and variations in the cover layer thickness, causing degradation in the quality of the playback signal. For this reason, the cover layer for the Blu ray discs is reduced to 0.1 mm. While HD DVD maintains compatibility with the traditional DVD having two substrate layers of 0.6 mm thick, a total of 1.2 mm, the BD has one 1.1 mm thick substrate and a 0.1 mm thick cover layer as illustrated in Figure 20.4.

A list of physical parameters of the red-laser DVD and the two blue-laser discs are listed in Table 20.2.

But while reduced cover layer thickness improves the manufacturability of the higher capacity disc, it also reduces the tolerance of the playback signal quality to the effects of dust and fingerprints compared to the conventional 0.6 mm cover layer design. To overcome this, a 2-nm hard coat is used at the readout surface to protect the disc from scratches and fingerprints. Alternatively, a cartridge may be used to protect the disc.

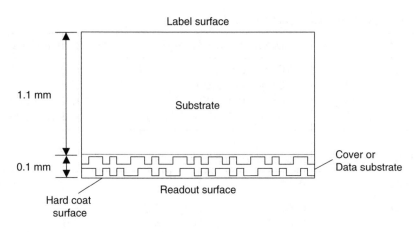

Figure 20.4 *The Blu-ray disc structure*

Table 20.2 *Comparision of the properties of DVD, HD DVD and Blu-ray technologies*

Parameter	DVD	HD DVD	Blu-ray
Pickup head laser wavelength (nm)	650	405	405
Focusing lens NA	0.6	0.65	0.85
Disc diameter (mm)	120	120	120
Cover layer thickness (mm)	0.6	0.6	0.1
Substrate thickness (mm)	0.6	0.6	1.1

Storage capacity

Capacity is given as so many Bytes in the same way as the capacity of a computer storage device such as that of a hard disk is specified. When the capacity exceeds thousands, millions or billions of bytes, as they often do, prefixes K (or k), M and G are used. In ordinary usage, prefix K = 1000 (10^3), M = 1000 K (10^6) and G = 1000 M (10^9). In specifying a computer data capacity, a 'binary' kilo, Mega and Giga are used which are based on powers of 2, namely 2^{10+} for Kilo, 2^{20} for Mega and 2^{30} for a Giga. This makes a computer or binary K = 1024 instead of the commonly used denary 1000 and so on. In DVD applications, using computer-based data capacities is not relevant. Nonetheless, it is not uncommon for it to be used in publications and specifications. For the purposes of this book, the computer binary prefixes will be used unless otherwise stated. To convert back to the denary prefix, i.e. the actual capacity, refer to Table 20.3.

Table 20.4 lists the capacity of the single-sided DVDs, HD DVDs and BDs.

Table 20.3 *Prefixes and their values*

Prefix	Name	Common use (denary)	Computer use (binary)
K or k	Kilo	$10^3 = 1000$	$2^{10} = 1024$
M	Mega	$10^6 = 1,000,000$	$2^{20} = 1,048,576$
G	Giga	$10^9 = 1,000,000,000$	$2^{30} = 1,073,741,824$
T	Tera	$10^{12} = 1,000,000,000,000$	$2^{40} = 1,099,511,628$

Table 20.4 *Capacity of different DVD formats*

Type	Name	SL/DL	Capacity (GB)
DVD	DVD-5	Single-layer	4.7
	DVD-9	Dual-layer	8.5
Blu-ray	BD-25	Single-layer	23.3
	BD-50	Dual-layer	46.6
HD DVD	DVD-15	Single-layer	15
	DVD-30	Dual-layer	30

Recording sectors

The DVD disc is divided into four areas. The main area is the *user* or *data area* where video and audio data are recorded. This is preceded by a *lead-in area* and *control data* areas and followed by a *lead-out* area. Before the video and audio data are recorded in the user area, the data bits are grouped into chunks known as sectors. Each sector commences with a header which includes, among other things, a unique ID number for the sector. The data is written on the disc surface, sector by sector. Reading the data off the disc involves directing the pickup head to the part of the disc surface where the required sector is recorded and once the sector is identified, data is extracted for further processing.

Rotation speed

In DVD applications, the disc may be made to rotate at a *constant linear velocity (CLV)*, in the same way as a CD-ROM or at a *constant angular velocity (CAV)*, similar to a hard disk. In the CLV technique, the track which the pickup head is reading must be moving across the head at the same speed regardless of the position on the disc. This means that the disc must rotate faster if the head is reading a part of the track which is nearer to the centre than if it was reading a part of the track that is nearer to the outer circumference. As the head follows the spiral track, the angular velocity of the disc must continuously change to keep the linear velocity constant. In the CAV technique, the rotation of the disc is constant resulting in the track moving faster across the head when a data that is being read is further away from the centre than nearer to it. To keep the bit rate constant, pits nearer the outer circumference must be spread out compared with pits nearer the inner circumference. The result is a lower average data density compared with the CLV technique. Compared with CLV, CAV has the advantage of faster access to the various sectors on the disc surface. This is because, when the pickup head moves to the required part of the disc, it does not have to wait for the angular velocity to change before reading the sector, as is the case with CLV.

CLV is used where the data on the disc is normally read sequentially, as in the case of a movie. Conversely, where the recorded data is not sequential with the pickup head having to move in search of the required sector, CAV with its fast access is used.

While CAV provides fast access to the recorded data, it is inefficient in terms of its use of the available recording surface of the disc. To maintain the advantage of fast access of CAV while at the same time improving the efficient use of the disc surface, the *zoned CLV* disc layout is used. In the zoned CLV technique, the disc is divided into multiple concentric rings, known as zones in which the speed of rotation remains constant. The only change in angular velocity takes place when the pickup head moves from one zone to another.

In the same way as CD-ROMs, DVD discs may be driven at multiple speeds: *single-speed X1*, *double-speed X2* and so on. *Multi-speed* drives spin the disc at multiples of the standard velocity to increase the rate at which data is read off the DVD disc.

Data transfer

The rate of data transfer from the disc to the playback processing unit determines the quality of application on offer such as SD or HD video resolution, audio mode, level of interactivity and games. Two transfer rates are specified: data transfer rate and video/audio transfer rate in maximum Mbps. For the traditional DVD format, the data rate is 11.08 Mbps and that for video/audio is 10.08 Mbps with a maximum video bit rate of 9.8 Mbps. The average video bit rate is in the region of 5 Mbps which is adequate for standard definition video ($720 \times 576i$ or $720 \times 480i$). The figures for HD DVD are 36.55 Mbps for both data and video/audio with maximum video bit rates of 28.0 Mbps. In the case of the Blu-ray format, the maximum data transfer rate is 36.0 Mbps and that for the video/audio at $1.5\times$ speed mode is a staggering 54.0 Mbps. This gives a maximum video bit rate of 40.0 Mbps. Both HD DVD and BD formats are thus suitable for high-definition resolution namely $1920 \times 740p$, $1920 \times 1080p$ and $1920 \times 1080p$. High-definition video is not necessarily confined to blue-laser-based discs. Using MPEG-4 AVC/AAC encoding, a X3 DVD-ROM has the capacity (8.4 GB) to store 120 min of HD video at a transfer rate of $3 \times 10.08 = 30.24$ Mbps, the same transfer rate as a single-speed HD DVD disc.

Table 20.5 compares the main properties of the three DVD formats: DVD, HD DVD and BD.

The symmetry between DVD and HD DVD formats enables the two formats to be combined on a single disc configuration designated as 'Combination' and 'Twin Format'. The purpose of these configurations so that consumers with a standard DVD player can buy the backward compatible discs and use them in full and when they are ready for an HD DVD player, they will could play the same discs, but this time making full advantage of the higher density data. Combinational discs are double sided, with HD DVD on one side and DVD on the other. Twin format configuration are DL SS discs with HD DVD on one layer and DVD on the other layer.

Triple-layer DVD

The latest development in DVD is the single-sided, triple layer illustrated in Figure 20.5 consisting of one DVD and two HD DVD layers. It is capable of either a DL 30 GB HD DVD and single-layer 4.7 GB configuration or single-layer 15 GB HD DVD and DL 8.5 GB DVD. This is a change from the original DVD twin standard that only allowed one HD DVD and one DVD, on the same side of the disc.

DVD encoding process

DVD technology employs a diverse range of advanced techniques including optics, data compression and processing, electro-mechanical servo control and micro processing as well as digital and television technology. Optical technology is used to write data onto a disc and to subsequently retrieve it from the disc. Retrieving information from a disc involves reading the data, bit by bit off a rotating disc using a laser beam that must be positioned accurately to follow a few micron spiral track on a rotating disc. The whole process of reading a DVD disc is controlled and all units are programmed by a powerful processing chip.

DVD video has three main components: *video, audio* and *sub-picture* (Figure 20.6). The latter include subtitles, captions, menus, karaoke lyrics, etc. These three components are stored in what is known as *streams*. In addition there are two other streams: *presentation control information (PCI)* and *data search information (DSI)*. Literally, a stream means a data flow, something akin to a track on a tape recorder. DVD-video supports one stream of video, up to eight streams of Audio and up to 32 sub-picture (SP) streams. HD DVD and BD formats support up to nine video streams, up to eight streams of Audio and up to 32 SP streams.

The streams are encoded separately to form video, audio and sub-picture packetised elementary streams (PESs) described in Chapter 4.

Table 20.5 *DVD, HD DVD and BD formats parameters*

Parameters	DVD	HD-DVD	Blu-ray
Storage capacity	4.7 GB (single-layer) 8.5 GB (dual-layer)	15 GB (single-layer) 30 GB (dual-layer)	25 GB (single-layer) 50 GB (dual-layer)
Laser wavelength	650 nm (red laser)	405 nm (blue laser)	405 nm (blue laser)
Numerical aperture (NA)	0.60	0.65	0.85
Disc diameter	120 mm	120 mm	120 mm
Disc thickness	1.2 mm	1.2 mm	1.2 mm
Protection layer	0.6 mm	0.6 mm	0.1 mm
Hard coating	No	No	Yes
Track pitch	0.74 µm	0.40 µm	0.32 µm
Data transfer rate (data)	11.08 Mbps (1×)	36.55 Mbps (1×)	36.0 Mbps (1×)
Data transfer rate (video/audio)	10.08 Mbps (<1×)	36.55 Mbps (1×)	36.0 Mbps (1×)
Video resolution (max)	720 × 480i – 720 × 576i	1920 × 1080p	1920 × 1080p
Video bit rate (max)	9.8 Mbps	28.0 Mbps	40.0 Mbps
Video codecs	MPEG-2 – –	MPEG-2 MPEG-4 AVC SMPTE VC-1	MPEG-2 MPEG-4 AVC SMPTE VC-1
Audio codecs	Linear PCM Dolby Digital DTS Digital Surround – Dolby TrueHD	Linear PCM Dolby Digital Dolby Digital Plus Dolby TrueHD DTS Digital Surround	Linear PCM Dolby Digital Dolby Digital Plus Dolby TrueHD DTS Digital Surround
DTS-HD	Dolby Digital DTS Digital Surround – – –	DTS-HD Dolby Digital Plus	DTS-HD
Interactivity	DVD-Video	Hdi	BD-J

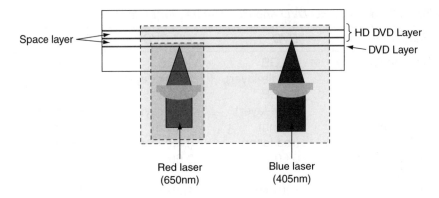

Figure 20.5 *Single-sided, triple-layer DVD disc*

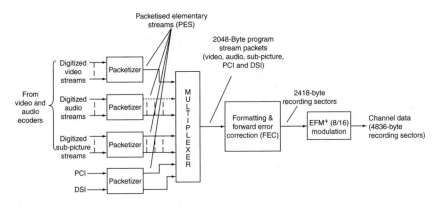

Figure 20.6 *DVD-video and HD DVD encoding*

In the case of DVD-video and HD DVD, before the packets are multiplexed they are broken into 2-KB (2048 Bytes) *user data packets* (Figure 20.6). Header and correction bits are then added to each packet resulting in 2046-Byte video, audio, etc. PES packets. The header includes ID bits to distinguish between the various types of packets: video, audio, sub-picture, PCI and DSI. The packets are then fed into the multiplexer to form the programme stream to be followed by the framing and forward error correction (FEC) to end up with 2418-Byte recording sectors. Before writing the data onto the disc, it is fed into an *8-to-16 modulator* (*EFM+*), which doubles the recording sector size to 4836 Bytes.

Each packet of data is fully identifiable in terms of data type, stream number and timing information for synchronisation purposes. The packets are then multiplexed to form what is known as a programme stream, which following framing and the addition of data correction bits, is stored on a disk. Unlike digital television broadcasting in which the bit rate of the programme stream is fixed and the data compression has to be adjusted to accommodate it, in DVD application, variable bit rate is used.

In the case of BDs, the video, audio, etc. PES packets from the encoders are chopped into smaller 188-Byte packets to form a transport stream using the same technique as that used for digital television.

Video encoding

Encoding the video information for all DVD disc formats follow the same lines as those for standard and high-definition digital television described in previous chapters. The traditional DVD-video format uses MPEG-2 data compression techniques including temporal and spatial compression as well as DCT and entropy coding. HD DVD and BD formats support MPEG-2 as well as MPEG-4, AVC (advanced video coding H.264) and SMPTE VC-1 (Microsoft Windows' media Video Codec). MPEG-4 and AVC (H.264) have been discussed in detail in Chapter 5.

VC-1 is a data compression procedure and is published and approved by SMPTE (Society of Motion Picture and Television Engineers) uses similar compression techniques to AVC. It was developed well after MPEG-2 and has benefited from the lessons learned. It employs many of the same advanced coding techniques as those used in AVC (temporal and spatial, block motion compensation, integer transforms and the use of a loop filter) and thus is able to achieve similar coding efficiencies. VC-1 is less complex than AVC and hence it is faster to decode with a *Windows-based software* (*WMV9*) that can be used on a 2–3 GHz personal computers. VC-1 has a number of profiles: simple (up to 384 Kbps), Main (2, 10 and 20 Mbps) and advanced (10, 20, 45 and 135 Mbps).

Audio encoding

DVD-video discs may incorporate up to eight audio streams; each one can convey mono or multi-channel audio using one of three coding techniques:

- MPEG-2 audio.
- Dolby Digital (also known as AC-3).
- Linear pulse code modulation (LPCM).

Both MPEG-2 (described in Chapter 6) and AC-3 coding involve audio compression to reduce the data bit rate. LPCM does not include any form of data compression and thus it has the highest quality and the greatest bit rate requirements.

HD DVD and Blu-ray formats support an enhanced set of audio encoding over and above those available for DVD-video. These include *digital theatre system* (*DTS*) which is optional in the DVD format but is mandatory for HD DVD and BD under a name DTS-HD constant bit rate (CBR). DTS-HD variable bit rate (VBR) is optional. Other audio-encoding techniques include *Dolby TrueHD* (previously called *meridian lossless packing, MLP*),

Dolby Digital Plus, AAC, MP3 and *WMA Pro* with sampling rates of 48, 96 and 192 kHz.

Dolby digital (AC-3) encoding

Dolby Digital, also known as AC-3 (AC for audio compression) supports all channel formats up to 5.1. It uses a sampling rate of 48 kHz with an average of 16 bits allocated to each sample. Its compression technique differentiates between short transient signals from long continuous sounds. It gives prominence to the latter in the form of long sample blocks compared to the transient sounds. AC-3 uses frequency-transform technique similar to DCT employed in MPEG video encoding. It provides smoother encoding compared with MPEG which creates arbitrary boundaries by its sub-band technique. Bits of up to 24 bits per sample are allocated dynamically to compensate for different listening environment, e.g. theatre, home or auditorium. The bit rate may be variable, although a fixed rate is normally used as is the case with MPEG audio encoding. Bit rates of 384 or 448 kbps are typical.

The Dolby Digital system is a high-quality digital sound system equipped with a channel dedicated to subwoofer output (LFE) in order to reproduce low frequency below 120 Hz in addition to the other five channels. Unlike the conventional 2-track Dolby Pro Logic system, Dolby Digital is a 5.1 track system with each channel discretely and digitally processed from the beginning, so they are independent and contribute to excellent channel separation. All five channels are reproduced at full bandwidth of 3 Hz–20 kHz. With these features, sound may be accurately reproduced in a home theatre context with the intended positioning of sound image and feeling along with surround sound with presence and power comparable to the movie theatre experience.

Linear PCM

LPCM is uncompressed and thus a lossless digital audio which has been used in CD and most studio masters. Analogue audio is sampled at 48 or 96 kHz with 16, 20 or 24 bits per sample (Audio CD is limited to a sampling rate of 44.1 kHz). As a result of the absence of compression and the high rate of sampling and quantisation, the bit rate could be excessively high and for this reason it is limited to 6.144 Mbps. The equivalent bit rates for MPEG and Dolby Digital are 448 and 384 kbps. LPCM supports up to eight channels, however, due to the limit of 6.144 Mbps, for five or more channels the lower sampling rate of 48 kHz must be used as indicated in Table 20.6.

The bit rate required by the non-compressed LPCM is normally too high to be accommodated by a DVD program (which may include other

Table 20.6 *Sampling rates quantisation for different number of audio channels*

Sampling rate (kHz)	Quantisation (bits/sample)	Maximum number of channels
48	16	8
48	20	6
48	24	5
96	16	4
96	20	3
96	24	2

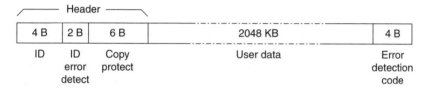

Figure 20.7 *PES construction*

elements such as video and other audio configurations) within the limits imposed by DVD specification on the maximum bit rate available for the various elements, namely 9.8 Mbps. For this reason, meridian lossless packing, MLP is used. MLP now named Dolby-HD compresses data, bit by bit, removing redundant data without any loss to quality. Dolby-HD achieves a compression ratio of 2:1 by using a combination of techniques namely *lossless matrixing, lossless waveform prediction* and *entropy coding.*

Multiple-language sound track may be included by dubbing each stream with a different language. At playback, a particular stream may thus be selected.

PES packet construction

A DVD/HD DVD PES packet contains 2 kB of user data such as video, audio and sub-picture information. The user data is preceded by a header and followed by a 4-byte error detection code (EDC) as illustrated in Figure 20.7 to form a packet, also known as a sector. However, before the header and the EDC bytes are added, the 2 kB user data bits are scrambled. Scrambling avoids long strings of zeros to ensure that energy is evenly spread.

The header consists of three sections: a 4-byte ID section (consisting of a 3-byte sector number and 2-bytes of error detection for the sector number) and 6-byte copy protection information.

As stated earlier, to improve error correction, BD format employs the MPEG-2 transport stream in which the PES packets are further divided into 188-Byte transport packets as is the case with digital television broadcasting.

Forward error correction—DVD/HD DVD

Digital signals, especially those with a high level of compressions, require an efficient error-correction capability. Apart from normal errors of propagation and processing, errors may occur as a result of dust or physical damage to the surface of the disc causing video or audio break-up or disruption. To avoid such errors, powerful FEC techniques are employed which can correct a burst error length of approximately 2800 bytes, corresponding to a physical damage of up to 6.0 mm. More stringent error-correction methods are used for the BD formats because of BD's higher sensitivity to scratches and fingerprints.

Data is structured into blocks of 32 kB (DVD/HD DVD) and 64 kB (DB) of user data arranged in rows and columns. Following row interleaving, error-correction rows and columns are added to form recording or physical sectors (Figure 20.8).

Non-return-to-zero encoding

Data is written on the disk using a *non-return-to-zero inverted* (*NRZI*) format in which a transition from pit to land represents a one and a non-transition represents a zero. The effect of this is to halve the number of transitions required to represent the bitstream. Consider the stream of ten pulses in Figure 20.9. With NRZI inverted, a change from land to pit or vice versa occurs only when a logic state of 1 is followed by a 1, or a logic zero is

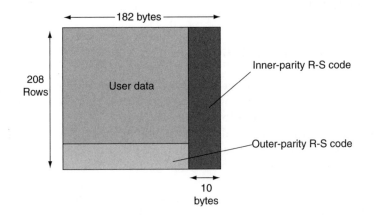

Figure 20.8 *DVD/HD DVD recording sector (5.3)*

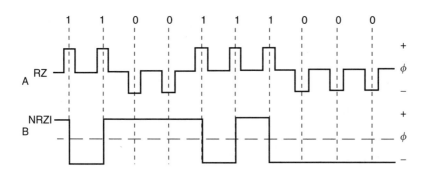

Figure 20.9 *Non-return-to-zero halves the number of transactions*

followed by a logic 1 as illustrated in Figure 20.9b. Otherwise, no change takes place. The result is five pulses or transitions instead of the original ten.

Eight-to-sixteen modulation (EFM+)

With NRZI encoding, it is possible to end up with a long stretch stream of zeroes or ones. Such a situation would result in a long stretch of a constant high or low voltage with no transitional edge. Since clock synchronisation depends on the regular occurrence of such a transitional edge, a long absence of such an edge would cause a breakdown of time synchronisation at the player. To avoid this eight-to-sixteen modulation is used.

Eight-to-sixteen (8/16) modulation converts each word of 8 bits (one-byte) of data into a 16-bit code selected from a conversion table. An 8-bit word has $2^8 = 256$ different combinations or words. A 16-bit code on the other hand, has $2^{16} = 65,536$ different codes. Out of these 65,536 codes, 256 are, carefully selected to minimise DC energy and reduce frequency. The selected codes are chosen in such a way that in each code at least two or at most ten zeros occur between any groups of ones. For example the 8-bit word 0000011 is converted into the 16-bit code 0010000001001000 as illustrated in Figure 20.10. The 16-bit code is then further encoded using NRZI resulting in a simple waveform shown which is then translated into two pits and a land. Eight-to-sixteen modulation is normally referred to as EFM+ (eight-to-fourteen Plus) because EFM (eight-to-fourteen) modulation) was originally used for modulating a CD audio channel data.

DVD playback system

The pick-up head reads the data on the disc, sector by sector (Figure 20.11). The speed of the disc, the position of the head as well as the beam focus is controlled by the servo control unit, which itself is controlled by the system control (sys con) microprocessor chip. The data from the pick-up head is fed into the decoding unit to re-produce the original audio and

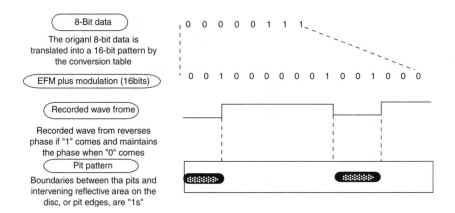

Figure 20.10 *Eight-to-sixteen modulation and pit pattern*

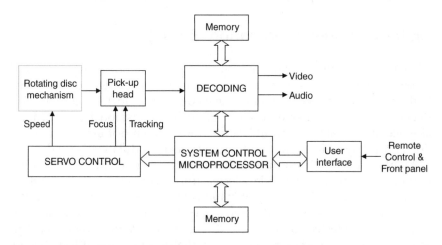

Figure 20.11 *DVD playback system*

video signals. The process of decoding involves error correction and data de-compression, both of which require the use of random access memory (RAM) store. Memory is also required by the sys con microprocessor to store the necessary processing routines as well as various configurations and settings. The user accesses the player's facilities using a remote control or front panel switches. Signals from these are sent to the microprocessor via the user interface.

The optical pickup head

Figure 20.12 shows a simplified diagram for a front end of a DVD player. The *optical pickup unit* (OPU), also known as the optical head, extracts the data from the DVD disc using a laser beam and then converting the

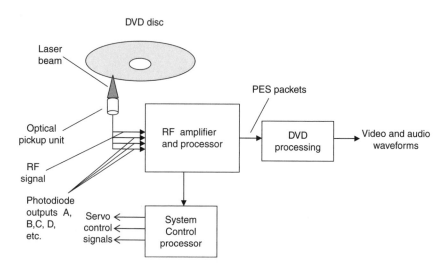

Figure 20.12 *The front end of a DVD playback system*

reflected laser beam into an electric waveform known as the *radio frequency* (*RF*) or *high-frequency* (*HF*) signal. The RF signal is an analogue signal which, following analogue-to-digital conversion, is processed by the RF processor on its way to the signal-processing unit to re-produce the original video, audio and other control and search information. The OPU also provides a number of other outputs, A, B, C, D, etc. to be used to ensure accurate focusing, tracking and speed control of the DVD disc. The RF processor also provides the necessary focus, tracking and other control signals.

The OPU is one of the most critical parts of a DVD player. It combines the LD which generates the laser beam, photodiode array to detect the reflected laser beam as well as all necessary lenses and mirrors in a single integrated device. It generates the laser beam, ensures that the beam focuses on and follows the recording tracks, detects the reflected waveform and produces streams of data for processing and control by the DVD player.

The OPU is mounted on an arm placed under the disc along which the pickup head can move in and out to follow a track or read different tracks. The OPU must be able to read the different types of DVD discs (DVD, HD DVD or BD, single-layer or DL) as well as conventional audio CDs. This means that it must have the capability to generate laser beams of different wavelengths: 780 nm infra-red for audio-CDs, 650 nm red laser for DVD-video and 405 nm blue laser for HD DVD and BDs. In addition, it must have the ability to focus at different depths to accommodate single and DL DVD discs.

The main components of a DVD pickup head are illustrated in Figure 20.13. It consists of a LD, a photo-detector array, splitting mirrors, a coil-controlled lens that can move up and down and sideways for focus and

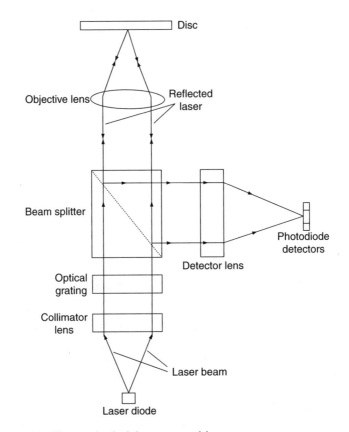

Figure 20.13 *The optical pickup assembly*

tracking control. The laser beam is generated by a low-power semiconductor laser diode, LD. The collimator lens forces the beam to follow a parallel path on its way to the optical grating lens which bends the laser to produce two beams: a main beam for the actual data stream and a side beam for tracking purposes. Before striking the disc, the beam is focused by the objective lens. The laser hits the disc surface and is reflected back towards the objective lens. Provided the beam is accurately focused, the reflected beam returns along the same path as the incident beam towards the beam splitter. The beam splitter turns the reflected beam by 90° to direct it towards the photodiode detector assembly via the detector lens.

Focus depth and numerical aperture

With a track pitch as small as 0.74 and 0.34 μm for DVD and HD DVD formats respectively and even smaller dimensions for the BD format, the laser spot size hitting the surface of the disc must be very small. It has to be small enough to distinguish between pits and avoid reading adjacent tracks. The spot diameter, W is directly proportional to the wavelength, λ and inversely proportional to the NA of the objective lens:

$$W = \frac{\lambda}{2 \times NA}$$

For audio-CD, the numeric aperture is 0.45 and the wavelength is 780 nm giving a spot diameter of

$$w = \frac{780}{(2 \times 0.45)}$$
$$= 867\,nm$$
$$= 0.867\,\mu m$$

For DVD-video, NA = 0.6 and λ = 650 nm, the spot width is W = 650/(2 × 0.6) = 0.541 μm.

Similarly for HD DVD, W = 405/(2 × 0.65) = 0.311 μm and for BD, W = 405/(2 × 0.85) = 0.238 μm.

The effect of larger NA is to reduce the focus depth causing the readout to be more sensitive to disc thickness and other irregularities. Although the thickness of the substrates may be reduced as in the case of the BD format, the optical head remains sensitive to disc warping which causes the disc to tilt. During readout, as the optical head moves across the disc, a changing tilt angle causes what is known as a *skew error*, which will disturb the reading of the data. For this reason, a *tilt* or *skew sensor* may be mounted on the surface of the optical unit to produce a skew error signal which is used to ensure that the optical head moves parallel to the disc surface.

The photodiode detector assembly

The photodiode assembly consists of a number of photodiodes which detect the strength of the reflected laser beam and produce a number of signals including the RF, focus error, FE and tracking error (TE) signals. The RF signal contains the actual video, audio and other information. Since the pits and lands reflect the beam with different strengths, the output level from the photodiodes will thus represent the pits and lands, i.e. a bitstream of zeros and ones.

The detector that monitors the main beam is divided into four quadrants or segments, A, B, C and D as shown Figure 20.14. The sum of the outputs of the four segments (A + B + C + D) represents the RF signal strength which is used for data processing. The same four-quadrant photodiode is used to produce the focus error (FE) signal.

Focus error

When the beam is correctly focused, the reflected beam forms a circular pattern on the photodiode (Figure 20.15a). However, if the beam is off focus, the beam forms an elliptical shape with a different aspect ratio as

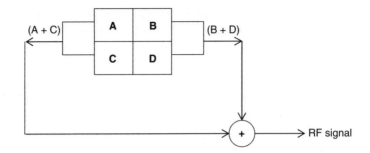

Figure 20.14 *The pickup head photodiode*

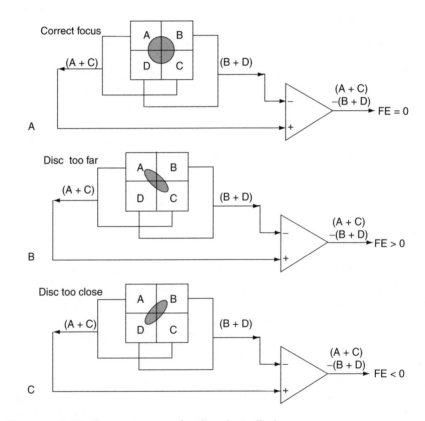

Figure 20.15 *Focus process for the photodiode*

shown. By comparing (A + C) with (B + D) a FE signal is produced (Figure 20.16b and c). When the beam is in focus (A + C) − (B + D) = 0. Otherwise, FE would be positive when the focus is too short and negative when the focus is too long.

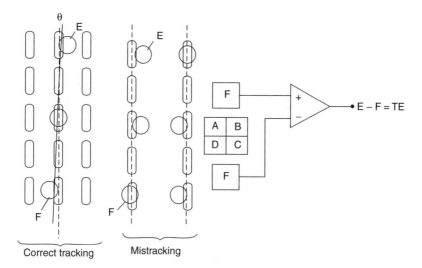

Figure 20.16 *Tracking (DVD players 6.5)*

Tracking error

Apart from the main beam which is used for the RF and FE signals, there are two side beams produced by the grating lens. These two side beams are directed to two separate photo-detectors (E and F) placed on each side of the main beam photodiode assembly as shown in Figure 20.16.

When the beam is on track, the outputs of diodes E and F are equal. However, if the beam leaves the track, it will energise one of the diodes more than the other producing a positive or negative tracking error. Other alternative techniques may also be used by different manufacturers, all of which involves the use of additional photo-diodes assemblies.

The tracking and focus two-axis device

The objective lens is contained inside two separate coils: one for focus and the other for tracking control, placed at right angles to each other. The coils are placed inside a magnetic field produced by a permanent magnet as shown in Figure 20.17. This assembly, known as a *μ2-axis actuator* will move the lens in the horizontal and vertical directions. Current through the focus coils will move the lens vertically to achieve focus and current through the track coil will move the lens horizontally (lateral movement) for tracking purposes.

Compatible pickup

The fact that DVD-video players could also play audio CDs has been and remains a major selling point. This is also true for the new generation blue-laser-based playback machines. HD DVD playback systems must now

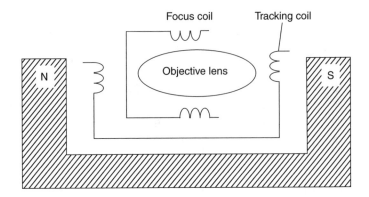

Figure 20.17 *Two-axis device*

be able to read not only HD DVD with a blue laser, it must also be able to read a DVD and a CD. Similarly for the BD playback systems, namely BD/DVD/CD. Therefore, an HD DVD/DVD/CD or BD/DVD/CD compatible pickup requires three different wavelength beam sources and three different NAs to meet the requirements of each disc standard. The obvious solution is to have three switchable pickup units, one for each application. The size of such a pickup system becomes large and in most cases impractical. Other techniques include a multi-beam laser unit in which three LD chips of blue, red and infra-red are mounted in one package. Each one of the three wavelengths can be emitted by applying a different voltage to the appropriate LD chip.

Signal processing and control

Figure 20.18 shows the principle units involved in the signal processing of a DVD/HD DVD playback system. The OPU, generates the laser beam with the appropriate wavelength and detects the reflected beam using a number of photodiodes to produce two types of signals: a varying signal representing pit and lands on the surface of the disc, known as the RF and a number of error signals used for the servo control system. All these signals are fed into the RF amplifier. The RF amplifier performs two distinct functions on the incoming signals. It amplifies the RF signal to a level suitable for processing by the next stage and carries out digital processing of the error signals for the servo DSP. The RF signal is fed to the RF processor which carries out the following functions:

- RF signal processing including de-multiplexing to extract the various video, audio and other PES packets.
- Clock sync extraction
- EFM+ (16/8) demodulation Frame/sector sync detection Error correction

The process of demultiplexing the PES packets requires a memory bank to store the PESs as they arrive to allow the RF processor to select and

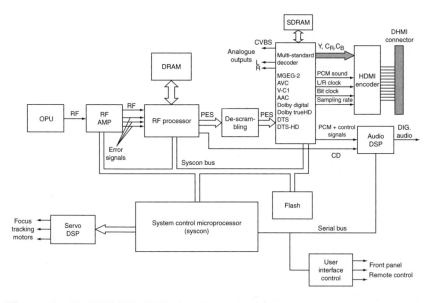

Figure 20.18 *DVD/HD DVD signal processing*

re-arrange them in the required order. Before audio and video decoding, the PES stream has to be descrambled to return it to the original bit sequence. Following that, the PES packets are sent to the multi-standard A/V decoder chip which carries out the decoding of the video, audio and sub-picture packets into digitised analogue video and pulse code modulated audio signals. Decoding compressed video involves data de-compression and the reconstruction of the video stream, picture by picture from the I frames and the P and B frames. The process of decoding thus requires a memory bank to store the I, P and B frames for reconstruction of actual frames to take place. A similar process takes place in the case of decoding compressed audio information. The memory bank also serves as a 1 s audio delay for lip-sync. This delay is necessary since the decoding of the video packets takes a longer time than the decoding of audio infor-mation. The A/V DRAM memory chip provides this memory bank. Digital video and audio PCM is fed to a HDMI encoder/transmitter for the HDMI output port. Analogue audio and video are also available for SCART, S-video and component video. Audio PCM together with control signals L/R, bit clock and sampling clock are fed to an audio digital signal processor (DSP) for conversion into a digitised audio output.

The whole process is programmed and controlled by a powerful micro-processor centred on a *system control* (*sys con*) chip with an embedded RISC microprocessor core. Sys con carries all necessary programming and control of signal and servo processing. It incorporates data, address and control bus structure together with a FLASH memory store. It provides a serial control bus for controlling such units as the audio DSP as well as the user interface.

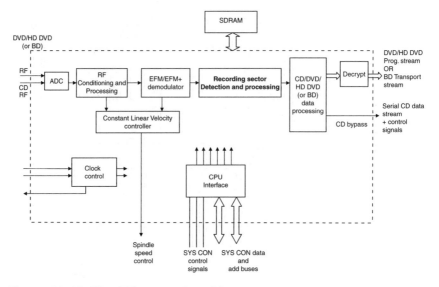

Figure 20.19 *The RF processing chip*

The FLASH memory is a non-volatile RAM chip used to store start-up and other initialising and processing routines. One of the functions of the sys con chip is to ensure that the correct track sector is read at the correct speed and with correct focus. This is carried out by the servo DSP chip. The final part of the player is the user interface which is managed by the interface control to provide the necessary interface between the player and font panel control buttons, front panel display and remote control handset.

The RF processor

The RF processor has two major tasks: signal processing and disc speed control. For this reason, this chip is also known as RF processor/digital servo chip (Figure 20.19). These tasks must be carried out for DVD/HD DVD as well as audio-CD applications. The input to the RF processor is a high-frequency waveform known as the RF signal. It is generated by the optical head as it reads the pits and lands on the surface of the disc. The waveform is an analogue format representing a digital bitstream. Since it is not repetitive, it cannot be observed on an analogue oscilloscope. On a storage digital oscilloscope, the waveform will be as shown in Figure 20.20. If observed on an analogue oscilloscope, the pattern shown in Figure 20.21 will be displayed. This pattern, known as the RF eye pattern, is generated because the 8/16 (EFM+) modulation results in restricting the number of ones or zeros that can follow in a sequence.

The frequency of the pattern is determined by the closeness, i.e. size of a pit. For blue-laser-based discs, the pit size is small and hence the pattern

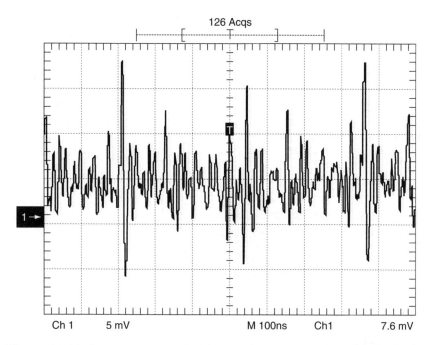

Figure 20.20 *Input waveform to RF processor captured on a digital storage oscilloscope*

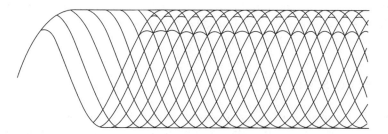

Figure 20.21 *The eye pattern produced by the RF signal on an analogue oscilloscope*

has a higher frequency requiring an oscilloscope with a minimum bandwidth of 100 MHz compared with that produced by a red-laser-based DVD disc which can do with a bandwidth of 40 MHz. Before processing can take place, the incoming analogue signal from the RF amplifier is converted into a digital format by the ADC. This is followed by the RF processor unit to produce a digital bitstream which is used for signal processing and spindle speed control. The bitstream from the RF processing unit contains 16-bit words which are converted back into their original 8 bits by the EFM/EFM (8/16) demodulator. This is followed by recording sector detection, error correction and RF data processing. For DVD, HD DVD and BD, the bitstream is de-crypted to re-produce the original program

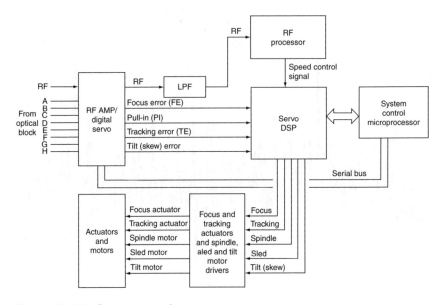

Figure 20.22 *Servo control*

stream for DVD/HD DVD or the transport stream or the BD. Audio CD data streams do not require de-crypting and thus they are fed directly into the next stage. The bitstream from the RF processor unit is also fed into a CLV controller which monitors the bit rate to determine the linear speed of the disc and a signal is sent to the servo controller to change the speed of spindle motor as appropriate.

Servo control

Figure 20.22 illustrates the main components of the servo control system of a DVD player. The RF amplifier receives the RF signal which it passes to the RF processor via a low pass filter (LPF). It also receives photo-detectors signals: A, B, C, D and so on from the optical head for infra-red, red and blue-laser discs formats. The photodiode signals are processed by the digital servo RF amplifier to produce three error signals: focus error (FE), tracking error (TE) and *pull-in* (*PI*) which are fed into the servo DSP for further processing. In addition, some players employ a fourth error signal, namely tilt (or skew) to detect and compensate for warped DVD discs.

The error signals

The servo DSP also receives a spindle control signal from the RF processor. The spindle control signal represents the data flow rate. A fast flow rate indicates a fast speed and vice versa. The servo DSP processes the servo input signals and produce focus, tracking, spindle speed, sled and depending of the manufacturer, tilt (skew) control signals which are fed to the appropriate drivers on their way to the relevant actuators and motors. The servo DSP is fully programmed and controlled by the sys con microprocessor chip as shown.

Playback start-up routine

When the tray is closed, or when 'playback' is selected, a start-up routine is initiated. The purpose of the routine is to determine the size of the disc, whether it is a CD or a DVD/HD DVD (or BD) disc and whether it is a single- or a DL disc and carry out what is known as a 'focus search'. The start-up routine will initiate the following steps:

- The spindle motor rotates the disk at a relatively high speed.
- The laser beam is directed onto the disc and the reflected beam is detected by the photodiodes to produce the RF signal.
- The sled motor moves the optical head across the disc from the centre towards the circumference and back again, to determine the diameter size and generate a TE signal.
- The two-axis actuator moves the objective lens up and down to obtain a focus, a process known as focus search.

CD/DVD detection

When the optical pickup head moves across the track, it generates a TE signal. If the TE voltage is low, then the inserted disk is a blue-laser disc; if it is high, it is a CD disc and in the middle, a DVD disc. Longer pits and wider track pitch produce a higher TE voltage.

Focus search

Once the type of disc is determined, focus search begins. At the beginning, the focus servo loop is kept open. The objective lens is moved up and down along an S-curve (Figure 20.23) crossing the full focus point. As it goes towards focus, the RF signal strength increases. When it reaches a specified threshold, determined by the type of the disc, a *Focus OK (FOK)* control line goes High and remains High. While the focus crosses the full focus (zero) line, the RF signal strength simultaneously goes to its maximum value and begins to fall again. At this moment, the servo loop is turned on and a focus is established.

Single/dual-layer detection

The difference between a single and a DL disc is the intensity of the reflected laser beam. A single-layer disc is 100% reflective while the first layer (layer 0) of a DL disc is only partially reflective. This is identified by

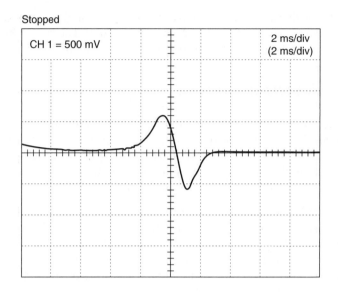

Figure 20.23 *S-curve for focus search*

what is known as the pull-in (PI) signal. The PI signal is obtained during the process of the focus search. A high PI signal indicates a highly reflective surface and thus a single-layer disc. Alternatively, a low PI indicates a DL DVD disc.

Tracking control

Tracking is ensured by the horizontal actuator of the two-axis device for small adjustments and by the sled motor for larger adjustments. The sled motor keeps the optical head moving along with the spiralling track and if necessary introduces a 'jump' when a new sector is selected for reading. The *sled error (SE)* is calculated inside the servo DSP which monitors the TE signal (Figure 20.24). The low-frequency component of the tracking error (TE) signal is caused by the gradual spiralling of the track towards the outside circumference. It is used to control the sled motor. The high-frequency

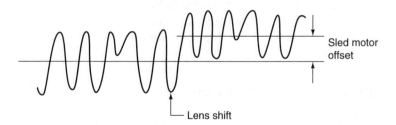

Figure 20.24 *Tracking error signal*

component, on the other hand, represented by spikes, is caused by the jerky track itself. It is used to control the tracking actuator.

Typical servo control chip

Focus and tracking error signals, FE and TE arrive from the RP amplifier/ digital servo at pins 29 and 23, respectively (Figure 20.25). They are processed by the servo DSP to produce a pair of positive and negative control signals at pins 1, 3 and 5, 7. These are then fed to the focus and track coils drivers to position the objective lens for correct focus and track. The system provides for three motor servo control: loading, sled and spindle. The loading motor drive is controlled by a single signal (pin 39) to open and close the tray. The purpose of the sled motor is to keep the optical unit moving along with the spiral track. The low-frequency component of the TE signal is used to calculate a sled error, SE. This error is then used to produce two pulse width modulated (PWM) signals that are fed into the sled motor drive. The drive mute is used to turn the sled motor off during such activities as focus search and pause.

The error signals for the spindle motor are generated by the RF processor. At the RF processor, the data bit rate emanating from the optical head is compared to a master clock. If there is a difference, the RF processor generates two error signals: a phase error and a speed error. These are fed into the servo DSP (pins 18 and 19) which processes them and produces a PWM signal for the spindle motor drive (pin 71) which in turn produces positive and negative signals to control the speed of the motor. These two signals are fed back via a summing amplifier to the servo DSP (pin 17) to complete the servo loop. The servo DSP is set up and controls the system control microprocessor via a 2-bit address bus (pins 80, 81), an 8-bit data bus (pins 82–89), Read, Write, IRQ and chip select (CS). The 27 MHz system clock is fed at pin 95 and a chip reset is at pin 68.

The user interface

Like other consumer equipment, the operations of the DVD player (play, pause, stop, rewind, etc.) are carried out by the user, either directly by pressing appropriate button on the front panel or by using a remote control handset.

At the heart of the user interface is the interface controller chip (Figure 20.26). The user interface controller is a dedicated microprocessor controller chip. It has its own individual clock and is powered by a permanent voltage, sometimes known as 'ever' voltage from the power supply. An ever voltage is a voltage that remains alive when the power supply is turned into the standby mode. The user interface controls the power supply mode of operation using two control lines: *Power Detect* and *Power*

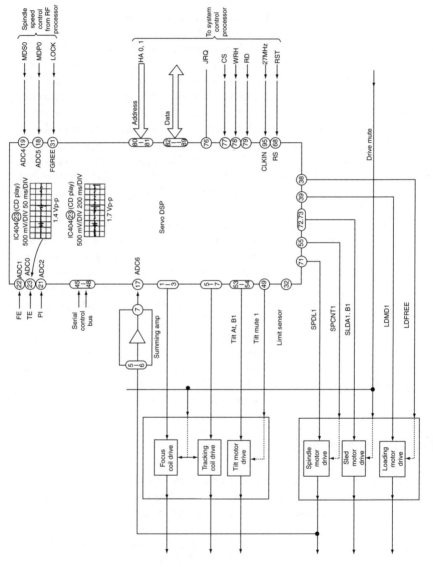

Figure 20.25 *Servo control chip*

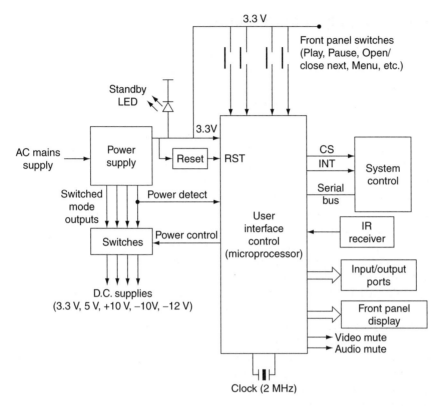

Figure 20.26 *The user interface controller*

Control. When the power supply is turned on from cold and following a short start-up routine, the power supply settles into the standby mode in which only the ever voltage (3.3 V in the diagram) remains switched on. The player will remain in this mode until a button on the front panel or the remote control handset is pressed or alternatively, the on/standby button is pressed. When this happens and provided the power supply is sound (indicated by Power Detect line High), the Power Control line from the interface goes high switching all the other voltages on and the player goes into the on mode.

The user interface has several other functions including:

- Receiving and decoding signals from the remote control handset
- Receiving and decoding instructions from the front panel button switches
- Controlling the front panel display
- Setting the appropriate switching of the input and output ports to set the video and audio outputs such as SCART and S-Video sockets.
- Providing video and audio mutes as necessary
- Communicating with the system microprocessor controller via a serial bus and other control lines such as CS and interrupt (INT).

Power supply requirements

Different elements of a DVD player require different DC power requirements in terms of voltage and power. Generally, processing and memory chips may require 3.3 V or 5 V, logic devices 5 V, fluorescent indicator tubes on the front panel –10 V, focus and tracking coils −12 V and spindle and sled motors 12 V.

A typical circuit diagram for a DVD player switched mode power supply is shown in Figure 20.27. FET Q101 together with transformer winding W2, C117 and zener Q102 form a self-oscillating circuit running at a rate determined by the time constant R105/C117. The capacitor charges up through R105 and when it reaches a certain voltage, it turns transistor Q101 on which discharges the capacitor turning itself off in the process and so on. Feedback control is provided by the network composed of opto-coupler PC101, transistor Q102 and zener D102. The opto-coupler monitors the voltage level of the 5 V rail. If the 5 V rail goes up, light emitted from the photo-diode increases, increasing the current through the photo-transistor, forward biases transistor Q102 and reduces the voltage at its collector. The effect of this is to increase the time it takes for capacitor C117 to charge up thus delaying the switching on of Q101. If the 5 V rail goes down, current through opto-transistor decreases, current through transistor Q102 falls, its collector voltage increases and capacitor C117 charges up faster. The effect of this is to bring forward the switching of Q101 and thus increasing the voltage. On the secondary side of the transformer, diodes D211 and D511 provide the +10 V and –10 V rails, D311 and D611 provide the Ever 5.5 and 3.8 V rails and transistor Q611 provides switched rail 3.3 V. When the poser supply is in the on mode, Q711 is turned on by a positive voltage from the power control line into its emitter, taking its collector voltage to 3 V. This voltage turns FET Q511 on switching the –10 V rail. The 3 V from the collector of Q711 turns transistor Q312 on and its collector voltage drops which turn transistor Q611 and FET Q211 on to switch the 3.3 V and the +10 V rails. With the 3.3 V available, Q621 and Q615 are switched on energising the green LED of double LED diode D615. When the power supply is turned into the standby mode, the power control voltage goes down switching the –10 V, +10 V and +3.3 V rails off, turning Q621/Q615 off and lighting the red part of the double LED diode.

Content protection

In the aftermath of the Content Scrambling System (CCS) used in DVD-video having been decrypted and widely distributed over the internet, AACS (Advanced Access Content System) designed a new protection system which enables authorised use and deters unauthorised copying and re-distribution. The system is based on what is called *'watermark'* which

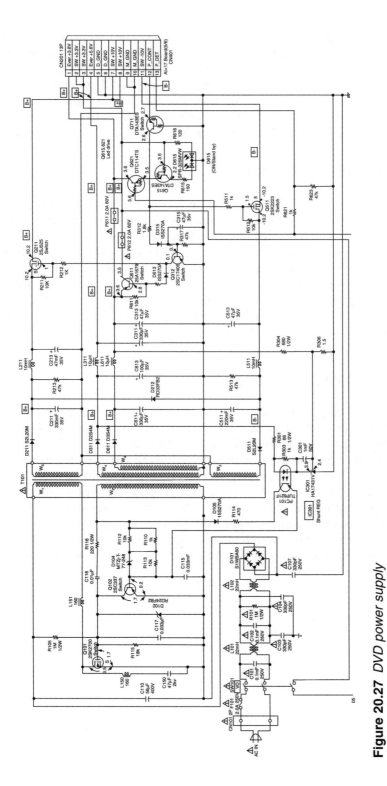

Figure 20.27 *DVD power supply*

the player would detect and enable playback. The absence of the 'water-mark' data means the disc is a copy causing the player to refuse to play it. The watermark may be set to a restricted number of copies, normally one after which playback is blocked. In addition, AACS provide guidelines pertaining the outputs over the analogue connections. This sets a flag to restrict the resolution for analogue outputs without the High-Definition Copy Protection (HDCP) to a max of 960 × 540.

Film grain technology

Another feature of HD DVD format, not available in the BD format, is the use of *Film Grain technology, FGT.* Film grain is perceived to be an important component of movie watching experience and many film makers chose specific film material because of their grain property. If DVD reproduction is to be true to the original intentions of the film makers, the film grain effect must be re-created. MPEG-4 AVC specifications include a standardised mechanism for transmitting film grain characteristics in the form of a *supplementary enhanced information (SEI)* message. HD DVD decodes this information and uses it to stimulate the original look of the film.

Interactivity

One of the main merits of both HD DVD and BD is enhanced interactivity and seamless connection to the internet. HD DVD use *iHD Interactive Format* while Blu-ray has opted for a format called *BD-J.* Unlike the DVD-video experience where a menu screen must first be selected by the user before the movie starts playing, the HD DVD and BD user is able to start a commentary from the main menu in the middle of the film without stopping it. Another enhanced feature is the ability to interact with the internet. For example a pre-recorded disc can be linked to a website to be constantly updated. Through this connection, studios can make additional content available to consumers, including clips, games, new releases and even enhanced subtitles.

21 Magnetic tape recording

History and development

The story of video tape recorders really began before the turn of the century with the experiments of Valdemar Poulson. By this time, the relationship between electricity and magnetism was well understood, and the idea of impressing magnetic pulses on a moving magnetic medium was sufficiently advanced in 1900 to justify the US patent on Poulson's apparatus, the Telegraphone. The medium was magnetic wire rather than tape; and without any form of recording bias and scant means of signal amplification, the reproduced sound signal was low, noisy, non-linear and lacking in frequency response. These problems of 'tape' and record/replay head performance are ones that have continually recurred throughout the history of sound and vision tape recording, as we shall see.

By the early 1930s, many advances had been made in the field. The d.c. bias or pre-magnetism of the recording wire had been tried with better results, then overtaken by the superior system of a.c. bias, as used today. The magnetic wire gave way to steel tape 6 mm wide travelling at 1.5 m/s, and performance became comparable with the contemporary disc recording system. Not 'hi-fi' by any means, but certainly adequate! The BBC adopted and improved the Blattnerphone system and in 1932 broadcast a programme of the Economic Conference in Ottawa, for which 7 miles of steel tape was used, edited by means of a hacksaw and soldering iron! This era also saw the first crude forerunner of a servo system in the Marconi–Stille machine of 1934. The earthed tape was arranged to contact insulated metal plates when it became slack, the plates being wired to thyratron control valves. Relays in the thyratron anode circuits modified drive motor currents to regulate tape speed.

A great impetus was given to the industry when it became possible to coat a flexible insulated base with a finely divided magnetic substance. This was achieved in Germany by Dr. Pfleumer and developed by Wilhelm Gaus under the auspices of AEG. This activity culminated in the successful demonstration at the 1935 Berlin Radio Exhibition of the first commercial sound tape recorder, the AEG Magnetophone, using a cellulose acetate tape coated with carbonyl iron powder. Performance of these sound recording machines steadily improved during the 1930s and 1940s to the point where at the end of the 1940s, much radio broadcast material was off-tape and indistinguishable from live programmes.

Once audio magnetic recording had become established in the radio industry, attention was turned towards the possibility of recording television images on tape. The problems were formidable, mainly because of the relatively large bandwidth of a television signal. Plainly, it would be necessary to increase the tape speed and one approach, by Crosby Enterprises in 1951, took the form of a 250-cm/s machine designed to record a monochrome picture whose frequency spectrum was split into 10 separately recorded bands with additional sync and control tracks. The same principle was embodied in a fearsome machine designed by RCA, in which the tape travelled at 600 cm/s to record on three separate tracks simultaneous R, G and B information for colour TV. Other designs involving longitudinal recording called for tape speeds approaching 1000 cm/s, amongst which was the BBC's video electronic recording apparatus (*VERA*) of 1956. Such machines were wasteful of tape and thoroughly frightening to anyone who happened to be in the room in which they were operating!

Already the seeds had been sown of a new system, one which was to hold the key to the modern system of TV tape recording. This was simply the idea of moving the record or replay heads rapidly over the surface of a slowly moving tape to achieve the necessary high 'writing' speed. Initially, the hardware consisted of a circular plate with three recording heads mounted near its edge at 120° intervals, their tips protruding from the flat surface of the faceplate. The 5 cm-wide tape was passed at 76 cm/s over the surface of the rotating plate, whose heads had an effective velocity of over 6000 cm/s, resulting in narrow accurate tracks across the width of the tape. It was a step in the right direction, but results were poor for several reasons. The bandwidth of the TV signal being recorded was difficult to get on and off the tape due to noise and head-gap problems and the valve technology of the time did not really lend itself to such requirements as an ultra-wideband, high gain and stable amplifier.

Two more factors were required for success, and these were engineered by Dolby, Ginsburg and Anderson, of Ampex between 1952 and 1955. The problems associated with the tape-track configuration were solved by the use of a horizontal head drum, containing four heads and rotating on a shaft mounted parallel to the direction of tape motion as shown in Figure 21.1a. The video heads lay down parallel tracks across the tape width, slightly slanted (see Figure 21.1b) with each head writing about 16 television lines per pass. The tape-to-head speed (i.e. the writing speed) at an incredible—to us today—40 m/s was most adequate, with response beyond 15 MHz. The final hurdle was cleared with the introduction of an FM recording system; this involves frequency modulating a constant-amplitude carrier with the picture signal before application to the recording head. The *Ampex Quadruplex* system was enthusiastically taken up by TV broadcasters, and rapidly became a world standard. For less exacting requirements and limited funds, a simpler recording system was really needed, and this gave birth to the helical scan system. The idea of a rotating video-head drum containing one or two heads, each laying down one complete television field per pass, was first mooted in 1953. The Japanese Toshiba company was foremost in the field in the early days, though several

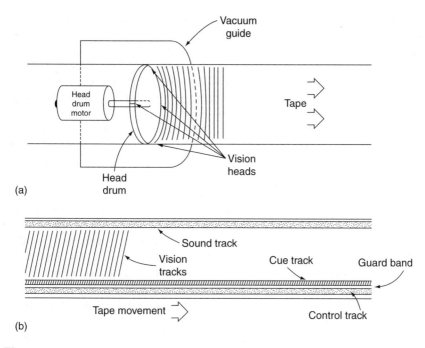

Figure 21.1 *(a) The transverse scan system. The vacuum guide draws the tape into the curved profile to match the contour of the head drum. (b) Resulting tracks on the tape*

other companies were working on the idea. By 1961, a handful of manufacturers were demonstrating helical machines, all with the open-reel system, each totally incompatible with all the others, and none having a performance which could approach that of the quadruplex system. The market for these machines was intended to be in the industrial and educational spheres, with very little regard as yet to the domestic market.

By 1980, the pace of machine development quickened with VHS ruled supreme in the UK and a wide range of software appeared alongside. In 1980, piano keys had given way to light-touch sensors and remote control, sophisticated timers and programmers had appeared and a form of freeze frame had been introduced. Another significant event of 1980 was the appearance of a new helical VCR format, the Philips/Grundig Video 2000 system, for use with the *video compact cassette* (*VCC*). The compact cassette designation indicated the intention that it should become as popular as the universally used audio compact cassette. In the event, V2000 format was not successful and production had ceased by early 1986. It used an advanced *dynamic track following* (*DTF*) tracking system, very similar to that of the later and more successful *Video-8 format*, whose ATF (automatic track finding) feature will be described later.

Specialised ICs had been making steady inroads into domestic VCR design, large-scale integration (LSI) and microprocessor devices becoming commonplace in contemporary machines. These made possible advanced remote control systems, comprehensive timers and programmers, trick-speed, still-frame and visual search features. They also led a trend away from mechanical complexity in the tape deck and towards electronic control of direct-drive systems; this considerably simplified the mechanics of the tape transport, threading and head drive systems, while retaining a relatively low electronic component count.

Advances had been made, too, in the field of portable VCR equipment. Purpose-designed battery operated machines became available in VHS, VHS-C and Video-8 format for outdoor location work. Those currently on offer incorporate the camera section and video recorder in one unit—camcorders. In this realm, there was strong competition between the VHS camp whose champions are JVC and Panasonic, and the Video-8 protagonists, led by Sony of Japan and having in its ranks many companies with backgrounds in the world of conventional and cine photography. Details of VHS, VHS-C and Video-8 formats will unfold throughout this chapter and the next, and the relative merits of the formats (including the high-definition [HD] variants, S-VHS and Hi-8) will be discussed later in this chapter.

The mid-nineties saw the advent of the digital format DV, primarily intended for camcorders, and taking advantage of MPEG compression techniques described in Chapter 4. For the first time here was a format backed by virtually all the manufacturers in the 'domestic' field, and DV-format products, with compatibility across the board, became available in a wide range of types and prices, with competition working well for the consumer. The late 1990s saw the introduction by Sony of Digital-8 format, a hybrid capable of dual-standard operation with Video-8/Hi-8 and digital recording and playback. JVC began marketing D (data)—VHS in 1999, a '*homedeck*' format for use with DTV television systems, and disc-based digital camcorders made their first appearance, from Hitachi, at about the same time. We shall look at the operation of digital tape systems later in this chapter. At the turn of the new century computer-type hard disc drives were incorporated into DTV set-top boxes (STBs) for recording (auto-recording) of TV programmes.

Magnetic tape basics

All magnetic materials, video tape coatings amongst them, may be regarded for practical purposes as consisting of an infinite number of tiny bar magnets, each with its own north and south poles. This is a simplification, but suits our purposes well. In the natural state, these bar magnets are randomly aligned within the material so that their fields cancel one another out, and no external magnetic force is present. To magnetise the material, be it a blank tape or a solenoid core, we have to apply an external magnetic force to align the internal magnets so that they sit parallel to one another, with all their N poles pointing in the same direction. When

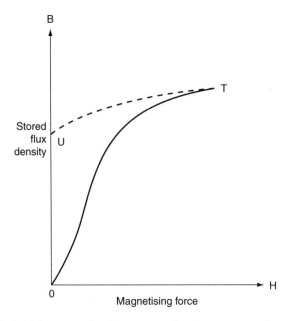

Figure 21.2 *Initial magnetisation curve of a ferro-magnetic material*

the external field is removed, most of the magnets remain in alignment and the material now exhibits magnetic properties of its own.

The relationship or transfer curve (Figure 21.2) between the external force and the retained magnetism on the tape is not linear. The transfer curve is a graph with the *magnetising force* (H) plotted along the horizontal axis, and the stored *flux density* (B) on the vertical axis. Our starting point is in the centre, point 0. Here, no external magnetising force is present, and the magnetic material's internal magnets are lying in random fashion—hence zero stored flux. Let's suppose we now apply a linearly increasing external flux. The flux density (strength of magnetism) in the material would increase in a non-linear fashion as shown by the curve 0–T. At point T, the material has reached magnetic saturation and all its internal magnets are rigidly aligned with each other—no increase in applied magnetic force will have any effect. If we now remove the magnetising force, bringing the H coordinate back to zero, we see that a lot of magnetism is retained by the specimen, represented by point U. This is the remanence of the material, the 'stored charge' as it were. For recording tape, it needs to be as high as possible.

Figure 21.3 is an expansion of Figure 21.2 to take in all four quadrants. We left the material at point U with a stored flux. To remove the flux and demagnetise the tape material, it is necessary to apply a negative magnetising force, represented by 0–V, whereupon stored flux B returns to zero. Further negative applied force (V–W) pushes the stored flux to saturation point in the opposite direction, so that all the magnets in the tape

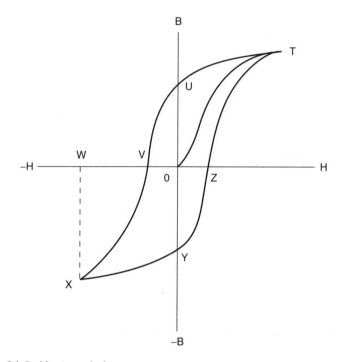

Figure 21.3 *Hysteresis loop*

are once again aligned, but all pointing the other way, represented by point X. Removal of the applied force takes us back to the remanence point, this time in the negative direction – point Y. To demagnetise the tape, a positive applied force 0–Z is required, an increase of which will again reverse the stored flux to reach saturation once more at point T. The diagram is called a magnetising *hysteresis curve* and one cycle of an applied magnetising waveform takes us right round it. It can now be seen how the 'degaussing' or erasing process works. Here, we apply a large alternating magnetic field sufficient to drive the material to saturation in both directions. The field is then allowed to decay linearly to zero, creating smaller and smaller hysteresis loops until they disappear into a dot at point 0, and the material is fully demagnetised.

Transfer characteristic and bias

Figure 21.4a is based on the previous diagram, but shows the initial magnetising curve in solid line (T0X) with the hysteresis curve in dotted outline. If we apply a magnetising force, 1H, and then remove it, the remnant flux falls to a value '1'. A larger magnetising force 3H will, if applied and removed, leave a remnant force '3', and so on, up to saturation point 8H and remanence point '8'. The same applies in the negative direction, and

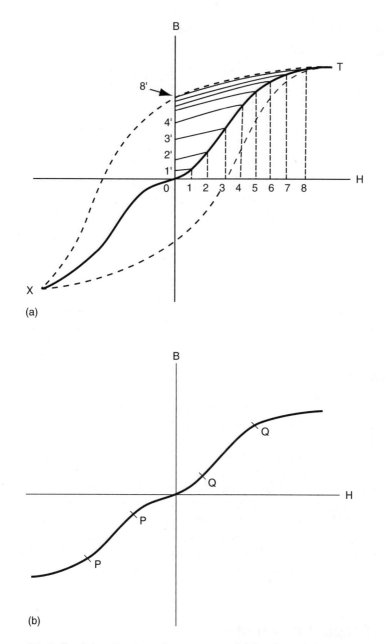

Figure 21.4 *Deriving the transfer curve: at (a) is plotted the remanent flux for eight linear steps of applied magnetising force; (b) shows the resulting transfer curve*

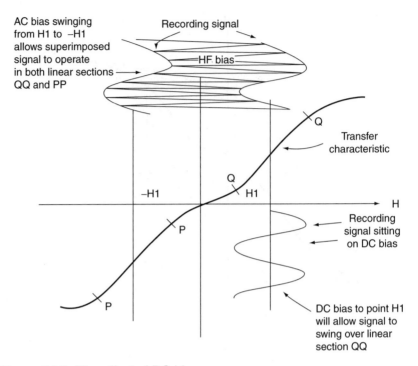

Figure 21.5 *The effect of DC bias*

plotting remanence points against applied force we get the curve shown in Figure 21.4b. This is the transfer characteristic. It is very non-linear at the middle and ends, but for a typical recording type, will contain reasonably linear sections on each flank, PP and QQ. If we can bias the recording head to operate on these linear parts of the transfer characteristic, the reproduced signal will be a good facsimile of that originally recorded. The d.c. bias (Figure 21.5) puts the head into one of the linear sections, but the recorded signal will be noisy and inadequate; a.c. bias, shown at the top of Figure 21.5, allows the head to operate in two quadrants, with superior results. We shall see that the chrominance signal in a VCR is recorded along with a bias signal, which is in fact the FM carrier for the luminance signal.

Head-tape flux transfer

The recording head consists of a ferrite 'ring', with its continuity broken by a tiny gap. A coil is wound around the ring, and when energised it creates a magnetic field in the ring; this is developed across the head gap. As the tape passes the gap, the magnetic field embraces the oxide layer on the tape and aligns the 'internal magnets' in the tape according to the electrical signal passing through the head. Provided that some sort of bias is

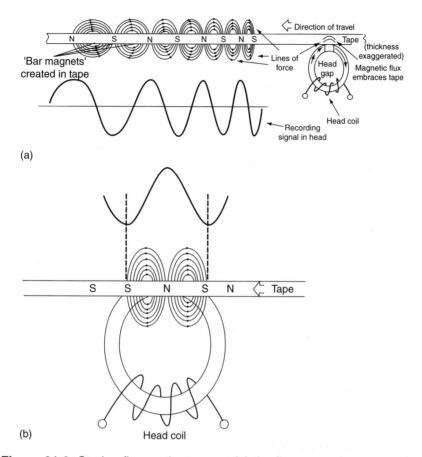

Figure 21.6 *Storing flux on the tape: at (a) the flux appearing across the head gap is penetrating the oxide surface to leave magnetic patterns stored on the tape; (b) shows the effect when one complete cycle of the recorded waveform occupies the head gap—no signal transfer will take place. This point represents f_{ex}, extinction frequency*

present (Figure 21.5) the relationship between writing current and flux imparted to the tape is linear, so that a magnetic facsimile of the electrical signal in the head is stored in the tape, as shown in Figure 21.6a. The tape passes the head at a fixed speed, so that low frequencies will give rise to long 'magnets' in the tape, and high frequencies short ones.

The linear relationship between head field strength and stored flux in the tape, described above, holds true when the wavelength to be recorded on tape is long compared with the width of the *head gap*. However, when the wavelength of the signal on the tape becomes comparable with the head-gap width, the flux imparted to the tape diminishes, reaching zero when the recorded wavelength is equal to the width of the head gap. This is illustrated in Figure 21.6b, where it can be seen that during the passage of a single point on the tape across the head gap, the applied flux has passed through one complete cycle, resulting in cancellation of the stored flux in the tape.

This point is known as the *extinction frequency* (f_{ex}) and sets an upper limit to the usable frequency spectrum.

For video recording, we need a large bandwidth and a high f_{ex}. This can be achieved by reducing the head-gap width or alternatively increasing head-to-tape or writing speed. There is a practical limit to how small a gap can be engineered into a tape head. The desirability for a narrow gap means that most practical heads are made by forming a narrow V-shaped groove in the back face of the core and grinding away the front face until the V-groove is just breached. In this way, gaps of the order of micrometres are achievable. With a typical gap of 0.3 μm used in domestic applications, a writing speed of about 5 m/s is required. Other HF losses also occur during recording. The head is by definition inductive, so losses will increase with frequency. Eddy currents in the head will add to these losses, as will any shortcomings in the tape-to-head contact. High frequencies give rise to very short 'magnets' in the tape itself, and it is the nature of these to tend to demagnetise themselves. For all these reasons, the flux imparted to the tape tends to fall off at higher frequencies, as in the solid line of the graph shown in Figure 21.7. To counteract this, recording equalisation is applied by boosting the HF part of the signal spectrum in the recording amplifier, as per the dotted line.

This is called *recording equalisation*, and its aim is to make the frequency/amplitude characteristic of the signal stored on the tape as flat as possible.

Replay considerations

As with any magnetic transfer system, the output from the replay head is proportional to the 'rate of change' of magnetic flux. Thus, assuming a constant flux density on the recorded tape, the replay head output will

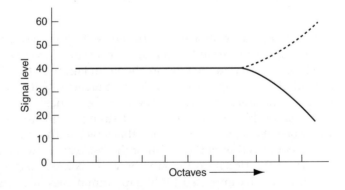

Figure 21.7 *Losses in the recording process. The dotted line shows a compensating 'recording equalisation' curve*

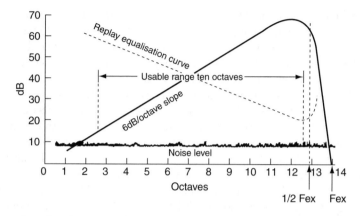

Figure 21.8 *Playback curve for a type with equal stored flux at all frequencies. For a 'level' output, the gain of the replay amplifier must follow the dotted replay equalisation curve*

double for each doubling of frequency. With a doubling of frequency known as an octave and a doubling of voltage being 6 dB, the relationship 6 dB/octave holds good until the extinction frequency, f_{ex}, is approached, when the head output rapidly falls towards zero. This is shown in Figure 21.8. The upper limit of the curve is governed by the level of the signal recorded on the tape, limited in turn by the tape's magnetic saturation point. At the low-frequency end, the replay head output is low due to the low rate of change of the off-tape flux. At some point, it will be lost in the 'noise' off-tape, and this will occur at about 60 dB down from peak level. Thus, even with playback equalisation (represented by the dotted curve in Figure 21.8), the dynamic range of the system is confined to 60 dB or so. With the unalterable 6 dB/octave characteristic, we are limited, then, to a total recording range of 10 octaves. This is inherent in the tape system and applies equally to audio and video signals. Ten octaves will afford an audio response from 20 Hz to 20 kHz, which is quite adequate. TV pictures, however, even substandard ones for domestic entertainment, demand an octave range approaching 18 and this is plainly not possible.

Modulation system

To be able to record a video signal embracing 18 octaves or more, it is necessary to modulate the signal onto a carrier and ensure that the octave range of the carrier is within the capabilities of the tape recording system. While the carrier could be amplitude modulation (AM), the frequency modulation (FM) system has been adopted because it confers other advantages, particularly in the realm of noise performance. An FM signal can be recorded at constant level regardless of the modulating signal amplitude, so that head losses and effects of imperfect head-to-tape contact are less troublesome. To achieve a picture replay with no perceptible background noise (snow), the S/N ratio needs to be over 40 dB, and this can be achieved by an FM recording system in a domestic VCR. Professional and broadcast machines can do much better than this!

An FM system, familiar to us in VHF sound broadcasts, starts with a *continuous wave (CW) oscillator* to generate the basic carrier frequency. The frequency of the oscillator is made to vary in sympathy with the modulating signal, audio for VHF sound transmitters, video for VCR recording systems and satellite broadcasts. For any FM system, the deviation (the distance that the carrier frequency can be 'pulled' by the modulating signal) is specified. In VHF sound broadcasting, it is ±75 kHz, giving a total frequency swing of the carrier of 150 kHz. In a VCR, carrier frequencies are specified for zero video signal amplitude—represented by the bottom of the sync pulse—and full video signal amplitude, i.e. peak white.

An FM system theoretically generates an infinite number of sidebands, each becoming less significant with increasing distance from the carrier frequency. Figure 21.9 shows the sidebands of a VHF-FM sound broadcast transmission. The modulating frequency is 15 kHz, and the sideband distribution is such that the first eight sidebands on either side of the carrier are significant in conveying the modulation information. Thus to adequately receive this double-sideband transmission, we need a receiving bandwidth of 240 kHz or so—in practice 200 kHz is sufficient, and this is the allocated channel width.

A bandwidth of 200 kHz for the transmission of a 15-kHz note seems very wasteful of spectrum space and certainly will not do for our tape system in which elbow-room is very limited! However, by making the carrier deviation small compared with the carrier frequency, the situation may

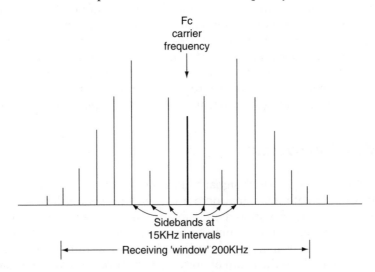

Figure 21.9 *Sidebands of a VHF-FM sound transmitter*

greatly be improved. This ration is known as the modulation index which is given by the formula

$$\frac{\text{carrier deviation}}{\text{modulation frequency}}$$

For the above example, the modulation index = 75/15 = 5.

If we can reduce the modulation index, the significant sidebands draw closer to the carrier frequency, and at modulation indexes below 0.5, the energy in the first sideband above and below the carrier becomes great enough for them to convey all the necessary information in the same way as those of an AM signal. In video tape recording we go a step further and use only one sideband along with a part of the other, similar to the vestigial sideband scheme used with analogue television broadcast.

By using a low modulation index, then, the sidebands of the FM signal can be accommodated on the tape. FM deviation has to be closely controlled and carrier frequencies carefully chosen to avoid trouble with the sidebands, which if they extend downwards from the carrier to a point beyond zero frequency, will not disappear, but 'fold back' into the usable spectrum to interfere with their legitimate fellows, leading to beat effects and resultant picture interference. The effect of a folded sideband is shown in Figure 21.10.

Pre-emphasis

Noise is the enemy of all recording and communications systems, and in FM practice it is common to boost the HF components of the modulating signal prior to the modulation process, a process known as pre-emphasis,

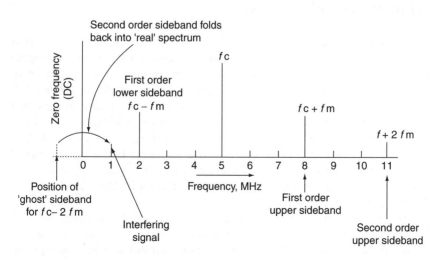

Figure 21.10 *The effect of folded sidebands. Carrier frequency is 5 MHz, modulating frequency 3 MHz*

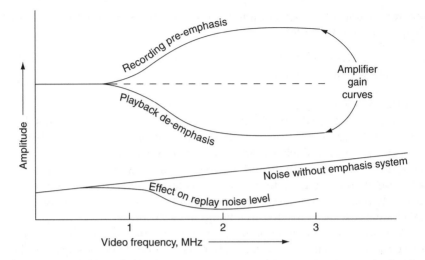

Figure 21.11 *The effect of pre-emphasis and de-emphasis on playback signal-to-noise (SNR) ratio*

and so do we in VCR FM modulation circuits. The effect of this after demodulation (in the post-detector circuit of a radio or the playback amplifier of a VCR) is to give a degree of HF lift to the baseband signal. In removing this with a filter, gain is effectively reduced at the HF end of the spectrum, with an accompanying useful reduction in noise level. The idea is shown in Figure 21.11, which illustrates the filter characteristics in record and replay, along with the effect on the noise level.

Principle of helical scan

We have seen that the necessary high head-scanning speed was first achieved by fast longitudinal recording techniques, then in the successful *Quadruplex* system by *transverse scanning* of the tape by a four-head drum with the video heads mounted at 90° intervals around its periphery. The transverse scan method has a lot going for it! The very high writing speed confers great bandwidth, enabling full broadcast-specification pictures to be recorded and replayed. The relatively wide tracks are almost at right angles to the tape direction so that any jitter or flutter in the tape transport has little effect on the timing of the video signal, merely causing slight momentary tracking inaccuracies, easily catered for by the wide track and the FM modulation system in use.

The cost of transverse scanning machines is high because of the need for a complex head drum system and precision vacuum tape guides. To maintain the necessary intimate contact between video head and tape, all VCRs have their head tips protruding from the surface of the drum so that they penetrate the tape and create a local spot of 'stretch'. Hence, the need

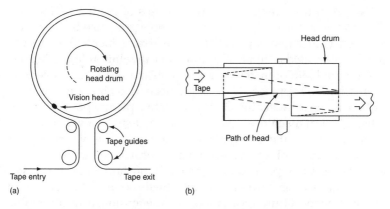

(a)

(b)

Figure 21.12 *(a) The omega wrap. This diagram shows the arrangement for a single-head machine; the path of the head is the heavy line in (b)*

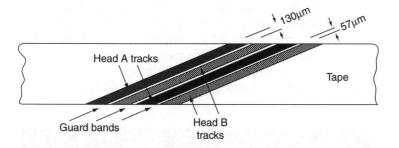

Figure 21.13 *The diagram of Figure 21.12 drawn from the point of view of the tape*

for a precise vacuum guide at the writing/reading point in a transverse machine to maintain correct tape tension. These techniques are not amenable to domestic conditions or budgets, and the alternative and simpler helical system has undergone much development. It is now used exclusively.

In a helical scan system, several problems are solved in one go, but other shortcomings are introduced. The idea is to wrap the tape around a spinning head drum with entry and exit guides arranged so that the tape path around the head takes a form of a whole or part-helix. The principle is shown in Figure 21.12 where a video head in the course of one revolution enters the tape on its lower edge, lays down a video track at a slant angle and leaves the upper edge of the tape ready to start again at the tape's lower edge on the next revolution. During the writing of this one track, the transport mechanism will have pulled the tape through the machine by one 'slant' video-track width, so that track 2 is laid alongside the first; each track is the width of the video head. Figure 21.13 shows how this works and gives an elementary impression of the track formation.

Let's look at the strengths and weaknesses of the helical format. Tape tension and hence head-tip penetration is now governed by the tape transport system rather than a precision guide arrangement. One or two complete television fields can be laid down per revolution of the head drum, so that the problem of matching and equalising heads (to prevent picture segmentation) disappears. In a single-head machine such as we have described, no head switching during the active picture period is necessary. Two-head helical machines have a simple head-switch or none at all; what price do we pay for these advantages? The main penalty is timing jitter in the recorded and replayed video signals. Because the tracks are laid at a small angle to the tape direction (about 5° off horizontal for VHS) they may be regarded as virtually longitudinal, so that the effects of the inevitable transport flutter, variations in tape tension, bearing rumble, etc. will be to introduce minute timing fluctuations into the replayed signal. This effect cannot be eliminated in any machine, and causes problems with colour recording and certain types of TV receiver.

In practice, two video heads are used in the drum of a helical VCR, pictured amongst others in the photo of Figure 21.14. The two-head system means that the tape needs only to be wrapped around half the video head drum perimeter, with one head joining the tape and beginning its scan as the other leaves the tape after completing its stint. To give a degree of overlap between the duty cycles of the two heads, the tape wrap is in fact

Figure 21.14 *Video head drum. This photograph of the underside shows five heads with a 'blank' (for mechanical balance) near the top. The heads at one o'clock and seven o'clock are double types for video, the others for hi-fi audio (two o'clock and eight o'clock) and flying erase near the bottom*

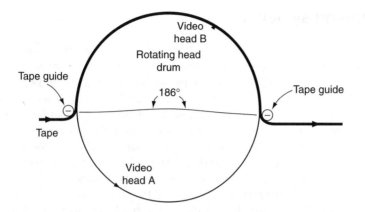

Figure 21.15 *Omega wrap for a two-head drum. The tape occupies rather more than half a turn of the drum*

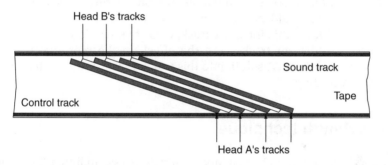

Figure 21.16 *Video tracks written by a two-head system, with each head writing alternate tracks. The relative positions of the sound and control tracks are also shown (VHS format)*

about 186°, slightly more than half a turn. This is known as an omega (ω) wrap outlined in Figure 21.15.

Track configuration

A typical track layout for a two-head helical VCR appears in Figure 21.16. Here we can see the video tracks slanting across the tape, shaded for head A, white for head B. At the edges of the tape, further tracks are present: the lower carrying a control track (serving a similar purpose to the sprockets in a cine film, and described later) and the upper carrying the sound track. Sound is recorded longitudinally in the same way as in an audio recorder, but with limited frequency response due to the low linear tape speed in the sorts of VCR we are dealing with. Tape is 12.65 mm (VHS) or 8 mm (Video-8) wide. It progresses at speeds between 22.4 and 10.06 mm/s, depending on format and mode.

Scanning systems

In the original VCR plan, a guard band was left between video tracks on the tape. This was true of the first machine to appear on the domestic scene, the Philips N1500. Linear tape speed here was over 14 cm/s, and the track configuration is shown in Figure 21.17. It can be seen that each video track is spaced from its neighbours by an empty *guard band*, so that if slight mist racking should occur, cross-talk between tracks could not take place. Each video track was 130 μm wide and the intervening guard bands 57 μm wide. This represents relatively low packing density of information on the tape, and it was soon realised that provided the tape itself was up to it, a thinner head could be used to write narrow tracks; if the linear tape speed was also slowed down the tracks could be packed closer together, eliminating the guard band. Using both ideas, tape playing time for a given spool size could be doubled or trebled. First, the problem of cross-talk had to be solved. Even if the mechanical problems in the way of perfect tracking could be overcome so that each head always scanned down the middle of its intended track, cross-talk would occur due to the influence of adjacent tracks, and the effect on the reproduced picture would be intolerable. A solution to this problem was found in the form of azimuth recording.

The azimuth technique

For good reproduction from a tape system, it is essential that the angle of the head gap on replay is exactly the same as was present on record with respect to the plane in which the tape is moving. In an audio system, the head gap is exactly vertical and at 90° to the direction of tape travel. If either the record or replay head gap is tilted away from the vertical, even by a very small amount, tremendous signal losses occur at high and medium frequencies, the cut-off point travelling further down the frequency spectrum as the head tilt or *azimuth error* is increased. If the same head is used for record and replay (as is usually the case in audio tape

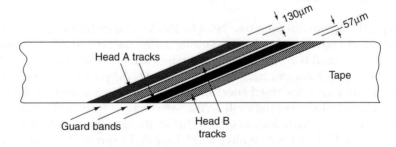

Figure 21.17 *Guard-band recording: the track formation for VCR format as recorded by the Philips N1500 machine*

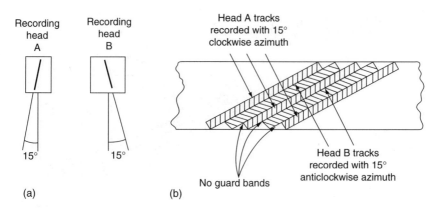

Figure 21.18 *Azimuth offset. The head gaps are cut with complementary azimuth angles so that the guard bands of Figure 21.17 can be eliminated*

recorders), the azimuth error will not be noticed, because there is no azimuth difference between record and replay systems.

This azimuth loss effect is the key to successful recording and replay of video signals without a guard band. Let's designate our video heads A and B, and skew A's head gap 15° clockwise and B's head gap 15° anticlockwise as in Figure 21.18a. This imparts a total 30° difference in azimuth angle between the two heads, and the result is video tracks on the tape like those in Figure 21.18b.

With the built-in error of 30°, head A will read virtually nothing from head B's tracks, and therefore the guard band can be eliminated. This was the modus operandi of the Philips VCR-LP format, using the same cassette and virtually the same deck layout as the original VCR format, but with linear tape speed reduced by 50% , and video track width down to 85 μm. It worked, and the 2-h machine was a reality. Subsequent formats use a smaller azimuth tilt: 6° for VHS and 10° for Video-8. The offset between heads is double this figure in each case.

The video track

Figure 21.19 shows four adjacent tracks on the tape. Track 1 is laid down by head A, and the 'phasing' of the spinning video head drum is arranged so that the A head enters on to the tape and starts to record just before a field sync pulse arrives. It will write about seven lines of picture before recording the field sync pulse, and then go on to write the rest of the lines in the field. By the time line 306 has been recorded, head A is leaving the top of the tape, having recorded one field of 312½ lines; head B has entered onto the tape and is about to record track 2, consisting of the next sync pulse and field; and the head drum has turned through half a revolution, or 180°. During this time, the capstan has pulled the tape through

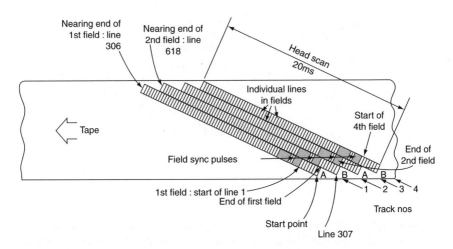

Figure 21.19 *The positions of TV lines and fields in the recorded tape track. Each head writes one field of video information*

the machine just far enough to ensure that track 2 lays alongside, and just touching, track 1. Track 3 is laid down by head A again, track 4 by head B and so on.

We can see, then, that signals recorded towards the lower edge of the tape correspond to those in the top half of the picture and vice versa. Thus, a tape damaged by creasing along the top edge may be expected to give a horizontal band of disturbance in the bottom half of the reproduced picture. Around the period of the head changeover point, both heads are at work for a brief instant, one just about to run off the top of the tape, and the other having just entered at the bottom. Thus, there is an overlap of information. All VCRs incorporate a head switch which electronically switches between the heads at the appropriate time, just before the field sync pulse. As should now be clear, this takes place at the very bottom of the picture, and any picture disturbance due to head changeover during record and replay is hidden by the slight vertical over-scanning which takes place in a correctly adjusted TV.

Miniature VHS head drum

The 62.5 mm drum diameter of the standard VHS specification is a great handicap in portable video equipment. To achieve a deck size small enough to be accommodated in a light camcorder, a small head drum is used: it is 41.3 mm in diameter. To permit record and playback of standard VHS tracks, some complexity in the mechanics and electronics of the machine is unavoidable, and Figure 21.20 shows the essence of the arrangement. The travel of the tape guides follows a longer and more sinuous path than before, to wrap the tape around 270° of the periphery of

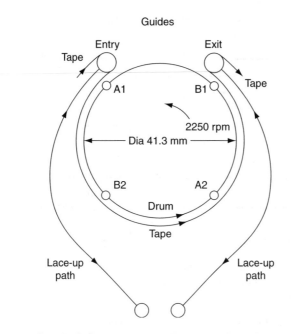

Figure 21.20 *Small VHS head drum. Four heads are required to read and write standard tracks*

the small drum, which rotates at 2250 rpm. This high speed is calculated to sweep a single drum-mounted head along the entire length of a standard VHS video track during its contact with the wrapped tape. The inclined and continuous ruler-edge around the lower drum assembly maintains the tape at the normal 5.302° angle to the head-sweep path.

Plainly, one pair of heads will not suffice to work this system. At the point when one head is leaving the tape wrap, there needs to be another just 90° ahead, the point where it is just entering the tape wrap. This ensures continuity of signal feed onto the tape, whose linear speed around the drum conforms to standard VHS specifications—2.34 cm/s for SP mode, 1.17 cm/s for LP mode. If the head which has just left the tape is writing or reading 'A' tracks, the one ahead of it and one behind it must be 'B' heads with azimuth angles cut accordingly. Hence, the A-B-A-B configuration of the four heads around the drum is shown in Figure 21.20. Each head scans every fourth track on the format-standard tape.

At any given moment only one of the four heads will be active in record or playback, and since two others will be in contact with the tape at this time a four-phase head-switching system is required during both record and replay. Figure 21.21 shows the switching sequence and the time relationship of the video signal to the active period of each head. The switching system is the same in record and playback modes, though of course the routing of the video FM carrier is opposite.

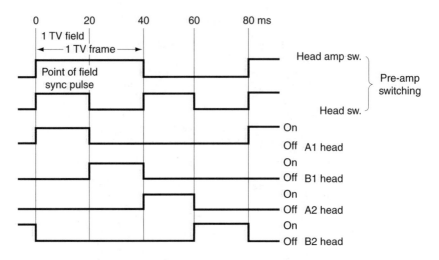

Figure 21.21 *Sequential switching for the heads in Figure 21.20. Switching is applied on record and playback*

It can be seen, then, that four heads are necessary in a 'small' VHS head drum to do the work of the two in a conventional drum. If separate heads are provided for SP and LP eight heads are required on the drum, though they can be mounted in four chips, each carrying two windings and two head gaps. Some VHS camcorders are additionally fitted with a flying erase head, giving an effective total of nine heads around the drum periphery, with a multi-winding rotary transformer to couple the recording signals to the heads.

Video-8 format was designed from the outset for a two-head 40 mm-diameter drum in a small camcorder. Even so, further miniaturisation has led to the use in Video-8 and Hi-8 camcorders of the same small-head technique, with a drum of 26.6 mm diameter.

Tracking

We have seen that during record, the phasing of the head drum position relative to the incoming video ensures that the field synchronising pulse is laid down at the beginning of each head scan of the tape. During replay, it is necessary to set and maintain the relative positions of the tape and head drum so that each head scans down the centre of its own track. The record phasing and replay tracking is carried out by the VCR's servo systems. A control track is recorded along the edge of the tape as a timing reference. This takes the form of a 25-Hz pulse train and is used on replay to set and maintain the relative position of the tape tracks and head drum.

Automatic tracking is achieved by sampling the amplitude of the off-tape FM carrier signal and applying it to the servo circuit to 'trim' the tracking for optimum results. Even so, a manual override is provided, most often used in replay of hi-fi sound cassettes which are 'difficult', perhaps due to slight head misalignment in the duplicating deck on which they were recorded.

Automatic track finding

Automatic track finding (ATF) is a more advanced form of tracking, in which head-guidance signals are continuously recorded in the video tracks. The concept was first introduced in the now-defunct Philips/ Grundig V2000 format, where it was known as DTF and was used with a piezo-bar mounting system for the heads, whose position could be set by applying d.c. deflection voltages to the piezo bars.

Video-8 format uses an ATF system, the essence of which is a pilot tone which is recorded with the picture throughout every video track. Four pilot tone frequencies are used: f_1, 101.02 kHz; f_2, 117.19 kHz; f_3, 162.76 kHz and f_4, 146.48 kHz. They are added to the luminance FM record signal and recorded in the sequence f_1, f_2, f_3, f_4 in successive head sweeps (see Figure 21.22). Relatively low frequencies like these are almost unaffected by any azimuth offset of the replay head, so pilot-tone cross-talk from adjacent tracks is easily picked up by the video heads during playback. The pilot-tone frequencies are chosen to have specific relationships as Figure 21.23

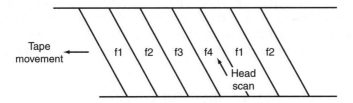

Figure 21.22 *Tone sequence laid on tape for ATF*

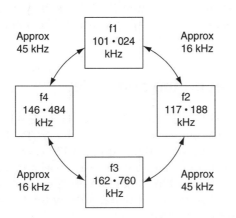

Figure 21.23 *ATF tones have a carefully selected frequency relationship*

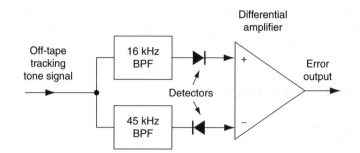

Figure 21.24 *Basic arrangement for derivation of an error voltage from inter-track pilot beats*

shows. The beat frequencies which arise when pilot tones from adjacent tracks are mixed are always 16 or 45 kHz. These beat products are used to steer the head path/tape track alignment for optimum tracking: when the levels of 16 and 45 kHz beat signal are equal, the replay head must be scanning along the dead centre of its video track indicating optimum tracking.

There are several ways in which the ATF pilot tones can be processed during playback. A simple one, illustrated in Figure 21.24, utilises two band-pass filters to pick off and separate the 16 and 45 kHz beat products so that they can be separately detected and measured. The d.c. outputs are applied to the differential inputs of an operational amplifier whose output forms the error signal. This error output can be used to phase-lock either the capstan or head drum servo to give accurate and continuous tracking correction with built-in compensation for 'mechanical' errors, tape-stretching, etc.

Video-8 tape-signal spectrum

The DTF pilot tones are recorded at the lowest part of the frequency spectrum as shown on the left of Figure 21.25. Here, they do not interfere with the 'signal' components of the tape recording. The other parts of the tape-signal spectrum will be dealt with later.

Luminance recording and replay

The luminance (or black-and-white) signal is dealt with separately from the chrominance (or colouring) signals in home VCRs. For luminance, the basic idea is to modulate the signal onto an FM carrier for application to the recording head and demodulate it to baseband during the replay process. As the simplified block diagram of Figure 21.26 shows, however, there are several other processes undergone by the luminance signal and these will be described in turn.

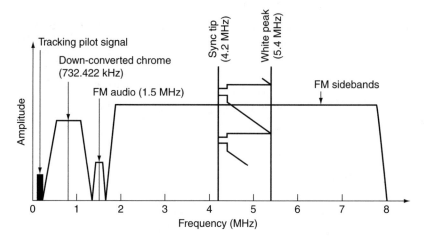

Figure 21.25 *The spectrum of signals on tape – Video-8 format. That for Hi-8 format is given in Chapter 25 of 3rd edition*

Figure 21.26 *The stages in the luminance-recording process*

The basic luminance signal that we wish to record may come from a TV camera or other local video source, or more likely, a broadcast receiver built into the VCR. In either case, it will be positive-going for white and will probably contain a chrominance signal modulated onto a 4.43-MHz carrier. It is important that the signal recorded on the tape is within the limits of the recording system, so the luminance signal is first passed through an AGC amplifier with a sufficiently wide range to compensate for signal inputs of varying amplitudes. This works in a similar manner to the AGC system of a radio or TV, by sampling its output level to produce a d.c. control potential, and applying this to an attenuator at the amplifier input. Thus, the output from the AGC stage will be at constant (say) 1 V amplitude.

In low-band (e.g. VHS) VCRs, luminance bandwidth is restricted to about 3 MHz, gives or takes a hundred kHz or so between the formats. If higher frequencies than this are allowed to reach the modulator, they will make mischief with sidebands, as explained earlier. A low-pass filter with a quite sharp cut-off around 3 MHz is incorporated in the record signal path, then, and this also eliminates all the chrominance components of the signal, which are based on a sub-carrier of 4.43 MHz. When recording in monochrome, more bandwidth can be allowed occupying the space normally reserved for the chrominance signal, and many machines have an automatically switched filter characteristic for monochrome and colour recordings as shown in Figure 21.27.

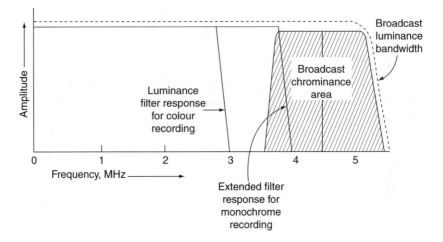

Figure 21.27 *VCR luminance-recording filter characteristics relative to the spectrum of the composite video signal as broadcast*

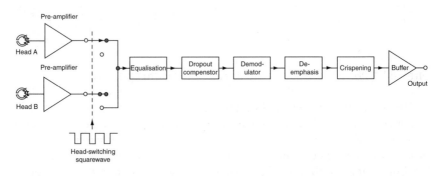

Figure 21.28 *The replay chain for the luminance signal*

During replay, the off-tape signal from the video heads is very small, and to maintain the necessary >40 dB S/N ratio in the reproduced picture, low-noise amplification is necessary. The main replay processes after the preamplifiers are head switching, equalisation, limiting, drop-out compensation, demodulation to baseband, de-emphasis, crispening and amplification, after which the luminance signal is resorted to its original form, usually 1 V peak-to-peak, negative-going syncs. A block diagram showing the order of the replay circuits is given in Figure 21.28 incorporating two heads (A and B) with a head switching arrangement. During replay, the signal from the tape head falls at the rate of 6 dB/octave. HF losses due to the approach of f_{ex} are largely compensated for by a resonance circuit associated with the replay head. To even-out the response on replay, further equalisation is provided, and in practice it takes the form of a 'boost' in the 2–3 MHz region, the major area for the lower FM sideband of the luminance signal. This is often catered for by the provision of a reactive component in the collector circuit of the preamplifier or a following stage within the IC. Equalisation is followed by dropout compensation (DOC).

Dropout

In a video tape, the magnetic coating is not completely homogenous and with video track widths smaller than the diameter of a human hair; even a microscopic blemish in the magnetic coating will delete some picture information. As the tape ages, slight contamination by dust and metallic particles and oxide-shedding effects will aggravate the situation. The effect of these tiny blemishes is a momentary loss of replay signal known as a *dropout*. Unless dropouts can be 'masked' in some way, a disturbing effect will take place in the form of little ragged black or noisy 'holes' in the reproduced picture.

In practice, the video information on one TV line is usually very much like that on the preceding line; so that if we can arrange to fill in any dropout 'holes' with the video signal from the corresponding section of the previous line, the patching job will pass unnoticed. What's required, then, is a delay line capable of storing just one TV line of 64 µs duration, so that whenever a dropout occurs we can switch to the video signal from the previous line until it has passed. This is known as DOC. For delay-line bandwidth reasons, this is difficult to achieve at video baseband frequencies, so it is carried out on the FM signal before demodulation. To avoid disturbance on the picture, the switching has to be very fast, virtually at picture-element rate. One form of the DOC technique outlined above breaks down when any dropout exceeds one TV line-duration, and this happens often – the tape area occupied by one line is microscopic. If the dropout period exceeds one line, the disturbance will be visible on the TV screen. To prevent this, further steps must be taken. Modern designs recycle the last 'good' TV line around a delay line and read it out continuously for the duration of the dropout.

Crispening

Following demodulation and de-emphasis, a technique known as picture *crispening* or *picture sharpening* is carried out. The 3 MHz capability of a standard VCR means that replayed pictures will lack the sharpness and definition of an off-air picture; and fine detail, for example the frequency gratings of a test pattern, will not be reproduced. Very little of the content of an average picture consists of fine repetitive detail and the subjective effect of limited HF response is a blurring of sharp vertical edges. The crispening circuit goes some way to compensate for HF roll-off by artificially 'sharpening up' vertical edges in the picture. The technique has been used in broadcast and CCTV for many years to compensate for the finite scanning spot size in cameras and fixed satellite services systems. In such applications, it is known as aperture correction.

Colour signals in the VCR

If we had a perfect video tape recorder, there would be no reason why we could not take the composite video signal, chrominance and all, and record it on the tape as an FM signal, demodulate it during replay and recreate the original video signal as described for luminance in the first part of this chapter. As we've already seen, a home VCR is far from perfect in its performance, and the direct recording idea will not work for two good reasons.

The bandwidth limitations of our VCRs have already been described, and with an upper cut-off frequency around 3.5 MHz, the colour signals at 4.43 MHz are left out in the 'cold' somewhat! Even in S-VHS format, whose signal passband approaches 5 MHz, it is still not possible to directly record an encoded colour signal. Because the VCR relies on mechanical transport, perfectly smooth motion cannot be achieved. Inevitable imperfections in bearing surfaces and friction drives the elasticity of the tape and minute changes in tape/head contact all combine to impart a degree of timing jitter to the replayed signal. In good audio tape systems, this jitter is imperceptible, and for luminance replay its effects can be overcome by the use of timebase correctors in professional machines, or fast-acting TV line-synchronising systems with domestic VCRs. The latter class of machine, in good condition, may be expected to have a timing jitter of about 20 μs over one 20 ms field period. Let's examine the effect of this on the chrominance signal.

We have seen that the hue of a reproduced colour picture depends entirely on the phase of the sub-carrier signal. 'Phase' really means relative timing, so if the timing of the sub-carrier signal relative to the burst is upset, hue errors will appear. One cycle of sub-carrier occupies 226 ns, and a timing error of this order will take the vector right round the 'clock' of Figure 21.28, passing through every other hue on the way! A timing error of 100 ns (one ten-millionth of a second) will turn a blue sky into a lime-green one, and the jitter present in a mechanical reproduction system will make nonsense of the colours in the picture. For acceptable results, sub-carrier phase errors need to be held within 5° or so, representing a permissible maximum timing jitter of ±3 ns as shown in Figure 21.29. The physical causes of jitter cannot be eliminated so an electronic method of sub-carrier phase correction is necessary. To achieve this, the luminance and chrominance signals are separately treated (Figure 21.30).

Let's first examine the problem posed by the high-colour sub-carrier frequency. Several solutions are possible, including decoding the chrominance signal to baseband (i.e. primary colour or colour-difference signals) before committing them to tape, then recoding them to PAL standard

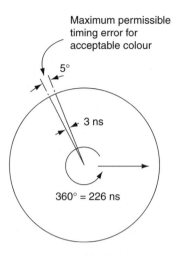

Figure 21.29 *Hue is dependent on phase angle, and only a very small phase change is permissible in a VCR system*

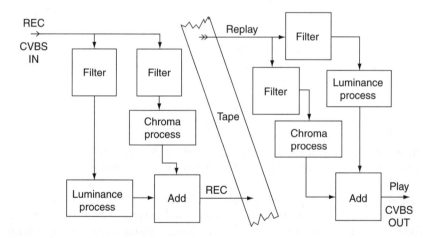

Figure 21.30 *'Streaming' of Y and C components of the video signal for tape recording*

during the replay process. This would require a lot of circuitry, and be vulnerable to hue and saturation errors. A better solution is to keep the PAL-encoded signal intact, bursts, phase modulation and all, and merely convert it to a suitable low frequency for recording. During playback, the composite chrominance signal can be re-converted to the normal 4.43 MHz carrier for recombination with the luminance component into the standard form of colour video signal, compatible with an ordinary TV set.

Space has to be found for the colour signal in the restricted tape frequency spectrum, however, and it is made available below the luminance

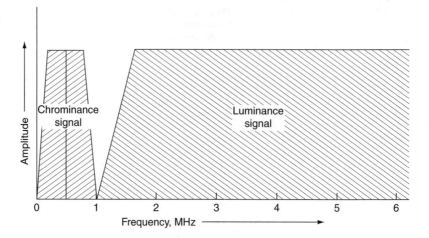

Figure 21.31 *Spectrum of luminance and chrominance signals on the tape. The luminance signal is now in FM form, and the chroma signal still at baseband, but with limited bandwidth and a lower carrier frequency*

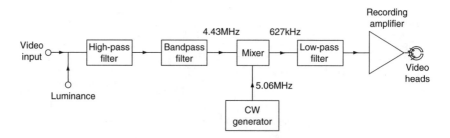

Figure 21.32 *The basic colour-under system. A heterodyne technique is used to convert the chroma signal to a new, low frequency*

sideband, where a gap is purposely left in the 0–1 MHz region. This area is kept clear by tailoring the luminance record passband and the FM modulator deviation to avoid luminance sidebands falling into it during colour recording. The colour signal, then, is down-converted and bandwidth-limited so that it occupies a frequency band of roughly 0–1 MHz. It is a double-sideband signal, so that the carrier frequency needs to be centred on about 500 kHz and the sidebands limited to 500 kHz or so. The spectrum of the down converted signal known as the *colour-under signal* is shown in Figure 21.31; compare this with Figure 21.26.

Down-conversion

The process for down-converting the chrominance signal is shown in Figure 21.32. The double-sideband signal is first separated from luminance information in a high-pass filter, then its bandwidth is limited to around 1 MHz by a bandpass filter. This curtailment of the sidebands of the modulated signal has the effect of restricting the chroma signal detail to around 500 kHz, less than half that of the broadcast signal. The chroma signal now passes into a mixer where it beats with a stable locally generated CW signal at around 5 MHz. This is the familiar heterodyne effect, and the mixer output contains components at two frequencies, those of the sum and difference of the two input signals. A low-pass filter selects the required colour-under signal, and rejects the spurious HF product.

The colour-under signal contains all the information that was present in the 4.43 MHz chroma signal except the finer colour detail; thus chroma amplitude and phase modulation are preserved, as well as the phase and ident characteristics of the burst, albeit based on a much lower sub-carrier frequency.

Up to now we have been quoting sub-carrier frequencies in round numbers. Format specifications quote the colour-under frequency very precisely, and taking VHS as an example, the local CW signal is at a frequency of 5.060571 MHz, which when beat against the 4.433619 MHz broadcast sub-carrier provides a colour-under centre frequency of 5.060571 − 4.433619 = 0.626952 MHz, or 626.952 kHz. Due to the broadcast sub-carrier modulation, sidebands extend for roughly 500 kHz on each side of this carrier.

The local CW signal at 5.06 MHz needs to be locked (i.e. have a fixed phase relationship) to the line syncs of the recorded signal and this is achieved by a phase-locked loop (PLL).

Chroma recording

The down-converted colour-signal based (for standard VHS format) on 627 kHz, and containing amplitude and phase modulation, is recorded directly onto the tape. Its amplitude is carefully controlled to achieve maximum modulation of the tape without non-linearity due to magnetic saturation. In this respect it is similar to audio tape recording, with the necessary HF bias being provided by the FM luminance record signal – the two are added in the recording amplifier, giving use to the waveform of Figure 21.32, where can be seen the recording-head signal for a colour-bar signal. The chrominance signal is superimposed on the constant-level FM luminance carrier, and this composite waveform is fed to the heads for recording on the tape.

During playback, the chrominance signal appears at the replay amplifier in the colour-under form described above and with a waveform like that in Figure 21.33. Impressed onto it are the timing variations, or jitter, introduced by the mechanical record and replay transport system and these have to be eliminated. First, though, the chroma signal is separated from the luminance FM signal in a low-pass filter, up-conversion to correct sub-carrier frequency (4.43 MHz) must also be carried out and we will take this first.

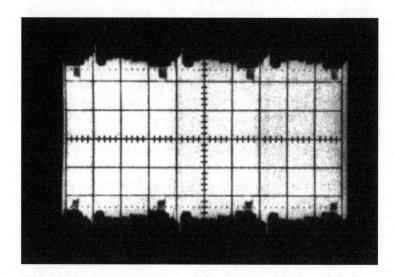

Figure 21.33 *The composite colour bar signal (FM luminance plus down-converted chroma) as applied to the recording heads*

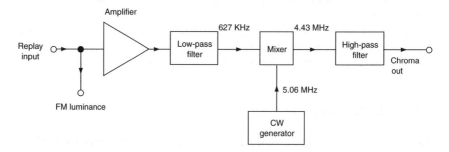

Figure 21.34 *Basic chroma replay process of up-conversion*

The 627 kHz is heterodyned with a local 5.06 MHz CW signal to produce a beat frequency of 4.43 MHz, selected by a suitable bandpass filter. The idea is shown in Figure 21.34. DOC and crispening are not relevant to the relatively coarse and *'woolly'* colour replay signal, and the recreated 4.43 MHz chroma signal is added to the luminance component and passed out of the machine.

Video-8 format chrominance

The basic principle of the colour system of Video-8 format is the same as for VHS already described, in that the chroma phase recorded on tape is manipulated to ensure that cross-talk signals come off the tape in antiphase over a two-line period. The colour-under frequency for Video-8 is $(47-^1/_8) \times$ fh, which is 732 kHz. It is locally generated, but in order to implement de-jittering during playback, is locked to incoming line sync in a PLL incorporating a $\div 375$ stage.

Signal processing chip

Throughout this chapter, we have used block diagrams to illustrate the conversion and conditioning of the luminance and chrominance signals for their recording on tape. In practice, dedicated chips are involved in providing all the necessary processing. A 2-chip solution incorporating a head drive/preamplifier chip mounted in or close to the head drum and a 'Y/C' chip on the main PC board. A typical chipset is shown in Figure 21.35 (a) for record and (b) for playback. Looking first at diagram (a) there are three video inputs, selected by a switch at chip pin 34 under bus control. Luminance noise reduction uses an electronic 1-line signal delay in IC302 at main chip pins 40/42, after which the Y carrier signal passes through three stages of emphasis on its way to the modulator. Emerging at IC pin 18, the carrier enters IC801 on pin 15 for application to the heads. The chrominance signal undergoes the processes described earlier to pass from pin 14 of IC301 to pin 16 of IC801, where it meets the luminance carrier en route to the video heads. IC801 also handles the drive to the hi-fi audio heads, whose operation we shall meet later in this chapter.

Turning now to the same chipset working on replay (Figure 21.35b), we can see on the left that this VCR is a dual-speed type with separate video head-pairs for SP and LP. The selected FM envelope signal passes into IC301 on pin 15 and splits two ways: up to the luminance limiter, demodulator and noise reduction stage (again using delay chip IC302); and down to the chroma up-converting stage, bandpass filters, etc. Luminance and chrominance are reunited in the Y/C mix block to emerge from the chip as a composite video/chroma signal at IC pin 38. The colour phase rotation switching pulses enter IC301 at pin 66 (diagram a); the colour crystal is hooked to pins 59 and 60, while an I²C bus passes user- and system-control commands into the IC on pins 63 and 64. Two separate SW25 pulse trains enter head amp/switch IC801, one for video and a second for audio, as will be described in the next chapter.

Audio signal processing

As we saw earlier in this chapter, the ingredients for success in recording TV pictures on tape are high writing speed and FM modulation. Within the immovable constraints of existing formats, it was obvious that for better sound performance these virtues must also be applied to the audio

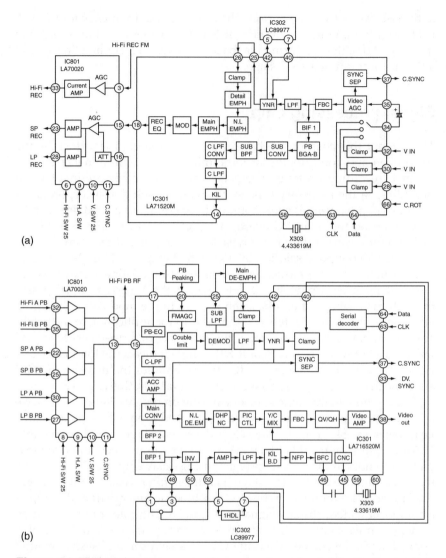

Figure 21.35 *Luminance processing, record mode at (a), playback mode at (b) (LG Electronics)*

signal. The obvious solution was to record an FM-modulated sound car-rier in or around the helical video tape tracks; but with the vision heads and tracks already chock-full of information, the problem was where to squeeze in the audio recording with regard to frequency-spectrum space and the 'magnetic' capacity of the tape track. Some form of multiplex system is required. Two alternative systems emerged for home VCR formats—*depth multiplex*, used in VHS models; and *frequency multiplex* in Video-8 format. The former is the most common, and will be covered first.

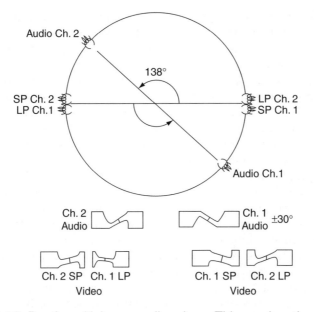

Figure 21.36 *Depth-multiplex recording drum. This one has the audio heads mounted 138° ahead of the double-gapped video heads*

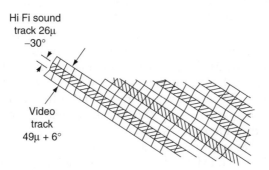

Figure 21.37 *Video/sound track relationship for VHS*

Depth-multiplex audio

This technique depends to some degree on the magnetic layer of the tape itself to discriminate between the video and audio signals. A separate pair of heads on the spinning drum is provided solely for the audio signal FM carriers. A typical layout of heads on the drum is shown in Figure 21.36. The two audio heads are mounted 180° apart and are arranged to 'lead' the video heads. The gaps cut in the audio heads have large azimuth angles: ±30°, sufficient to prevent cross-talk from adjacent tracks at the carrier frequencies (around 1–2 MHz) involved. The mounting height of the audio heads is set to place the hi-fi audio track in the centre-line of the corresponding video track. This track layout is shown in Figure 21.37, which also gives an idea of the relative audio track widths; the audio tracks are half the width of the vision tracks they accompany. The audio heads have a relatively wide gap resulting in a deep magnetic pattern penetrating some microns into the magnetic coating of the tape. Shortly following the audio head comes the regular video head, whose gap is in the region of 0.25 μm. Writing (for the most part) at higher frequencies, it creates shallower, shorter magnetic patterns which penetrate less than 1 μm into the magnetic surface. Thus, the recorded tape contains a 'two-storey' signal: a buried layer of long-wavelength audio patterns under a shallow top layer of video patterns.

During replay, the same heads operate on the same tracks as before. The video head picks off its track with little impairment, and with minimum cross-talk from sound tracks because of the large disparity (36° for VHS-SP) between the azimuth angle of recorded track and replay head. The audio head during replay is handicapped by the barrier presented by the 'video layer' on tape, but the resulting 12 dB or so of attenuation – thanks to the use of bandpass filters and the noise-immunity of the FM carrier system—does not prevent noise-free reproduction of the baseband audio signal as long as the tracking is reasonably correct. For hi-fi sound, the tracking performance is critical, especially in LP modes where narrow tracks tend to magnify any errors.

As with the video heads, switchover between the two rotating audio heads is carried out during the overlap period when both heads are momentarily in contact with the tape, one just leaving the wrap and one just entering. Because of the angular offset between video and audio heads on the drum (Figure 21.36), the head switchover point for the latter is set by a second, delayed head flip-flop square wave triggered from the drum tacho-pulse. The FM carrier signals to and from the audio heads on the drum require their own rotary transformers, which may be concentric with those for the video signals under the drum or mounted above.

The frequencies used for VHS stereo hi-fi in depth-multiplex systems are 1.4 MHz (L) and 1.8 MHz (R); the FM modulation sidebands extend for about 250 kHz on each side of the (unsuppressed) carriers. Each audio head deals with both FM carriers throughout its 20 ms sweep of the tape—during replay, the carriers are separately intercepted by bandpass filters for processing in their own playback channels.

Frequency-multiplex audio

An alternative approach, used in Video-8 VCRs, to helical sound recording is to use the video heads to lay down on the tape an FM audio soundtrack. The baseband frequency response is limited to about 15 kHz. As Figure 21.26 earlier shows the audio FM carrier is based on 1.5 MHz and has a deviation-plus-side-band width of about 300 kHz. It is possible to insert this 'packet' between the outer skirts of the upper chrominance and lower luminance sidebands, permitting it to effectively become part of the video signal so far as the heads and tape are concerned. The FM audio signal carrier is added to the FM luminance signal in the recording amplifier, and laid on tape at a low level – some 18 dB below that of the luminance carrier. This suppression of sound carrier level helps prevent mutual interference between sound and vision channels, whose outer sidebands overlap to some degree.

The original AFM recording system made provision only for monophonic sound. The advent of hi-fi stereo recording in the competing VHS format led to the adoption of a stereo AFM plan for Video-8 formats, using a second sound carrier at 1.7 MHz. Unlike VHS hi-fi, however, it is not possible here to simply use one carrier for each channel because compatibility with mono AFM equipment must be maintained. The stereo-AFM system is illustrated in Figure 21.38. Incoming L and R audio signals are brought together in an adder, and the result halved in an attenuator to derive an $(L + R)/2$ signal for FM modulation and recording on tape at 1.5 MHz. This forms the compatible mono signal. The L and R record signals are also routed to a matrix in which they are subtracted to produce an $L - R$ signal, which is now halved to render an $(L - R)/2$ signal. It is this which is processed and frequency modulated onto the 1.7 MHz carrier. During replay, the recovered $L + R$ and $L - R$ signals are added and subtracted in separate matrices to derive L and R signals. Readers who are familiar with VHF-FM analogue stereo sound broadcast/reception techniques will recognise this 'stereo-difference' technique – in radio broadcasts, the difference signal is conveyed in a sub-carrier at +38 kHz.

In replay mode, the separation of the various off-tape signals in Figure 21.26 is carried out by four separate filters: a low-pass type with cut-off around 180 kHz for ATF tones; a low-bandpass filter centred on 732 kHz for colour-under signals; a narrow bandpass one tuned to 1.5 MHz for interception of the audio-FM carrier; and finally a high-pass (roll-on about 1.7 MHz) acceptor for the FM vision signal with its sidebands.

PCM audio

All newly developed hi-fi audio systems use digital encoding: examples are compact disc, the NICAM TV stereo plan, DAB radio broadcasting and digital TV. Although digital transmission and recording requires more bandwidth than other systems it has the advantages that the two-state signal is more robust than its analogue counterpart, and (so long as the quantising rate is high) is capable of superb S/N ratio and wider dynamic range.

The Video-8 format has provision for stereo PCM in addition to the monophonic AFM sound facility described above. The most expensive Video-8 camcorders and homebase VCRs are fitted up for PCM operation. It is impractical to give full details of the system in this book, and what follows is a basic outline of the technique used.

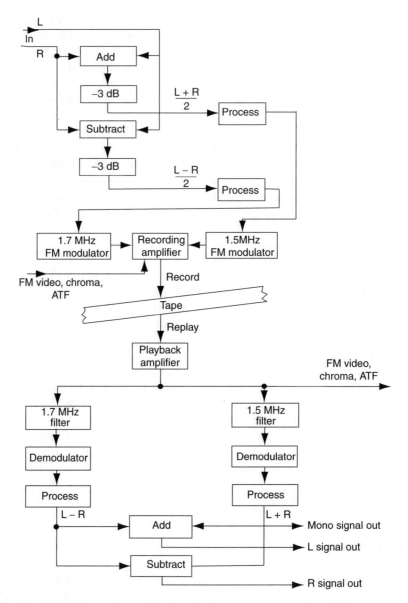

Figure 21.38 *Record and playback signal processing for AFM stereo audio in Video-8 format*

Figure 21.39 shows the functional blocks in the PCM sound processing. Initially, the audio signal for record is amplitude compressed in a compander. Next follows a quantisation process in which it is sampled at 2 fh, (31.25 kHz) to 10-bit resolution. Since the tape system here cannot cope with 10-bit data, a conversion is carried out to 8-bit: the process is a non-linear one, in which low-level signals are in effect given 10-bit (1024-level)

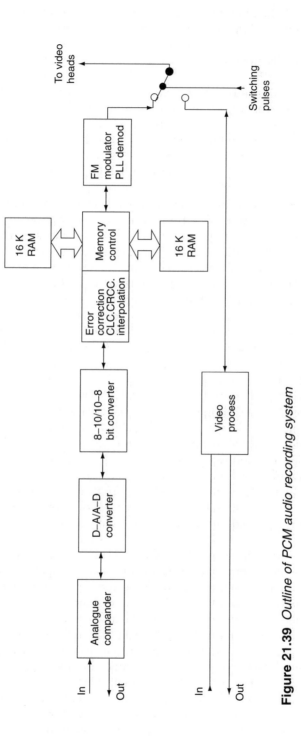

Figure 21.39 *Outline of PCM audio recording system*

descriptions, falling in three stages – as signal level increases—to 7-bit data (128-level) for the largest signal excursions. This has the effect (during playback) of concealing quantising noise in the loudest sound peaks, where they go virtually unnoticed; indeed the overall S/N ratio is subjectively equivalent to 90 dB.

Unless their transmission/recording media are very secure, digital data systems need error-correcting artifices to repair or conceal corruption of the data by noise and distortion. About 38.5% redundancy is imparted to the PCM audio data by the addition to it of further data in the form of a cyclic redundancy check code (CRCC) – used during playback for error detection and correction. The data rate is high, and the effect of a tape dropout would ordinarily blow a hole in the information stream; to prevent this the data is 'scattered' on tape according to a cross-interleave code (CIC), part of the Video-8 format. The effect of a tape dropout thus becomes distributed during replay, and the 'frayed edges' can be repaired by use of the CRCC and an additional parity check system.

The 8-bit data words are now temporarily stored in a pair of 16K RAM memories. Writing to memory is performed in real time. Readout from memory is much faster: all the data (which contains information on both stereo channels) is clocked out in less than 3 ms at 20 ms intervals. The effect of this 7:1 time compression is to push up the data-rate to about 2 Mbps, but to confine the audio transmission period to a small time-slot. As we shall see in a moment, a separate place is found on the video tape for this data-burst. Frequency shift keying (FSK) is used in which the bits are tone modulated at 2.9 MHz for 0 and 5.8 MHz for 1.

At this point, the PCM signal is ready to go onto tape. The switches on the right of Figure 21.39 change over once per field period, in synchronism with TV field rate and RAM readout. This feeds data to the video heads alternately, during a period when each is scanning a 'forward extension' of the helical vision track on tape – this is illustrated on the right of Figure 21.40. The conventional video tracks (which also contain AFM audio information, duplicating the PCM sound track) are recorded over 180° of the head track, but the head/tape angle is such that they occupy about 5.4 mm of the 8 mm tape width. The extra 30° or so of tape wrap shown on the right of Figure 21.40 is devoted to PCM recording: while one head is writing PCM data, the other (diametrically opposite on the drum) is recording the last lines of the TV picture at the top of the video tracks on the right of the diagram.

During playback, head switching ensures that PCM data read off the tape is routed to the audio section for the appropriate 30° scan/3 ms continuing to the left in Figure 21.39, the data, still in time-compressed form, is read into the same pair of 16K memories as was used during record.

Memory readout takes place in 'real time', expanding the data to give continuous data at a lower bit-rate – that at which the memories were loaded during record. Readout sequence is governed by the cross interference code (CIC) mentioned above in order to de-interleave the data, scattering and fragmenting errors in the process. In the memory control

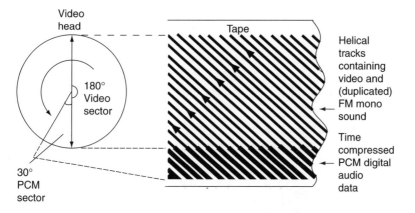

Figure 21.40 *The PCM data in recorded on a 'forward-extension' of the video tape tracks; an extra 30° of head rotation is reserved for this*

chip is also carried out error correction by means of the CRCC and parity checks mentioned earlier. For dropouts too severe to be repaired by these means, interpolation provides a 'patching' system which in the face of sustained and continued corruption, devolves to a PCM mute action, switching the audio output line back to the AFM sound track. If this is also corrupted, silence will ensue!

The 8-bit data now passes to the 10-bit conversion stage to make ready for D-A conversion. As is common in these designs, the record A-D converter is used for this, now switched to perform D-A operation. The analogue signal reconstituted at the D-A converter output is still in amplitude-compressed form, and is now expanded to full dynamic range in a logarithmic compander.

The complete VCR

Figure 21.41 shows a block diagram of a VCR. To avoid regular plugging and unplugging operations, the VCR is permanently connected 'in series' with the aerial lead to the TV receiver. When the machine is off or recording a programme other than the one being viewed, it is important that normal TV reception is not affected by the machine's presence, so a loop-through facility is provided in the aerial booster, a small RF amplifier which is permanently powered. Its modest gain cancels the losses incurred in the extra RF plugs, sockets and internal splitting of the RF signal within the VCR.

UHF tuner and IF amplifier

Most recordings made on a home VCR come via broadcast transmissions, so the machine needs a tuner and receiver built-in to select and demodulate broadcast programmes. An effective AGC circuit is provided to ensure

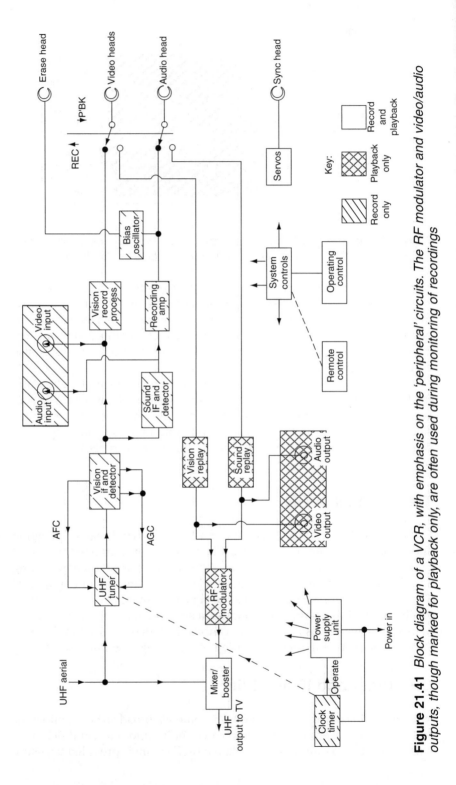

Figure 21.41 *Block diagram of a VCR, with emphasis on the 'peripheral' circuits. The RF modulator and video/audio outputs, though marked for playback only, are often used during monitoring of recordings*

a constant signal level to the recording section, and AFC (automatic frequency control, sometimes known as AFT, auto fine tuning) feedback maintains correct RF tuning. Some VCRs have auto-set routines, in which they tune themselves, working from a program in the control section, and set their own clock from teletext.

Sound

An inter-carrier IF amplifier and detector follow the vision detector to provide a sound signal for recording on a tape. After the detector, the sound recording processes follow audio cassette practice, with the audio signal being added to an HF bias source before recording on a longitudinal track on one edge of the tape as described earlier. The slow speed of the tape and narrow audio tape track do not make for ideal recording conditions, and noise reduction systems are commonly used to improve performance. Foremost among these is the Dolby system, which involves a form of non-linear and frequency-dependent pre-emphasis during the recording process, and complementary de-emphasis in replay.

Stereo machines use one of the hi-fi systems, VHS hi-fi, AFM-stereo or PCM. Virtually all hi-fi stereo VCRs have built-in NICAM decoders described in Chapter 17. Most pre-recorded cassettes offered for sale or rent also incorporate good stereo and surround sound tracks.

Bias oscillator

To provide a suitable a.c. recording bias for the 'longitudinal' audio signal and to generate an erasing signal to wipe out video tracks, a power oscillator is used. Usually consisting of a discrete transistor oscillator built round a feedback/driver transformer, it operates at about 60 kHz, and is used only in record mode. Besides the audio head, it powers a full-width erase head to wipe all signals, sound, picture and control track, off the tape on its journey towards the head drum during recording. As described in earlier chapters, no recording bias signal is required for the video heads as this function is performed for luminance and chrominance by the constant-level FM luminance recording signal.

RF modulator

When the machine is playing back, a TV signal at baseband is produced, and this cannot be applied to a TV receiver whose only signal input facility is via its aerial socket. To cater for this, an RF modulator is provided within the VCR, working as a tiny TV transmitter. Baseband audio and video signals are applied to its input for modulation onto a UHF carrier, whose frequency is chosen to fall into a convenient gap in the broadcast spectrum. A rear-access preset control swings this frequency to avoid beat effects with other RF signals in the vicinity. VCR modulators are normally programmable for any channel in the broadcast band. Output level is set to be 1–3 mV (the optimum for a TV receiver), and this modulated RF carrier signal is added to the booster output for application to the VCR's RF output socket.

During record, regardless of the signal source which may be off-air, TV camera, STB, cable, another VCR or whatever, the signal being recorded on tape is applied to the RF modulator so that it may be monitored on the TV set if desired. This is called the E–E (electronics to electronics) mode to distinguish it from off-tape playback mode. The E–E signal is taken off from the record electronics as late as possible, and several of the record and playback circuits (such as the luminance/chrominance adding stage) are usually included in the loop.

Clock back-up

Although most types of EEPROM can retain stored data (for 10 years or more) without the need for a sustaining voltage, it is necessary to keep the clock oscillator—but not its display—ticking over during power cuts and while the VCR is temporarily disconnected from the mains. It saves having to reset the time, and retains stored timer-recording data. Clock back-up supply comes from a small battery or electrically large capacitor which becomes isolated from the main +5 or +3.3 V supply by a diode when mains power is lost, and can sustain the oscillator and timer-programme memory for a period varying from a few minutes to several hours, depending on the design of the machine.

Video tape formats

There are four major analogue home video tape formats in the consumer field: VHS, S-VHS, Video-8 and Hi-8. VHS owes its origin to JVC, while the Video-8 formats are later developments, using more advanced features. Over the years, the competing formats have shaken down into well-defined categories. Video-8 is favoured for camcorder use, while VHS is the system for general-purpose homedeck use, and virtually the only one in which pre-recorded cassettes are produced for sale and rental. The high-band variants Hi-8 and S-VHS find a relatively small market, even though they are capable of producing better pictures than the others.

VHS stands for video home system: this format was designed by the JVC company in Japan. It was released in Europe in 1978. A clever adaptation of the standard tape and cassette package is VHS-C (compact) which permits a small and light VHS camcorder. The cassette is a small

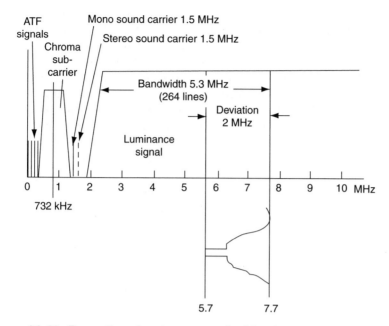

Figure 21.42 *Recording signal spectrum for Hi-8 format*

(92 mm × 59 mm × 23 mm) housing containing up to one hour's worth of standard 12.7 mm VHS tape. It fits a small camcorder weighing less than 1 kg, and incorporating a small head drum, thin direct-drive motors and a solid-state image sensor. Back at home the small cassette is loaded, piggyback style, into a normal-size adaptor shell for replay or editing in the standard VHS machine.

Super-VHS is an advanced 'high-band' variant of the established format. It uses a high FM carrier, with deviation from 5.4 to 7 MHz. To permit this, an advanced video tape formula is used in a cassette of conventional size, shape and running time. A very small video head gap, and 'fine grain' magnetic tape permits a baseband video frequency response approaching 5 MHz, and on-screen resolution better than 400 lines. Noise ratio is also better than that of the low-band formats.

The *Video-8* format uses 8 mm-wide tape, from which it gets its name. Unlike the others, this system was developed by a consortium of companies and accepted for use by over 127 of the world's major audio/video manufacturers. It was the first domestic video system to use digital audio recording. Its other advantages are a small cassette (92 mm × 58 mm × 20 mm), facilitating miniaturisation of the equipment, be it portable or home-base type; a flying erase head for good edits, also featured in some VHS equipment; and the exploitation of new tape and head materials and techniques for better performance. Although offered in home-base form for tabletop use, Video-8 is seen mainly as a camcorder format, whose primary advantages are excellent sound, light weight and high performance.

As with VHS, development of the basic Video-8 format led to the production of a high-band variant, with picture resolution exceeding 400 lines under favourable circumstances. The *Hi-8* variant, has higher FM record frequencies and a greater sideband spread than Video-8, as shown in Figure 21.42. It is used with metal evaporated (ME) and high-grade metal powder (MP) tapes. In camcorders, which is the form in which nearly all Hi-8 hardware is produced, the lens and CCD image sensor are upgraded to accommodate the higher resolution capability, and the bandwidth of the luminance path widened to suit. As with S-VHS, provision is made for separate handling of Y and C video components throughout the chain, and for their passage through S-links to suitably equipped TVs, monitors and copy-VCRs. In other respects, Hi-8 is similar to Video-8 format, with the same cassette dimensions, sound systems, tracking and transport arrangements.

22 Digital recording and camcorder

Digital recording may be carried out using a magnetic tape, an optical disk (CD or DVD) or a hard disk. In all cases, the analogue video and audio are sampled and compressed to form digital data bitstreams that are stored on the recording medium. Recording on optical discs was fully explained in Chapter 20. We shall therefore start with digital magnetic tape recording.

Digital video recording on a magnetic tape was launched in 1994 under the name of DV employing a tape cassette known as *digital video cassette* (*DVC*) and its smaller form factor *MiniDV*. The DV specification defines both the type of compression and the tape-recording format. Though DV quickly became the standard for home and semi-professional recording, it spawned a number of variants, most notably Sony's *DVCAM* and Panasonic *DVCPRO* formats targeted at the professional end of the market. Sony's consumer Digital8 format is another variant, which is similar to DV but recorded on Hi8 tape. A high-definition version of DV has also been developed under the name of HDV. While it uses the DV and MiniDV tape form factor, it employs the full MPEG-2 compression toolkit including inter-frame compression which is not used in the standard definition DV format.

Other formats have also been introduced including the S-VHS and the D-VHS but they did not get general acceptance by the public, especially as DVD recording started to gather pace. S-VHS improves VHS's luminance resolution by boosting the frequency deviation of the luminance carrier. D-VHS (D for Data) is a digital video format developed by JVC, in collaboration with Hitachi, Matsushita and Philips. The main departure from S-VHS D-VHS is that D-VHS records the packetised MPEG-2 transport stream packets, the same data format used for digital television applications.

The advantages of a digital video recording system are several. The format itself affords higher picture resolution, better in practice than a colour TV broadcast. The digital signal can be transferred to another (digital) tape without any loss of quality, and edited and manipulated likewise, wholly in the digital realm. Both video and audio signals can be 'dubbed' because they occupy different areas on the tape. The colour-bleed and -smear which arises from low chroma bandwidth in analogue tape systems disappears, as do the Y/C band-sharing artefacts due to PAL and other colour-encoding systems. Because the picture is built up 'from scratch' as it were, in the playback electronic system, there is no potential for noise or interference on picture or sound, though of course when the digital bitstream from the tape gets too corrupt the sound disappears and the picture becomes progressively 'blocky'; intermittently 'frozen' wholly or in segments; and then disappears altogether, just like the 'digital cliff' phenomenon in digital television broadcasting.

Digital video (DV) uses two types of cassettes: the *L-size* (120 ×90 × 12 mm) and *S-size* (65 × 48 × 12 mm) known as MiniDV. Both are much smaller than an audio or Video-8 type. These are capable of storing 11 GB of data equivalent to 1 h of high-quality sound and vision in SP mode, or 1.5 h in LP mode on 6.35-mm-wide tape, with excellent sound quality and 500-line picture definition, a good match for large-screen home-TV sets. We shall look more closely at the tape, cassette and system specifications later.

Signal processing: record

DV makes use of the same MPEG-2 redundancy removal techniques as those used for digital video broadcasting described in Chapter 4. Recall that MPEG-2 included two main data-compression techniques: spatial or intra-frame and temporal or inter-frame. In DV applications, only intra-frame compression is used resulting in a much lower, but adequate compression ratio of about 5:1. By using intra-frame compression only, we end up with one type of frame, namely the I-frame which is amenable to editing and 'trick-play' operations.

Figure 22.1 outlines the initial processes in the DV record system. Sampling rates are 13.5 MHz for luminance and 6.75 MHz for chroma. 8-bit quantisation is used on Y, Cb and Cr signals, making an initial data rate of 214 Mbps: it is reduced to 25 Mbps (25.146 Mbps to be exact) for recording by the use of discrete cosine transform (DCT) spatial redundancy coding and variable-length coding (VLC). It should be noted here that for PAL a 4:2:0 sampling structure is used while for NTSC, 4:1:1 is used. This does not

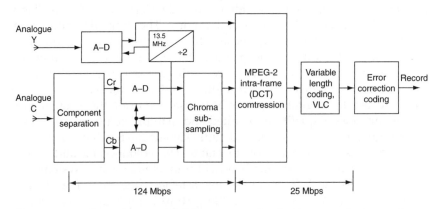

Figure 22.1 *Sampling and DCT compression in the record stage of a DVC tape*

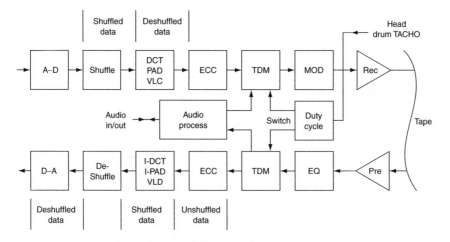

Figure 22.2 *DVC data processing system throughout*

affect the bit rate as was explained in Chapter 3. The bit-reduction process is the same as that used for DVB being based on blocks and macro-blocks – indeed these blocks can sometimes be seen during replay if the data becomes corrupt, e.g. by mistracking.

The record and replay system for DV is shown in block diagram form in Figure 22.2. Following A-D conversion in the record mode, the data is 'shuffled' by bringing together 16 macroblock samples from widely separated picture areas and assembling them in what appears to be a random sequence to reduce statistical distortion due to compression ratio bias, and to give best-possible 'trick' and still-frame reproduction during playback. Later in the chain, the data is de-shuffled to regain correct order for recording on tape. Next comes the data-compression stage, and then ECC, a form of Reed–Solomon protection, primarily to combat the effects of tape mistracking and dropout. The video data is then multiplexed with audio data and other control and housekeeping data by the *time-division multiplex* (*TDM*) switch. The video data only occupies about one-half of the duration of a head scan. The remainder is used for the audio and control/housekeeping data streams. Finally, the data is modulated and fed via a drive amplifier to the recording heads.

DV audio processing

As for the audio channel, the user can select either 16 or 12-bit sound mode. The former containing high-quality stereo signals, while the latter offers four separate channels suitable for dubbing. The sound data is sampled and packetised, stored and released to the modulator in time with the TDM switch changeover, itself synchronised by the head-drum PG/tacho pulse. The bit rate for the audio bitstream is fixed at 1.25 Mbps and that for the control/housekeeping at 8.716 Mbps. Given that the video bit rate is 24.146 Mbps, the total bitrate for the DV recording is 35.382 Mbps (normally referred to as 36 Mbps).

Signal processing: replay

The first stages in DV replay are signal amplification and equalisation, the latter not to give best bandwidth and noise performance as in analogue tape systems, but to optimise the demodulator's discrimination between 0 and 1 symbols. Thus furbished, the off-tape signals are gated by the TDM switch to the audio or video stages in synchronism with their readout. The video data is checked and – where possible – errors are repaired in the ECC stage; any data beyond repair is deleted and replaced by good data from the same position in the previous scanning line, a process akin to the action of a dropout compensator in an analogue VCR.

Now the data is shuffled once more, using the same look-up table as was used during the record phase, for its passage to the inverse DCT (I-DCT) where data decompression is carried out. After de-shuffling the picture is built up, frame-by-frame and stored in a SRAM memory chip: it is read from here into the D-A converter whose output provides the reconstituted video signal. The analogue audio signal is reconstituted likewise, using the inverse processes to those applied during the record phase. Both video and audio can be conveyed from the machine in digital form (IEE 1392/Firewire/i-Link) into other digital equipment (see Chapter 25).

DV track configuration

DV head drums have a diameter of 21.7 mm, and rotate at 9000 rpm to give a writing speed of about 10 m/s. Two video heads are fitted, with azimuth angles of ±20°: this amounts to a 40° offset between them, with a correspondingly low level of cross talk pick-up from adjacent tracks. The heads are very narrow to give a track pitch of just 10 μm in SP, 6.67 μm in LP mode, corresponding to tape speeds of 18.8 and 12.6 mm/s respectively. The tape's helical wrap embraces 186° of the drum's periphery; the signal envelope is clipped at the 180° point by the head switch.

A complete picture frame (with corresponding sound signals) takes 12 tracks for PAL and 10 tracks for NTSC. Figure 22.3 shows the track arrangement for NTSC. Each 32.89-mm-long track consists of four segments, written sequentially by one head as it scans across the tape at an angle of 9.167°. First comes the insert and tracking information (ITI) segment, containing reference data for track height/positioning and to provide a tracking signal for audio and video dubbing and inserts. The ITI data is used purely for 'housekeeping' purposes by the VCR itself during playback. The next two segments are the audio and video data we have already met; in the video sector are recorded additional data such as date and time of recording, camera settings,

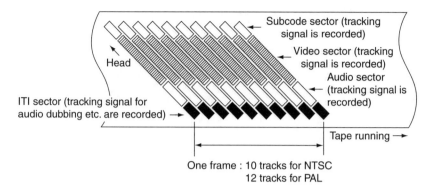

Figure 22.3 *Track layout on DVC tape: the physically separate video and audio sectors permit dubbing of either*

widescreen mode flag, etc. These can be displayed during replay if required. The final sector in the tape track shown in Figure 22.3 contains sub-code data, primarily a time code which can be used to identify every frame, useful for editing. Also in the sub-code are cue flags for location of each new shot, and for marking still-frame/photo shots when the camera is set to that mode.

DV tracking system

As with the Video-8 format described in the previous chapter, automatic head-tracking during replay is carried out using '*tones*' recorded along the tracks. In DV the system is simpler than for Video-8, as Figure 22.4 shows. Here two tone frequencies are used, 465 and 697 KHz, alternating with no recorded tone in a four-track sequence. When scanning a track with no pilot tone the off-tape F1 and F2 cross talk signals are made equal by the tracking-servo system, signifying a track-centre trajectory of the head; this condition corresponds to maximum pilot-tone amplitude during replay of an F1 or F2 track.

System parameters

The specifications for DVC-SD format are set out in Table 22.1 for the electrical and coding sections and in Table 22.2 for the mechanical and tape-track characteristics.

DV tape and cassette

A new type of tape was developed for DV recording, having five layers as shown in Figure 22.5a. It is 6.35 mm wide and seven microns thick. The third layer is the magnetic one, having two high coercively evaporated

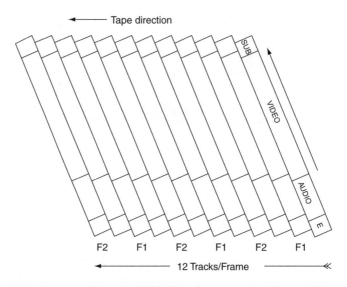

Figure 22.4 *Auto-tracking for DVC. Two frequencies F1 and F2 are used as shown at the bottom*

Table 22.1 *Data and coding specifications for DV*

Error-correction code	Reed–Solomon code
Recording clock (Fclk)	41.85 MHz
Data transfer rate	Video: 24.9 Mbps
	Audio: 1.55 Mbps
Video signal	Digital component recording
Resolution	NTSC: 720 × 480, 4:1:1
	PAL: 720 × 576, 4:0:2
Sampling scheme	NTSC: 4:1:1
	PAL: 4:0:2
Video sampling frequency	Y: 13.5 MHz
	C_R and C_B: 6.75 MHz
Digital video compression system	DCT intra-frame; variable-length coding (VLC)
Audio signal recording system	PCM digital recording
Audio Channels	2 ch, 4 ch
Audio Sampling frequency	48; 44.1; 32 kHz

metal coats for good reliability, high output and low noise. The back coating has very low surface friction, while the overcoat hard carbon layer is durable, flexible and 'tough' in terms of abrasion resistance.

Mini-DV cassettes come in 30 and 60 min types, DV-M30 and DV-M60 respectively. They are capable of 50% more running time in the LP mode,

Table 22.2 *Mechanical and tape-track specifications for DVC*

Item		525/60 system	625/50 system
	Recording system	Rotary 2 heads azimuth recording	
	Drum rotation speed (rps)	150/1.001	150
	Drum diameter (mm)	21.7	
	Writing speed (m/s)	10.202/1.001	10.202
Tp	Track pitch (μm)	10.00	
Ts	Tape speed (mm/s)	18.831/1.001	18.831
Qr	Track angle (deg)	9.1668	
Lr	Effective track length (mm)	32.890	
Qe	Effective wrap angle (deg.)	174	
Wt	Tape width (mm)	6.350	
He	Effective area lower edge (mm)	0.560	

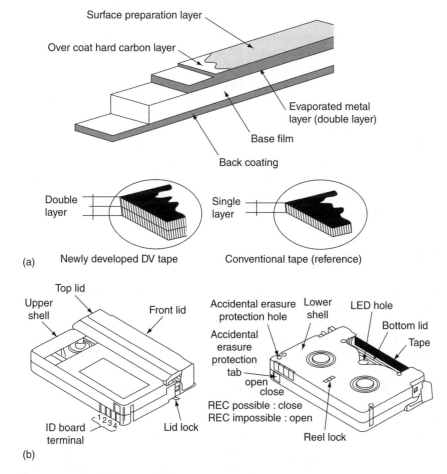

Figure 22.5 *DVC tape (a) and cassette (b)*

whose picture is just as good, but whose tracking tolerance is 'tighter', with a greater risk of data errors. The mini-DV cassette package measures $66 \times 48 \times 12.2$ mm, and is illustrated in Figure 22.5b. ID terminals 1–4 may merely connect to internal resistors to indicate to the deck's syscon details such as tape type, capacity and grade. Some MiniDV cassettes have a small memory chip known as *memory in cassette (MIC)*. The chip is an I_2C-controlled EEPROM that functions as serial data ports for read and write. The non-volatile memory chip typically 1, 8 or 16 kB capacity stores record dates and photo time/date (stills) along with DV format data like ID, mode, and size and tape type.

Digital-8 format

The Video-8 and Hi-8 analogue formats are much better suited to portable (camcorder) operation than VHS ones; the smaller cassette makes for a smaller and lighter machine, even lighter than those *VHS-C* (C for compact) cassette. The Video-8 formats were developed specifically for portability, while VHS was adapted from the basic large cassette/home deck concept developed by JVC in the 1970s. In an attempt to maintain their market lead, and to provide a 'bridge' between analogue and digital systems for existing users, the inventor of Video-8, Sony, introduced a new 'dual' D-8 (Digital 8) format. D-8 decks (virtually all of them in camcorder form) can replay analogue Video-8 and Hi-8 recordings as well as their own digital recordings, using the same head pair, drum and tape wrap, and the same type of tape. During playback, the type of signal on tape is automatically detected and the replay circuits switched accordingly. If required an analogue signal off tape may be made available in digital form, using an A-D converter inside the camcorder.

D-8 recording system

For compatibility the Digital-8 head drum must necessarily conform to the 40 mm diameter of the Video-8 format. Figure 22.6 compares D-8 and DVC-format tape-track configurations. While the PAL–DVC system records each TV frame in 12 tracks, D-8 lays down a frame in six tracks, each divided into two consecutive sub-tracks. To achieve this the D-8 drum speed is 4500 rpm. Figure 22.7 shows at the top a comparison between the NTSC and PAL variants of analogue Video-8 format, with one track per frame in each case: for NTSC the tape speed is slower and the tracks narrower than for PAL. The lower section of the diagram illustrates that tape linear speed is virtually the same in Digital 8 for both TV systems, with five (NTSC) or six (PAL) tracks per frame, all tracks being 16.34 microns wide. Both trace 150 tracks/s, 30 five-track frames for NTSC and 25 six-track frames for PAL. Digital-8 video heads have a width of 19 microns, a little wider than the D-8 track pitch (each

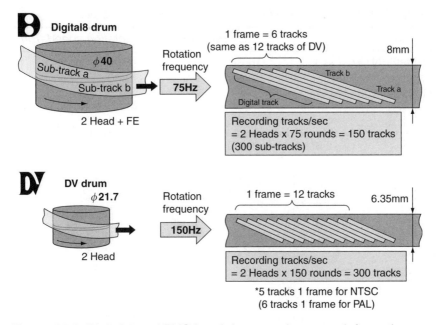

Figure 22.6 *Digital-8 and DVC head drums and tape-track formations compared*

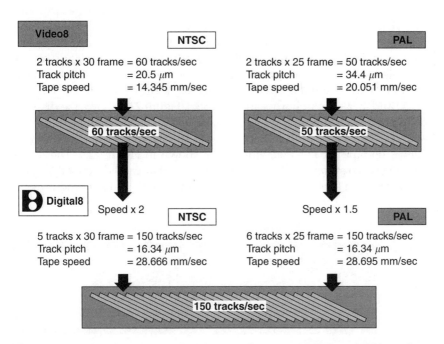

Figure 22.7 *Analogue and digital 8 mm formats: NYSC and PAL variants*

Table 22.3 *Track-scanning in three formats using 8 mm tape*

Format	Track Pitch (μm)	Rotation speed (rpm)
Digital8 (NTSC/PAL)	16.34	4500
Hi8/Standard8 (NTSC SP)	20.5	1800
Hi/Standard8 (PAL SP)	34.4	1500

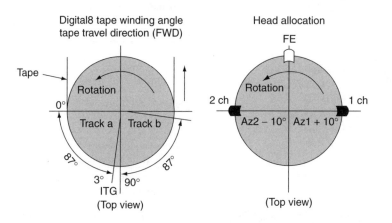

Figure 22.8 *Head disposition and azimuth cuts on Digital-8 head drum*

new head sweep erases the excess of its predecessor during record), and somewhat narrower than the Video 8/Hi-8 tracks of any analogue tape which may be presented. In the latter case, the heads sweep along the centre of the pre-recorded tracks, the signal they produce displaying slightly higher S/N ratio than would come from a correct-width head. Table 22.3 shows the track pitch for D-8 and the variants of video-8: the drum is designed to rotate at 1500 and 4500 rpm in European D-8 decks. Figure 22.8 shows the disposition of the three heads on the 40 mm-diameter drum, and their azimuth settings, ±10° for compatibility with Video-8 and Hi-8 recordings. The track patterns on tape for D-8 and Video-8 are shown in Figure 22.9. The track-following servo system for D-8 is the same as that already described for DV: 465 and 697 kHz tones are recorded in alternation with no-tone head-sweeps

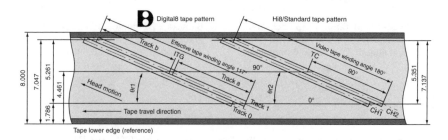

Figure 22.9 *Track patterns for D-8 and Video-8 formats*

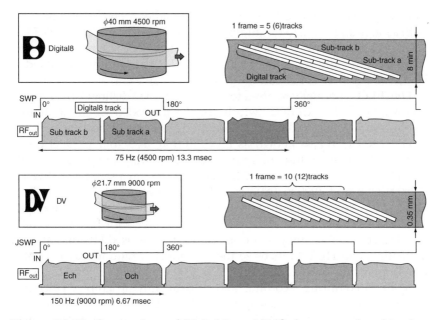

Figure 22.10 *Comparison of Digital-8 and DVC drum speed and track readout. Figures in brackets are for PAL/625*

in a four-track sequence, ATF error being detected only in playback of no-tone tracks. When replaying an analogue recording the servo system switches to Video-8 ATF operation as described in Chapter 21.

Because of the two sub-tracks per head sweep of the D-8 tape pattern, and the fact that drum rotational speed is half that of a DVC drum, the off-tape RF signals during digital replay are identical to those of DVC as Figure 22.10 shows.

Other DV formats: DVCAM

Sony's DVCAM is a semi-professional variant of the DV standard that uses the same cassettes as DV and MiniDV, but transports the tape 50% faster. This leads to a higher track width of 15 µm. Video and audio encoding is the same as regular DV, namely MPEG-2 intra-frame and VLC. However, the greater track width lowers the chances of dropout errors. All DVCAM recorders and cameras can play back DV material. DVCAM tapes have a shorter recording time than DV of almost third.

DVCPRO (professional)

Specifically created by Panasonic for the professional end of the market, the DVCPRO family is directed towards electronic news gathering (ENG) use. It provides better linear editing capabilities and robustness. It has a track width of 18 μm and Metal Particle instead of Metal Evaporated tape with a longitudinal analogue audio cue track. Audio is only available in the 16 bit/48 kHz variant and there is no EP mode. *DVCPRO* uses 4:1:1 colour sampling structure for both NTSC and PAL. Apart from that, standard DVCPRO is otherwise identical to DV.

DVCPRO is normally referred to as *DVCPRO25* to distinguish it from the high-value EGN *DVCPRO50* which doubles the coded video bitrate from 25 Mb/s to 50 Mb/s, and uses 4:2:2 sampling structure instead of 4:1:1.

DVCPRO HD, also known as *DVCPRO100*, uses four parallel codecs and a coded video bitrate of approximately 100 Mbps, depending on the format. DVCPRO HD is also 4:2:2. *DVCPRO HD* downsamples native HD (1280 × 720p and 1920 × 1080i) signals to a lower resolution as follows: 1280 × 720p downsampled to 960 × 720p and 1920 × 1080i downsampled to 1280 × 1080i for NTSC frame rate and 1440 × 1080i for PAL frame rates. This is a common technique, utilised in most tape-based HD formats such as *HDCam* and HDV to increase the compression ratio to approximately 6.8. To maintain compatibility, DVCPRO100 equipment internally down-samples video during recording, and subsequently upsamples video during playback. A camcorder using a special variable-frame rate (from 4 to 60 frame/s) variant of DVCPRO HD called *VariCam* is also available. All these variants are backward compatible but not forward compatible. There is also a *DVCPRO HD EX* format, which runs the tape at slower speed, resulting in twice as long recording times.

DVCPRO cassettes are always labelled with a pair of run times, the smaller of the two being the capacity for DVCPRO50. The colour of the lid indicates the format: DVCPRO tapes have a yellow lid, longer "L" tapes made especially for DVCPRO50 have a blue lid and DVCPRO HD tapes have a red lid. The formulation of the tape is the same, and the tapes are interchangeable between formats. The running time of each tape is 1 × for DVCPRO, 2 × for DVCPRO 50, 2 × for DVCPRO HD EX and 4 × for DVCPRO HD, since the tape speed changes between formats. Thus a tape made 126 min for DVCPRO will last approximately 32 min in DVCPRO HD.

A more recent high-definition magnetic tape format is high-definition video (HDV) introduced in 2003. HDV is high-definition, wide screen version of the DV magnetic tape format. Like DV, HDV recording stores data on the same DV and MiniDV tapes as SD camcorders.

The HDV codec is based on MPEG-2 video compression employing both intra-frame (spatial DCT) and inter-frame (temporal) techniques. As a result of the inter-frame compression, HDV frames vary in size depending on whether they are I, P or B frames. HDV is native 16:9 using 4:2:0 sampling, recording YCbCr frames in 1280 × 720p, 1980 × 1080i or 1980 × 1080p resolutions at frame rates of 60i 30p, 50i or 25p. All HDV variants record to existing standard DV format DVCs, the most popular form factor of which is the Mini-DV shell. The transport stream interface conforms to IEEE 1394 (FireWire). The HDV specifications are listed in Table 22.4

Table 22.4 *High definition video (HDV) Specifications*

Storage medium	Same as DV format, DV or MiniDV cassette
Video signal	720/25/30/50/60 lines/Hz progressive or 1,080/50/60 lines/Hz interlaced
Aspect ratio	16:9
Video compression	MPEG-2 video (profile and level MP@H-14)
Luminance sampling frequency	75.25 MHz with progressive scanning, 55.7 MHz with interlaced scanning
Video sampling format	4:2:0
Video quantisation	8 bits for both luminance and chrominance
Video bit rate after compression	About 19 Mbit/s with progressive scanning, 25 Mbit/s with interlaced scanning
Audio compression	MPEG-1 Audio Layer II
Audio sampling frequency	48 kHz
Audio quantisation	16 bits
Audio bit rate after compression	384 kB/s
Audio mode	Two-channel audio
System data format	MPEG-2 system
Stream type	Transport stream with progressive scanning, packetised elementary stream with interlaced scanning
Stream interface	IEEE 1394 (MPEG-2TS)

Hard disk recording

Data recording on a *hard disk drive* (HDD) is carried out in the same way as recording on an electromagnetic tape in which a thin magnetic surface is magnetised by a flux produced by an electromagnetic head gap. The disk rotates at a constant speed and data in the form of 0 and 1 bits are written along concentric tracks as shown in Figure 22.11. To retrieve the data from the disk, a read head is used and the process is reversed. Early hard disks used a single read/write head. Later, *metal-in-Gap* (MIG) heads were used, and today thin film heads are common. With these later technologies, the read and write head are separate units, but are placed on the same actuator arm.

The hard disk consists of a number of rigid disks called platters made of aluminium covered with a thin layer of magnetic coating and placed in a sealed container. A separate read/write heads are provided for each active

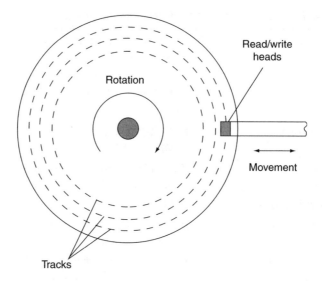

Figure 22.11 *Hard disk platter*

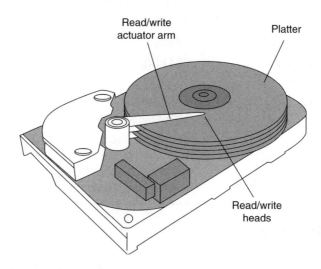

Figure 22.12 *The hard disk drive, HDD*

platter surface as shown in Figure 22.12. Each active surface is formatted into a number of tracks known as cylinders and each cylinder is divided into a number of 512-Byte sectors. The disks rotate at a very high angular speed of up to 10,800 rpm and can have a capacity of over 300 GB. The sealed container keeps dust and other contamination away from the surface allowing the read-write heads to float on a cushion of air very close to the disk surface which helps to increase the potential size of the disks. The hard disk could be part of a personal computer for general use or dedicated to a single task as in the case of Sky+, *personal video recorders* (*PVRs*) and *twin tuner terrestrial DTV receivers*. In all cases, the video and audio must be digitised and compressed before it is stored on the disk.

Digital video recorder

The first consumer *digital video recorders* (*DVRs*) were launched at the 1999 Consumer Electronics Show in Las Vegas. Apart from improved picture quality, DVRs provide advanced features in terms of time shifting, pausing live TV, instant replay of interesting scenes and chasing playback where a recording can be viewed before it has been completed. Most DVRs use the MPEG-2 for encoding analogue video signals.

Video signals may be available in a digital format directly from a digital source via an HDMI or a DVI port. In such cases, the MPEG packets are recorded directly on the hard disk. Alternatively, where the video signal is analogue RGB or YC_RC_B signals, analogue-to-digital conversion and data compression must first be performed by the DVR before recording can take place.

Many satellite, cable and terrestrial DTV decoder manufacturers are incorporating an in-built DVR in the form of a hard disk into their set-top box such as the Sky + box and the twin-tuner terrestrial box. In these cases, encoding in not necessary as the signal is already digitally encoded. The in-built DVR simply stores the digital stream directly to the disk. With this arrangement, interactive functions can continue to be provided on recorded programs. Today, Satellite DTV set-top boxes feature two separate LNB inputs allowing two separate recordings to take place simultaneously while a recorded program is being played back. The same facility is also available in twin-tuner terrestrial set-top boxes without the need for a dual aerial sockets.

Recording files

Data (video, audio, etc.) are recorded on the had disk in the form of files. The interleaved audio and video data forms a container (*AVI*) file with extension .avi which can be written to the hard disk. A container is a computer file format that can contain various types of compressed data. Simpler container formats can contain different types of audio data, while more advanced container formats can support audio, video, subtitles, chapters, and control/information data known as *meta-data* or *tags* along with the synchronisation information needed to play back the various streams. Windows Media Video (WMV) is another *VC-1/MPEG-2* derived codec specification which may be combined with the *Windows Media Audio* (*WMA*) into an AVI with extension *.wmv* or *Advanced Systems Format* (*ASF*) container (extension *.asf*) or a *Matroska* container (extension *.mkv*). WMV is a popular codec for video streaming on the internet together with its high-definition companion, *WMV HD*.

Audio and video capture

Audio and Video capture is a process of storing AV on the hard disk of a personal computer using AV capture cards. AV capture cards are normally PCI card or *AGP* (Accelerated or Advanced Graphic port) graphic card that allow a computer to capture composite, component video or directly as UHF modulated signal from a terrestrial or a satellite antenna. Video capture cards must be distinguished from video editing cards; the latter have dedicated hardware for processing video beyond the analogue-to-digital conversion.

Video capture starts with a fast A-D converter, providing YUV 4:2:2 data. Its sampling rate (hence final picture quality) can usually be selected in software, primarily to suit the resolution available from the source; and the computer/HDD capabilities. Typical resolution settings are 384×288 pixels (low band/VHS/Video-8 standard) and 640×480 (NTSC) and 768×576 (PAL), corresponding to standard television with zoom and other facilities. High-definition capture cards are also available with $1920 \times 1080i$ picture resolutions. Following sampling and quantisation, data compression takes place to reduce the bit-rate; this is essential to reduce the amount of hard disk space for the program. Most analogue capture cards work to Motion-JPEG (M-JPEG) standard, with which the data rate can be reduced by a factor of up to 100, though of course there's a trade-off between compression rate and moving-image quality. At compression ratios up to 5:1 the results are very good; between 5:1 and 15:1 some deterioration in quality is perceptible, depending on motion rates and picture 'busyness'. For compression ratios higher than 20:1, MPEG-2 algorithms are used: Main Profile@ Main Level MP@ML for SDTV and Main Profile @ High Level (MP@HL) for HDTV. The compressed video data is now applied to the computer's data bus, along with audio data, ideally produced and processed on the same card to avoid lip-synchronisation problems.

Editing

The editing process is conducted by a special software program with which the user's requirements are specified in terms of trimming and combining clips, adding transitions, effects, titles and so on. Sophisticated editing programs offer many effects and facilities: *batch capture*, animation, *'morphing'*, *'paints'*, chroma-key, filters, image re-sizing etc. The final picture/sound programme schedule is built up on a timeline, a series of horizontal on-screen rows, each representing a video, audio or effects track, and progressing in time from left to right. This timeline can be scaled as required, ranging from the entire required 'movie' to just a few frames, seconds or minutes. All these instructions are stored with frame/timecode markings, but not yet executed. Editing is followed by a process called rendering.

Rendering

The *rendering* process is a long one, during which all the instructions are carried out on a frame-by-frame basis, pulling sound and vision data off the hard disc, processing it as required, and then progressively reassembling it back onto the HDD, which ideally is a separate one from that fitted as a part of the computer's basic system. The programme-time capacity of a disk depends on the capture resolution and data-compression rate chosen by the user in software. During replay the edited material comes off the hard disk as an .avi or .wmv files for de-interleaving, reverse DCT processing and data expansion: these take place on the capture card (codec) where the data is finally D-A converted and produced as composite video or S-video and baseband audio, a process called printing to video.

DV Editing

The A-D and data-processing involved in off-line computer editing of analogue video necessarily introduces degradation of picture and sound. Where the footage is captured in a camcorder to DV tape the data-compression ratio is fixed by the system at 5:1, with an excellent, tailor-made data-reduction algorithm. So long as this data is fully preserved during the editing and storage phases, it becomes possible to carry out 'transparent' editing, entirely in the digital realm, and with no degradation at all of picture and sound quality, no matter how many generations of dubbing takes place. This is achieved with a DV capture card having a Firewire/i-link/IEE 1394 input/output port. It contains no A-D/D-A converters or data-compression systems, acting merely as a buffer/interface between the serial input line and the computer's data buses. Of course it carries out many other functions in the realm of software/instruction implementation. Once editing is complete, it is exported back to the DV tape in a digital format using a Firewire facility.

Recordable DVD

The original DVD specification related to the read-only (non-Recordable) DVD-ROM explained in chapter 21. Recording on a DVD requires a blank disc capable of being written on. Today, there are several competing recordable formats for DVD applications. They fall into two groups:

- *Recordable* (write once) format: *DVD-R* and *DVD + R*, R for recordable and
- *Re-writable* or *re-recordable* (write several times) format: *DVD-RAM*, *DVD-RW* and *DVD +RW*.

DVD-R, DVD-RW and DVD-RAM formats are supported by Panasonic, Toshiba, Apple Computer, Hitachi, NEC, Pioneer, Samsung and Sharp. They are also supported by the DVD Forum. DVD+R and DVD+RW formats on the other hand are supported by Philips, Sony, Hewlett-Packard, Dell, Ricoh, Yamaha and others.

The various recording formats have different features: DVD-R is compatible with over 80% of all DVD players and DVD-ROM drives. DVD-RW was the first DVD recording format released that is compatible with almost 70% of all DVD players and DVD-ROM drives. It supports single side 4.7 GB and double-side 9.4 GB disc capacities. DVD +R is compatible with over 80% of all DVD players and DVD-ROM drives. DVD+RW has better features than DVD-R such as lossless linking and both constant angular velocity (CAV) and constant linear velocity (CLV) writing. Its greatest advantage is that it is compatible with almost all DVD players and DVD-ROM drives. It supports single-side 4.7 GB and double-side 9.4 GB disc capacities. DVD-RAM has the best recording features but unfortunately it is incompatible with most DVD players and DVD-ROM drives. DVD-RAM discs are used more as a removable storage device than a recording medium for audio/video information.

Dual layer (or double-layer) technology is supported by a range of manufacturers including Dell, HP, Verbatim, Philips, Sony, Yamaha and others. Also known as DVD + R DL or DVD-R DL, they provide two individual recordable layers on a single-sided DVD disc. They have a capacity of 8 GB and normally referred to as DVD-R9.

There are two versions of DVD-R format: DVD-R(G) for General and the DVD-R(A) for Authoring. It is also possible to create a hybrid disc which is partially read only and partially recordable, sometimes known as DVD-PROM. This would normally be a disc with two layers, one of which is a read only and the other a recordable part.

While DVD-ROM disc such as a DVD-video disc have pits stamped permanently onto its surface, writable or recordable discs use other techniques to produce the same effect of a pit. In the case of a write-once DVD discs, a *photosensitive dye* is used to cover a reflective metallic surface. When the dye is heated by a high-power (6–12 mW) pulsating laser beam, it becomes darker and hence less transparent resulting in a weaker reflection. A pulsating laser is necessary to avoid overheating the dye creating oversized 'pits'. The disc is manufactured with a *wobbled groove* which is moulded into the substrate. The wobbled groove provides a self-regulating clock to guide the laser beam as it burns the disc. The wobbled track is pre-divided into sectors and each sector is identified by pre-stamped header. DVD-R is available in two versions. Version 1 has a

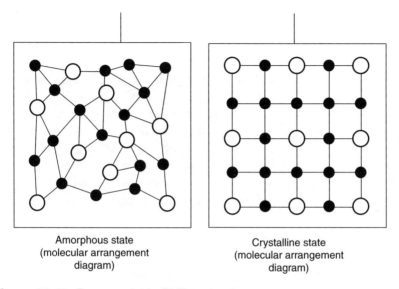

Amorphous state	Crystalline state
(molecular arrangement	(molecular arrangement
diagram)	diagram)

Figure 22.13 *Re-recordable DVD technology*

capacity of 3.95 billion bytes (3.68 GB) and version 2 has a capacity of 4.7 billion bytes (4.37 GB).

Re-writable DVD discs including DVD-RAM employ a different technology, known as *phase-change*. Phase-change technology uses a metal compound that changes its reflectivity as it moves between a crystallised and an amorphous state (Figure 22.13). When the compound is heated by a low-power laser, it melts creating a crystallised spot with a high reflectivity known as 'eraser'. Alternatively, if it is heated by a high-power laser, it melts and then cools down rapidly to form an amorphous (non-crystalline) spot of low reflectivity, known as *'mark'*. The marks can be read by a low power laser to retrieve the data from the disc. Phase-change uses the same wobbled groove as DVD-ROM, but it writes its data bits in both the groove and the land between the grooves as shown in Figure 22.14. The purpose of the wobbled grove is twofold:

- The generation of a spindle motor control signal
- The generation of a gate signal used in detection of the land pre-pits

The *pre-pits* on the other hand ensures high precision when writing the data and it provides the recording address and other information that are necessary for writing on a DVD disc.

Layout of recordable DVD

The DVD disc is divided into four areas or zones as shown in Figure 22.15. The main zone is the data or information zone where video and audio data are recorded. This is preceded by a burst cutting area which is between 44

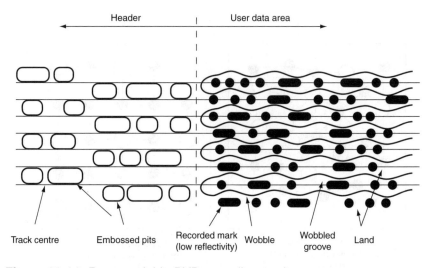

Figure 22.14 *Re-recordable DVD recording tracks*

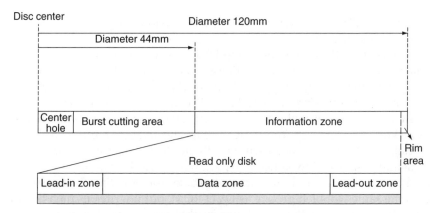

Figure 22.15 *Disc layout for DVD-ROM*

and 47 mm from the centre of the disc. The information zone itself commences with lead-in zone and followed by a lead-out area.

Before the video and audio data are recorded in the user area, the data bits are grouped into 2048-byte chunks or sectors. Each sector commences with a header which includes, among other things, a unique ID number for the sector. The data is written on the disc surface, sector by sector. Reading the data off the disc involves directing the pickup head to the part of the disc surface where the required sector is recorded and once the sector is identified, data is extracted for further processing.

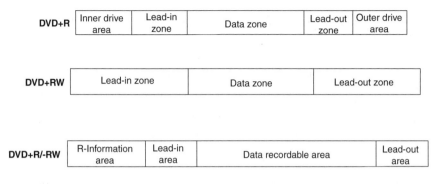

Figure 22.16 *Layouts of information zone for recordable and re-recordable DVD discs*

Burst cutting area

The burst cutting area is invariably used to stamp up to 188 Bytes of information related to the individual disc such as ID codes and serial number or any other information that may be used for inventory purposes or storage systems such as jukeboxes for quick identification of the individual disc. It may be read by the same pickup head that is used for reading the user data off the disc.

Recordable and re-recordable discs have the same burst cutting area organisation as DVD ROMs. Where they differ is in the layout of the information or data zone as illustrated in Figure 22.16.

Write strategies

Laser current waveforms needed to record on different media (CD-R, CD-RW, DVD±R, DVD±RW, DVD-RAM, HD-DVD, Blu-ray) vary, depending on the media and on the data transfer rate used. The shape and timing of these laser current waveforms for these different media and different data transfer rates are referred to as '*write strategy*'. A write pulse generator (WPG) is used to feed the appropriate current into the laser driver to produce the required laser current waveform. An example of a laser current waveform for a typical DVD-RAM write strategy is shown in Figure 22.17.

When a logic one follows a logic zero, the laser current is increased to the "first write" (fw) level. If the number of consecutive ones is only three (3 T mark from right to left in Figure 22.17), the shortest allowed mark length for DVD-RAM, the laser current pulse consists only of the "*first write*" and the "*cool*" (c) levels, before going back to the erase level. For a 4 T mark, the laser current pulse is already a little bit more complex: A "*bottom*" (b) and a "*last write*" (lw) pulse are inserted between the "first

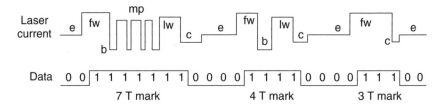

Figure 22.17 *Write strategy for DVD-RAM where: e = erase, fw = first write, b = bottom, mp = multiple pulse, lw = last write and c = cool*

write" and the "cool" level. For marks longer than 4 T, as shown for 7 T, *multi-pulses* (*mp*) are inserted between the "first write" and the "last write". Typical HD-DVD strategies are very similar to the DVD-RAM strategy shown in Figure 22.17, but the number of multi-pulses is higher by 1, e.g. a 7 T mark has 4 instead of 3 multi-pulses and the shortest allowed mark length is 2 T.

While the laser current can have up to six levels (erase, first write, bottom, multi-pulse, last write, cool), the differential output current of the WPG is only a four-level signal. Some relatively simple mapping from the four current levels of the WPG output signal to the six levels of the laser current needs to be done by the laser driver IC. Having only a four-level signal rather than a six-level signal increases robustness of the data transmission. The main task of the WPG is to generate the precise timing for the laser driver.

Other strategies used for standards other than DVD-RAM use simplified variants of the patterns shown in Figure 22.17. Such as the "bottom" level made equal to the "erase" level and dispensing with the "cool" pulse and so on.

DVD recorder

A DVD recording system consists of two major sections: AV playback/record (PB/REC) and servo-control. The servo-control and AV playback are described in detail in chapter 21. In the video playback mode, RF signals from the optical pickup unit (OPU) is amplified and processed to produce MPEG-2 PESs (packetised elementary streams) that are decoded by the MPEG-2 decoder to produce analogue video signal in the form of component video or Y/C a shown in Figure 22.18. Composite video (CVBS) is also available from the PAL/NTSC encoder. In the record mode, analogue video from a tuner via a colour decoder, or direct component video are first MPEG-2 encoded to produce video PESs that are fed into the write strategy pulse generator via an ATAPI interface and hence forth to the blue laser driver. *ATAPI* which stands for AT Attachment Packet Interface (AT is advanced technology) is a standard interface between a PC or other computer-controlled devices on the one hand and CD/DVD or tape

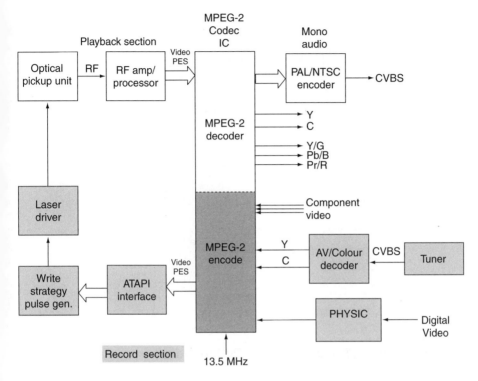

Figure 22.18 *Video processing in a DVD recorder*

backup drives on the other. It is based on the *IDE* (Integrated Drive Electronics interface) specifications used for HDDs with some additional commands needed for controlling a CD/DVD or tape drive. Where the video input is digital, a *PHY* (pronounced fi and stands for physical layer in the OSI communication model, see Appendix A5) IC is used. The PHY chip is a fast (400 Mbps) Transreceiver which acts as an interface between digital and analogue domains.

A similar process is applied to the audio signals in the playback and record modes as shown in Figure 22.19. The playback is the same as that described in Chapter 20. In the record mode, mono or stereo audio is MPEG-2 encoded by the MPEG-2 codec chip. PCM sound from the MPEG-2 encoder goes to the ATAPI interface via sound serial bus I$_2$S on its way to the write strategy pulse generator where it is multiplexed with the video signal before going to the laser driver.

HDD recorder

Video processing activities similar to those used in DVD recording are required for recording A/V signals on a HDD where audio and video PESs are stored as files on the disk. A block diagram of a DVD/HDD

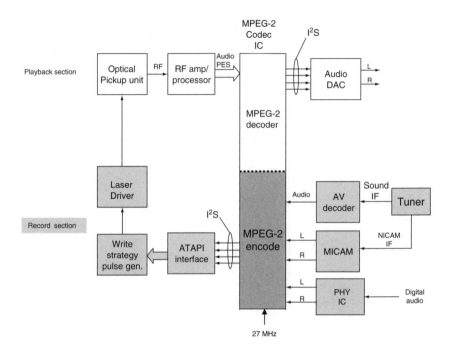

Figure 22.19 *Audio processing in a DVD recorder*

recorder using system processor made by LSI Logic is shown in Figure 22.20. DMN 8603 is a system-on-chip IC which provides comprehensive solutions to DVD recorders including DSP servo control and video and audio encoding/decoding with in-built microprocessor. Two ATAPI interfaces are available, one for the DVD drive and the other for the HDD drive. The HDD itself is the normal IDE-type hard disk with a capacity ranging from 80 GB upwards.

In DVD HD recording, two modes may be used: *real time* in which the DVD is written to directly and *non-real time* in which the data bitstream is first recorded into an HDD first and transferred to the DVD at a later stage.

Camcorders

Camcorders (camera recorders) are the most advanced form of 'domestic' technology. Combining the elements of camera, encoder, VCR electronics, deck- and lens mechanics and computer control, they draw on the systems, technologies and techniques described in most of the chapters of this book.

Figure 22.21 shows the functional block diagram of a digital camcorder system. Light from the optical lens assembly projects an image onto the

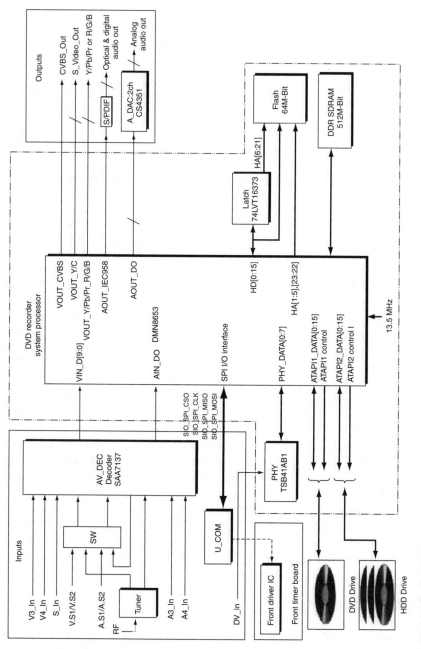

Figure 22.20 *DVD/HDD recorder*

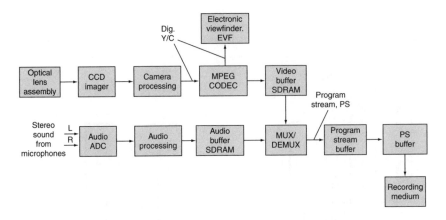

Figure 22.21 *Functional block diagram for a camcorder system*

charged coupled device (CCD) imager. The CCD is a photosensitive array which is charged by the light falling on it. The charge is then converted into a continuous analogue voltage when the CCD charged elements are scanned line by line. After the scan is completed, the CCD elements are reset to start the exposure process for the next video frame. Embedded within the CCD is an analogue-to-digital converter to produce a digital output for further processing by the camera processing block ready for data compression by the MPEG codec. The camera processing chip carries out such functions as *'steady shot'*, zoom and focus motor control and digital picture effects. The MPEG-coded data are fed into a video buffer. Digitised Y/C data are also fed into the *electronic viewfinder* (EVF) for monitoring by the user. Stereo sound from audio microphones are A/D converted and the PCM audio data placed into an audio buffer. The MUX/DEMUX receives the compressed video and PCM audio streams from the corresponding buffers, packetises and multiplexes them into a standard MPEG-2 program stream (PS) to be stored in a PS buffer. Data in the PS buffer are then used to write on the recording medium which could be a DVD disc, an HDD or a magnetic tape. In the playback mode, the process is reversed and this is the reason for using an MPEG codec chip instead of just a coder and MUX/DEMUX instead of just a MUX. In the playback mode, data from the recording medium are demultiplexed and decompressed and fed into the EVF for display.

Optical section

The lens assembly is a complex ensemble of precision glassware, with adjustment for focus and zoom by sliding members within the barrel. Both focus and zoom rings are driven by miniature electric motors; where manual (hands-on lens ring) adjustment is provided for these, the motors drive through slipping clutches. Zoom motor control is provided for in a body-mounted rocker switch, sometimes offering two zoom speeds. The auto-focus system consists of a servo loop with *through-the-lens* (*TTL*) picture-sharpness sensors. It is controlled by a microprocessor, and many models offer a choice of picture zones for auto-focus operation.

Also inside the lens assembly is a multi-bladed iris, whose operation is similar to that of a moving-coil meter. It too is part of a servo loop, this time controlled by the level of luminance signal coming from the image sensor. The higher the light level in the televised scene the smaller the iris opening, with a consequential improvement in depth of focus field. There is also an AGC system in the video amplifier: between these two control loops the signal level—and hence contrast in the reproduced picture—is held constant over a huge range of ambient scene brightness. In very low light situations, however, noise (grain, snow, confetti) intrudes on the picture.

Interposed between the lens and the image sensor are two optical filters. An infra-red cut filter prevents most infra-red radiation reaching the sensor so that heat and similar energy sources have little effect on the picture, while a crystal filter takes out the finest detail in the incoming scene to prevent 'beat' and patterning effects due to the dot-matrix structure of the sensor IC and its colour filter and to minimise cross-colour effects.

CCD image sensor

The CCD image sensor consists of a mosaic of photodiodes, each of which acquires and stores a charge proportional to the intensity of light falling upon it. The charges are transferred by horizontal and vertical CCDs to the image-sensor's output terminal, whence they emerge (after processing in a sample-and-hold stage) as a serial video signal. The colour components of the image are captured by a coloured translucent dot-matrix overlay bonded to the face of the pick-up sensor chip. Digital colour cameras generally use a Bayer mask over the CCD. Each square of four pixels has one filtered red, one blue and two green (the human eye is more sensitive to green than either red or blue). The result of this is that luminance information is collected at every pixel, but the colour resolution is lower than the luminance resolution.

Better colour separation can be reached by three-CCD devices and a dichroic beam splitter prism, that splits the image into red, green and blue components as was explained in Chapter 12. Each of the three CCDs is arranged to respond to a particular colour. Some semi-professional digital video camcorders (and all professionals) use this technique.

Image stabilisation

A problem with very small camcorders is *shake and wobble* in the reproduced image due to the natural tremor in the human hand which holds it. It is especially troublesome at extreme zoom settings. To mitigate this, *optical image stabilisation* (*OIS*) is used. There are different techniques that have been developed. One brought to the market by Sony, involves a vari-angle refractive prism, consisting of two silicon–oil filled plate-glass panels linked by bellows, mounted in the lens assembly. The prism's angle, relative to the pick-up sensor's image plane, is varied by drive coils under the control of a piezo-electric sensor of pitch and yaw of the camcorder's body. Another technique use DSPs to analyse the image on the fly and then move the sensor appropriately.

A third electronic technique depends on the use of an A-D converter in conjunction with a large DRAM digital memory in which one whole field of picture information is temporarily stored. The image for recording is read from the centre section of the memory bank, using approximately 85% of its contents, area-wise. The pixel data corresponding to the outer periphery of the picture is selectively used by the electronic *image stabiliser* (*EIS*) processor with reference to four motion-detection zones in the picture and an algorithm which distinguishes between camera shake and natural movement in the picture.

Digital image stabilisation is used in some video cameras. This technique shifts the electronic image from frame to frame of video, enough to counteract the motion. It uses pixels outside the border of the visible frame to provide a buffer for the motion. When the camera moves, the whole frame changes position on the chip, but it still registers in its entirety. The camera's computer picks a fixed point in the picture, like the outside edge of the frame, then stabilises that edge as a fixed point on the recording.

Digital picture zoom

In addition to the optical zoom facility provided on all camcorders, those fitted with a digital field store are able to offer a further 'zoom' feature in which only the memory data corresponding to the central section of the picture is read out at normal field and line scanning rates. The effect on screen is of fewer but larger pixels rather than real image magnification, so it counts more as an 'effect' than as a true extension to the zoom range.

Electronic viewfinder

Electronic viewfinder (EVF) systems invariably include colour LCD screens of the sort described in Chapter 11, some types of which are on a side-mounted flip-out panel, and others built onto the camcorder's backplate, in either case removing the need for an eyecup magnifier system. They have the advantage of giving the operator some idea (though not as accurate a one as a conventional colour monitor) of the colour balance of the pictures being recorded, and can be viewed by more than one person at a time. On the debit side is their relative lack of picture definition; and their vulnerability to being 'washed out' by sunlight. Some camcorders offer the best of both EVF systems with both a monochrome tube and a colour LCD screen.

Audio section

All camcorders have an integral microphone, which may be a mono or—with hi-fi camcorders—a stereo type. Many have external microphone sockets for greater versatility and an opportunity to eliminate the motor- and handling noise which is almost inevitable in quiet situations where the *automatic level control* (*ALC*) drives up the gain of the audio amplifier. The action of the ALC circuit is necessary to enable the camcorder to cope with the huge range of sound levels it may encounter, from the rustle of leaves to the roar of a jet aircraft. Where a hi-fi sound recording system is provided in the more expensive models, the dynamic range for sound recording is much greater and the ALC action can be less harsh. The camera sound is passed out of the camcorder during record along with the composite video signal for monitoring if required: very often a headphone socket is also provided for audio monitoring 'on the hoof'.

D-8 signal processing

As stated earlier, the recording medium for modern camcorders may vary. Figure 22.22 shows the various stages of a typical Digital-8 recording system. The CCD imager's output is cleaned up, gain-controlled and then A/D converted within IC502 into a 10-bit bit stream, sampled at 13.5 MHz. IC251 separates Y and RGB data to derive Y, Cb and Cr values, all at 8-bit depth and the latter two at 3.375 MHz sampling frequency. Also within this chip is carried out the 'steady-shot' process in which the image is stabilised by a fast pan/tilt reading operation from a picture memory bank with a larger effective area than the useful picture size. The next chip in line, IC351, performs blocking, data shuffling, digital picture effects and D-A conversion for use in the viewfinder and to produce an *electronics-to-electronics* (*E-E*) output monitoring signal.

IC301 is the MPEG codec concerned with the bit-reduction process. Here the data undergoes DCT, quantisation, VLC and framing to achieve data compression to about one-fifth of the original content. The compressed video is joined by the audio bitstream inside IC301 for passage into IC302, where error correction ECC, deshuffling and encoding takes place into a form suitable for recording on tape. Further processing takes place inside IC104, primarily to convert the data from 5-bit parallel form to a serial (one line) data stream for application via recording amplifier IC102 to the video heads. For this application an FE (flying erase) head is also present on the drum to wipe off previously recorded data.

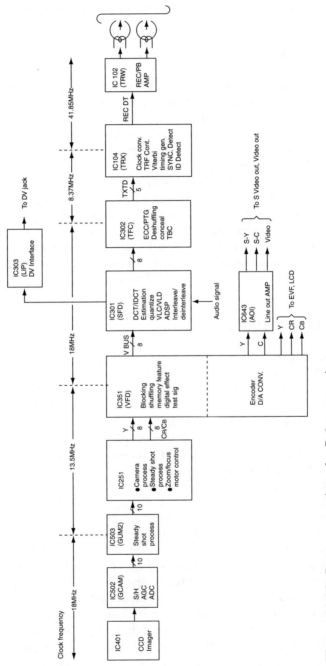

Figure 22.22 *Data processing of a D-8 camcorder*

D-8 playback

During replay of digital video tapes, many of the same ICs are used, with their roles reversed to perform the inverse functions to those carried out during record; the same applies to memory chips. Thus in Figure 22.22, IC102 now operates as head preamplifier; IC104 carries out clock conversion, synchronous detection etc.; IC302 shuffles the data and provides error-correction and concealment; IC301 takes care of I-DCT, de-interleaving and data expansion; and IC351 de-shuffles the data, D-A converts it and encodes the Y and C components into PAL (or other analogue) form to produce signals for the viewfinder and the AV-out ports. The digital in/out port interface is IC303, shown at the top of Figure 22.22.

23 Cable and on-line television

Originally the philosophy of cable television was simple: to bring television to those who could not receive it by means of conventional RF aerials for reasons of local topography, the impracticability of installing an aerial, or aesthetics as judged by the local council or planning authority. Many *communal-aerial television* (*CATV*) systems had their origins in the dark ages of 405-line television on VHF, with few and scattered transmitting sites. With the spread of the UHF terrestrial broadcasting network (transmitters have been installed for communities of fewer than 300 souls) the necessity for CATV as an alternative to listening to the radio has all but disappeared; blocks of flats, hotels and similar domiciles can be served by *master-aerial TV* (*MATV*) which is a small-scale cable system, working from a master aerial and distributing signals to tens, rather than thousands, of TV sets.

In some cases, out-of-area programmes were made available on a community cable network as a 'bonus' to subscribers, although the advent of Channel Four sometimes meant dropping this facility, as many cable systems (especially wired-pair HF and some co-axial VHF networks of long standing) had a maximum capability of four vision channels. Because of government policy, cable operators could, in general, only distribute the programmes of the national broadcasters, and this (in the UK at least) tended to limit the popularity of cable systems. Further problems for old cable networks were the propagation of teletext signals through the network, the difficulty of maintaining good bandwidth and delay characteristics for colour signals, and the incompatibility of commercial home VCR machines with the special receivers (called terminal units) used with some cable systems.

In the more recent past, interest in cable networks has revived, with encouragement from the UK government, and many networks are now in place. An incentive to viewers is the provision of cheaper telephone calls and broadband connection. Another attraction of new cable franchises is that alongside 'off-air' material they are permitted to broadcast exclusive programmes, not obtainable except over the cable, and thus create a demand from the viewing public and a financial incentive for the cable operators. The advent of satellite services opened further prospects for the cable system; those viewers who are unable to accommodate a receiving dish, or who wish to view programmes intended for other European countries (with or without English soundtrack) can be catered for, as the installation of receiving dishes for fringe satellite and direct broadcasting by satellite (DBS) reception is most economically done on a 'community' basis.

Provided the demands (and hence finance) were there, the cable scheme would enable television to become as locally based as the current BBC and *independent local radio* (*ILR*) district services, particularly relevant in the provision of text and data transmissions of local interest only; and in the potential for local advertising.

Cable types

The early cable transmission system consisting of twisted pairs carrying HF vestigial sideband TV signals is now obsolescent, though used in some districts for satellite programme relay.

The choice for multi-channel TV carrying videotext and other data channels lies between co-axial, standard telephone and glass-fibre optic cables, and each has various advantages and disadvantages. Co-axial cable is currently well established, and scores on the counts of easy interfacing with existing equipment and lower initial cost, at least on comparatively short runs. The *optical-fibre* technique requires more complex terminal equipment but has advantages in the areas of data-handling capabilities, immunity to electrical interference, security against 'tapping' and a small physical diameter, enabling a greater number of services to be laid in existing ducts, and a reduction in the cost of routing. Optical fibre has advantages over co-ax in terms of transmission efficiency, too; for a given data rate the repeaters ('boosters' to overcome transmission losses) can be spaced at greater intervals than with the co-ax system. Regarding cost, glass is intrinsically much cheaper than copper and in volume production the glass-fibre technique may well show an overall cost advantage over copper cables.

It has been shown that for trunk lines optical transmission has much to recommend it, and British Telecom currently operates many glass-fibre optic links for transmission of television, audio, data and telephone traffic. Some can operate at very high data rates (140 Mb/s) which confers the simultaneous ability to handle thousands of phone calls, or many broadcast-quality digital TV channels. It may be that the most economical way to implement a large cable system will be to adopt a 'hybrid' solution, with optical-fibre trunk routes to local distribution points, whence co-axial cables will '*spur off*' to individual dwellings grouped around the fibre cable head. Alternatively, connection can be made to existing subscribers' telephone lines for conveyance of picture, data and sound.

Transmission modes

In the same way as air or space can be used to carry virtually any radio frequency using different modulation systems so is it with co-ax and fibre-optic systems. Obviously the '*launching*' and '*interception*' methods differ, and for fibre the basic carrier is light, rather than an electrical wave. Thus we can use baseband, AM, FM, PSK or PCM in cable systems, at such carrier frequencies as are appropriate to the signal, the distance between terminals and the transmission medium. Fibre-optic cables will work from analogue baseband to the 140 Mb/s PCM mode. It should be remembered that the light-carrier in a fibre system (usually infra-red rather than visible light) is itself an electromagnetic wave with a frequency of the order of 3×10^8 MHz or 300 THz so that the upper limit on the rate of data throughput in an optical fibre, is perhaps, limited by technology rather than physics! The main restriction on bandwidth in glass fibre links using modulation of a sub-carrier (sub-, that is, to the frequency of the light wave itself) is the effect of fibre-dispersion, which tends to slightly *'blur'* in time the sharpness of received pulses, giving an integration effect to their shape. Fibre-optic cable is a very efficient carrier: a typical loss figure at 1550 nm light wavelength is 0.23 dB per km.

The network

There are two basic methods of cable distribution. The more traditional is the tree-and-branch system (see Figure 23.1a) in which all available programmes are continuously sent over the network in separate channels,

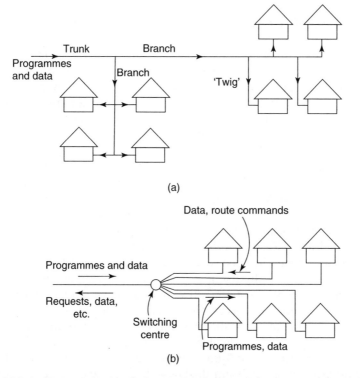

Figure 23.1 *Two methods of signal distribution: (a) the "passive" trunk-and-branch system; (b) the "interactive" switched-star network. These are discussed in the text*

with user selection by means of some form of switch at the receiving point. This was the modus operandi of the original radio and TV cable system, in which each household may be regarded as being on the end of a 'twig'.

The alternative and better system is termed a 'switched-star network', shown in Figure 23.1b. Here the available programmes are piped to a 'community-central' point analogous to a telephone exchange. The subscriber can communicate with the 'exchange' and request the desired programme(s) to be switched to his line for viewing, listening or recording. The advantage of the switched-star network is its interactive capability, whereby the subscriber can 'talk-back' to a local or central exchange; this opens the possibility of the 'single-fibre' household in which all communications services (radio, TV, telephone, internet, banking, public-utility meter reading, text and data, etc.) come into the dwelling via a single link which can, by means of recording devices at either end, be utilised during off-peak and night hours. The switched-star configuration lends itself well to Pay-TV (either pay-by-channel or pay-by-programme) because security is more easily arranged. It is simpler to deny a programme to a non-subscriber by a central 'turn-key' process than by expensive signal scrambling and 'authorised unscrambling' systems of the sort described in the previous chapter.

Propagation modes in glass-fibre cables

Since the basic transmission 'vehicle' in fibre-glass cables is light energy we must now see how the signal is launched into the cable and intercepted at the receiving end. Depending on the distance to be covered, the sending device may be an LED or low-power semiconductor laser operating on a wavelength (infra-red) around 850 nm. The radiant energy in the sending device is surprisingly small, typically 200–300 µW for an LED and 1–3 mW for a laser. The fibre termination is an integral part of the light-source encapsulation for maximum coupling efficiency, permitting virtually all the light to be concentrated in the cable.

The receiving device is a light-sensitive diode, again intimately coupled to the fibre end. For low noise and highest possible sensitivity this pick-up device is 'tuned' to the light wavelength in much the same way as a radio set is tuned to an RF transmission. For short-haul reception a silicon *PIN* photodiode is generally used; with long-distance fibre cables greater sensitivity can be obtained by using an avalanche photodiode which combines the property of light detection with an internal amplification process. A PIN diode (p-type, intrinsic, n-type diode) with a near perfect resistance at RF and microwave frequencies that depends on the DC current through it.

The glass-fibre core, made of silicon dioxide (SiO_2) glass, is of very small diameter, typically 50–100 µm, surrounded by an intimately bonded cladding layer of about 20 µm thickness. Further layers give strength and environmental protection, the outer jacket consisting of a tough waterproof polyurethane cover. The transmission of light along a glass fibre depends on the phenomenon of total internal reflection in which the light, when it

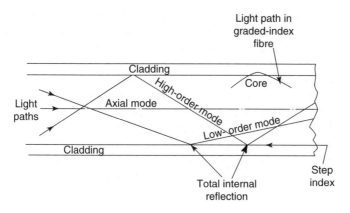

Figure 23.2 *Propogation modes in fibre-optic lightguides*

encounters the inner surface of the fibre wall, is *'bounced'* by the mirror-like wall surface back into the fibre. Light can enter the fibre end at any angle and the bounce path in transmission can thus take several forms, as shown in Figure 23.2. Large bounce angles give rise to a long path length (in terms of the passage along the entire cable) and are called high-order transmission modes; at lesser angles the total path length is shorter, described as low-order transmission mode. A beam which travels down the axis of the fibre takes the shortest possible path length in what is known as axial mode.

The nature of light propagation down a fibre depends on its diameter and on the difference in refractive index between the fibre core and its cladding material. Where a sudden change of refractive index is present at the fibre wall we have a *step-index* fibre in which several modes (high- and low-order) are taken by the light. An alternative form of construction is the *graded-index* fibre, where the interface between 'core' and 'cladding' represents a more gradual change in refractive index. This has the effect of making the light rays turn less sharply when they encounter the fibre wall and thus reduces reflection loss, as shown at the top right of Figure 23.2. Low-order modes predominate in such a graded-index fibre cable, and such high-order modes as are present travel faster along their longer path, reducing the fibre-dispersion effect described earlier. Graded-index fibres offer a low transmission loss and greater bandwidth than step-index types, at the expense of higher production costs and greater coupling losses at the junctions between the fibre and the sending and receiving devices.

Both step- and graded-index fibres operate in what may be called mult-imode with many possible light path angles within the fibre. If we can arrange a fibre to concentrate on the axial mode we shall significantly reduce the transmission loss. In this monomode system a high-grade glass core is used, with a small diameter in the region of 5 μm (about one-tenth of diameter of multi-mode cable cores). The light wavelength used here is longer, around 1.35 μm, and the much straighter light path gives very good transmission efficiency. *Repeaters* (regenerators) are therefore required at much longer intervals than in conventional fibre (and particularly co-axial) systems, and in a typical monomode transmission system repeaters can be as much as 30 km apart; this economy in equipment easily outweighs the cost disadvantage of the monomode fibre cable itself.

For repeaters generally, the operating power can be sent down the cable in co-axial systems, as described earlier for domestic masthead amplifiers. Fibreoptic cable plainly cannot carry DC, but conductive members can if required be incorporated in its protective sheath, or local power sources can be used, in view of the long intervals between repeaters made possible by fibre-optic technology. Many thousands of kilometres of fibre-optic cable are in use in the UK, primarily by British Telecom.

Multimedia 'cable TV'

Cable TV now offers interactive services, video-on-demand, cheap telephone connection and internet access as well as the normal TV channels. For internet connection, a 2-way communication is necessary and this is obtained by allocating a relatively small part of the bandwidth with a bit rate of 500 kbps–1.5 Mbps to what is known as *'upstream'* data communication from the subscriber to the provider and a larger bandwidth with a bit rate of up to 35 Mbps to what is known as *'downstream'* data commu-

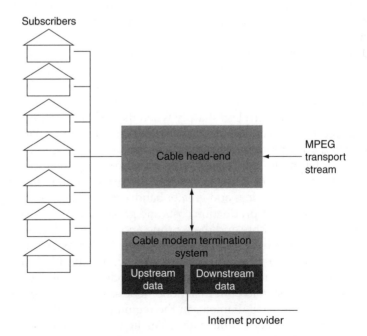

Figure 23.3 *Multimedia 'cable TV'*

nication from provider to subscriber. This is the same technique used in the telephone ADSL system described in the next chapter. Two modems are used: a *cable modem* at the subscriber's end and *cable modem termination system* (*CMTS*) a the provider's headend.

At the provider's head-end, the CMTS provides many of the same functions provided by a broadband internet connection. The CMTS takes the traffic coming in from a group of subscribers on a single channel and routes it to an *Internet Service Provider* (*ISP*) for connection to the Internet (Figure 23.3). The cable providers will also have servers for accounting and logging, *Dynamic Host Configuration Protocol* (*DHCP*) for assigning and administering the IP addresses of all the cable system's users, and control servers. In the narrower upstream, time division multiplexing, TDM is used. The bandwidth is divided into '*bursts*' of time of few milliseconds, in which users can send commands, queries, etc. to the Internet.

At the subscriber's end, the cable modem which includes a tuner, can be either internal or external to the computer or it can be part of a set-top cable box, requiring that only a keyboard and mouse be added for Internet access.

The two-way communication is also used to provide other services including video-on-demand as will be described in the next chapter.

24 Multimedia convergence

In the digital world, a 'bit' is a 'bit' whether it represents audio, video, data, software, etc. Consequently, in principle, there is no reason for separate networks for broadcasting and telecommunications. Of course, the reality is slightly more complex. In practice, there are some fundamental distinctions between broadcasting and telecommunications. Broadcasting essentially delivers one-way, one-to-many services, whereas telecom operators provide two-way, one-to-one services. Nevertheless, the boundaries between broadcasting and telecom have become increasingly blurred since the early 1980s; some analogue TV services have been used to deliver limited 'telecom-like' services, such as delivery of encrypted teletext services for individuals or closed groups of users. The introduction of digital radio and TV networks opens up new opportunities for data services, particularly for delivery to portable or handheld devices. Demand for short 30-s video clip (e.g. a football goal) on mobile phones is increasing. Large-scale on-demand video services via *Universal Mobile Telecommunication System (UMTS)* proved to be uneconomic. Instead, attention has turned to the benefits of one-to-many services: rather than sending individual video streams to each consumer, it would be much more efficient to transmit the same material simultaneously to all those interested. Of course, this 'new idea' is actually 'broadcasting'!

The over-used concept of convergence can apply to incorporating internet services onto a broadcasting platform, known as *Digital Multimedia Broadcasting*, or conversely, incorporating broadcasting services onto the internet platform, known as *on-line convergence*, or a mixture of the two.

Digital multimedia broadcasting

Arising from the hype surrounding the Internet, there has been increasing interest in offering multimedia services to mobiles. First, there was the big bang of selling frequencies for UMTS all over Europe, but it turned out that UMTS will not offer the huge bandwidth that modern streaming internet applications, such as TV, require. This means that, apart from point-to-point applications, there is an increasing requirement for point-to-multipoint, wireless, internet access technologies. Hence the terrestrial broadcast systems coming into focus, as a means of streaming multimedia content to mobile, portable and handheld receivers.

There are two different technical solutions that could meet these requirements:

- *DVB-H* (H for handheld), the latest terrestrial standard from DVB.
- Digital audio broadcasting (*DAB*), adapted for multimedia delivery.

A third solution, ISDB-T from Japan is not to be deployed in Europe.

Although the fact that all three components: audio, video and data are presented in digital form makes it possible for them to share the same transmission medium, it is not sufficient for piratical convergence. What makes the whole enterprise a practical possibility is their adherence to a standard network communication model, the *open system interconnect* (*OSI*) model (refer to Appendix A5).

DVB-H

The terrestrial version of the DVB system (DVB-T) was fully described in Chapter 8. Developed in the mid-1990s, it was primarily intended for portable and stationary reception using roof-top antennas. The design of the system was strongly influenced by the cost of the receiver.

To make the receivers cheaper, *time interleaving*—which would have benefited mobile reception—was not implemented; instead, the same error correction as the satellite system, DVB-S, was used. DVB-T can effectively be used for mobile and portable reception provided the multi-antenna diversity receiver is available to enable high-speed mobile reception of DVB-T. However fast, such varying channels are error prone. The situation is worsened by the fact that antennas built into handheld devices have limited dimensions and cannot be continuously pointed at the transmitter if the handheld terminal is in motion. This is just one of the problems of using DVB-T. The stumbling block for the use of straight forward DVB-T for mobile devices is however, the very practical problem of battery life. Power consumption of DVB-T front ends is too high to support handheld receivers that are expected to last from one to several days on a single battery charge.

To make DVB-T suitable for mobile multimedia services, a dedicated standard for handhelds, based on DVB-T, was necessary. It is called DVB-H (DVB-Handhelds). The aim is to provide an efficient way of carrying multimedia services over digital terrestrial broadcasting networks to handheld terminals.

DVB-H specifications were drawn up with the following objectives:

- To power off some part of the reception chain to increase the battery useful lifetime.
- EASY access to services and seamless transition from one service to another;
- Sufficient flexibility/scalability to allow reception of services at various speeds, while optimising transmitter coverage.
- To mitigate against the effects of high levels of man-made noise such as car ignitions interference.

- To provide a generic way to serve handheld terminals in various transmission bands and channel bandwidths in various part of the world.
- To receive multimedia services using a single antenna in the portable, mobile and indoor environments.
- To maintain maximum compatibility with existing DVB-T networks and systems.

These requirements were drawn up after much debate and with an eye on the emerging convergence devices providing video services and other broadcast data services to third generation, *3G* handheld devices.

Handheld screen resolution

Broadcasters initially supported the *Quarter VGA* (QVGA, 320×240 pixels) standard, while cellular phone carriers supported Quarter Common Intermediate Format (*QCIF*, 176×144 pixels). DVB-H solves the rift between them. When the user watches a program on a mobile phone, there will be two types of content on the mobile phone screen: a broadcast program (such as a sport or drama) by a broadcast service provider, and *custom data* relevant to the program (such as online shopping information) prepared by a telecom carrier.

DVB-H system properties

The main properties of DVB-H are: *time-slicing*, IP interfacing, enhanced signalling and *in-depth interleaving*. In order to save power, a power-saving algorithm based on time division has been introduced. The technique, called time slicing, results in a large battery power saving. In order to provide a common platform with Internet services, and for reliable transmission in poor signal-reception conditions, IP interfacing with an enhanced error-protection scheme was developed. This scheme is called multi-protocol encapsulation—forward error correction (*MPE-FEC*). It employs powerful channel coding on top of the channel coding included in the DVB-T specification and offers a degree of time interleaving. Furthermore, the DVB-H standard features an extra network mode, the 4K mode, offering additional flexibility in designing single-frequency networks (SFNs) which still are well suited for mobile reception, and also provides an enhanced signalling channel for improving access to the various services. Convergence with Internet services is accomplished by internet protocol (IP) encapsulation of Internet services prior to the transport multiplexing stage.

Time-slicing

A special problem for DVB-H terminals is the limited battery capacity caused by the relatively high power consumption of a DVB-T front end which is in the region of 600–1000 mW. You will recall from Chapter 8 that before any one of the multiplexed elementary streams of the selected pro-grammes can be accessed, the whole data stream has to be decoded first. A large part of the power consumed by the front end is therefore unnec-essary. The power-saving made possible by DVB-H is derived from the fact that essentially only those parts of the transport stream which carry the data of the service currently selected have to be processed.

In order to do this, the data stream needs to be reorganized in a suitable way for that purpose. With DVB-H, several services are multiplexed using pure time division. The data of one particular service are therefore not transmitted continuously but in compact periodical bursts with interrup-tions in between. At the transmitting end, several services with differ-ent bit rates are multiplexed and a continuous, uninterrupted transport stream at a constant bit rate (CBR) is maintained.

To indicate to the receiver when to expect the next burst, the time to the beginning of the next burst is indicated within the burst. Between the bursts, data of the elementary stream is not transmitted, allowing other elementary streams to be transmitted using the remaining bandwidth. Time slicing enables a receiver to stay active only a fraction of the time, while receiving bursts of a requested service.

Bursts entering the receiver have to be buffered and read out of the buffer at the service bit rate. Practically, the duration of one burst is in the range of several hundred milliseconds whereas the power-save time may amount to several seconds. Depending on the ratio of on-time/power-save time, the resulting power saving may be more than 90%.

Time slicing offers another benefit for the terminal architecture. The comparatively long power-save periods may be used to search for chan-nels in neighbouring network cells offering the same service but better reception. This is important as the handheld receiver movement may take the user from one network cell to another. In this way, a channel handover can be performed at the border between two cells which remains imper-ceptible for the user.

IP interfacing

In contrast to other DVB transmission systems which are based on the DVB transport stream adopted from the MPEG-2 standard, the DVB-H system is based on IP. The IP operates at Layer 3 (Network) of the seven-layer OSI model (for details of the 7-layer model, refer to Appendix A5). In the preceding layer (Transport Layer 4), two types of protocols are available: a unicast (one-to-one) *Transmission Control protocol* (*TCP*) and multi-cast UDP. TCP is a 'reliable' connection-orientated service which ensures that a connection is made and an acknowledgment is received before data is exchanged. In contrast, *user data protocol* (*UDP*) is an 'unreliable' connectionless service which sends out messages regardless of a connection being established. For the purposes of DVB-H,

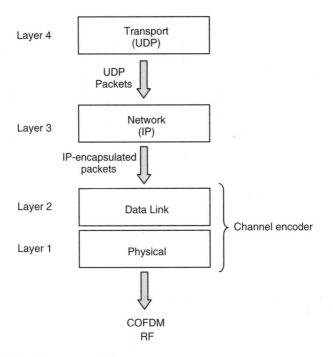

Figure 24.1 *IP encapsulation*

UDP is used which is sent datagram packets to Layer 3 for IP encapsulation. Layer 3 precedes the bottom two layers (*Data Link* and *Physical layers*) which incorporate the channel decoder (Figure 24.1). The IP encapsulated packet from Layer 3 is fed into the channel decoder as just another elementary stream to be multiplexed with other elementary streams from MPEG-2 broadcast services to form the MPEG-2 transport stream.

The IP interface allows the DVB-H system to be combined with other IP-based networks. This combination is one feature of the IP Datacast system. The manner in which the IP data is embedded into the transport stream is carried out by means of data piping technique known as *multiprotocol encapsulation (MPE)*.

Enhanced FEC

One of the main problems facing mobile TV transmission is the low signal-to-nose, S/N (or carrier-to-noise, C/N) ratio and the effect on the received radio frequencies caused by the *Doppler* effect. The Doppler effect, named after Christian Doppler, is the apparent change in frequency of a wave received by a handset that is moving relative to the transmitting source. This is the same effect on sound waves when the source, e.g. an ambulance

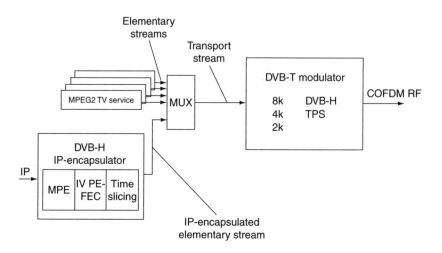

Figure 24.2 *DVB-H transmitter*

moves towards or away from a stationary person. To overcome this problem, enhanced FEC is employed (Figure 24.2).

On the Data Link layer that precedes the Physical layer, an additional stage of forward error correction (FEC) is added. This technique, called MPE-FEC, is the second main innovation of DVB-H besides the time slicing. MPE-FEC complements the physical layer FEC of the underlying DVB-T standard. It is intended to reduce the S/N requirements for reception by a handheld device by as much as 7 dB.

The MPE-FEC encoder creates a specific frame structure, incorporating the incoming data of the DVB-H codec. The FEC frame consists of a maximum of 1024 rows and a constant number of 255 columns (Figure 24.3). The frame is separated into two parts, the application data table (191 columns) and the RS parity data table (64 columns). The application data table is filled with the IP packets of the service being received. After applying the RS error to the coding, the IP packets are read out of the application data table and are encapsulated in IP sections. This is followed by the

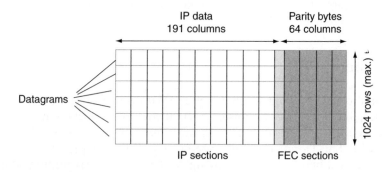

Figure 24.3 *Multi protocol encapsulation (MPE)*

parity data which are read out of the RS data table column-by-column and are encapsulated in separate FEC sections. Each MPE-FEC frame is contained within one time-slicing burst.

By adding parity information calculated from the datagrams and sending this parity data in separate MPE-FEC sections, error-free datagrams can be output after MPE-FEC decoding at the receiver despite a very bad reception condition.

The effect of MPE-FEC overhead in terms of increased redundancy can be fully compensated by choosing a slightly weaker transmission code rate. This MPE-FEC scheme should allow high-speed single antenna DVB-T reception using 8K/16-QAM or even 8K/64-QAM signals. In addition MPE-FEC provides good immunity to impulse interference.

Separating the IP data and parity data of each burst makes the use of MPE-FEC decoding in the receiver optional, since the application data can be utilised while ignoring the parity information. MPE-FEC-ignorant (but MPE capable) receivers will be able to receive the data stream in a fully backwards-compatible way.

The enhanced FEC, the MPE and the time-slicing technique are directly aligned with each other (Figure 24.2). The IP input streams provided by different sources as individual elementary streams are multiplexed according to the time-slicing method. The MPE-FEC error protection is calculated separately for each individual elementary stream. This is followed by encapsulating the IP packets and embedding them into the transport stream. As can be seen, all relevant data processing is carried out before the transport stream interface in order to guarantee compatibility to a DVB-T transmission network.

Enhanced signalling

At the physical layer, DVB-H introduces three extensions to satisfy the demands of handheld sets, namely an enhanced signalling, a new 4K mode option and in-depth interleaving.

As was described in Chapter 8, in DVB-T one of the COFDM channel is reserved for transmitter parameter signalling (TPS) to carry information such as the mode (2K or 8K), guard interval length, modulation type and code rate within the transmitted frame. The TPS information is carried by specified carriers spread over the entire COFDM frame. One carrier in each symbol is allocated to carry one bit of TPS using simple BPSK (Binary PSK). With DVB-H, additional information related to time slicing and whether MPE-FEC protection is used is included as an extension to the TPS channel. The purpose of the DVB-H signalling is to provide a robust and easy-to-access signalling to the DVB-H receivers, thus enhancing and speeding up service discovery. TPS also provides a faster way to access signalling than demodulating and decoding the Service Information (SI) or the MPE-section header.

Table 24.1 *Parameters of the three OFDM modes supported by DVB-H*

OFDM Parameter	Mode		
	2K	4K	8K
Overall carriers	2048	4096	8192
Modulated carriers	1705	3409	6817
Useful carriers	1512	3024	6048
OFDM symbol duration (μs)	224	448	896
Guard interval duration (μs)	7,14,28,56	14,28,56,112	28,56,112,224
Carrier spacing (kHz)	4.464	2.232	1.116
Maximum distance of transmitters (km)	17	33	67

The 4K transmission mode

As was stated in Chapter 8, DVB-T provides two OFDM modes; 2K and 8K (2048 and 8192 carriers respectively). DVB-H specifications provides for an additional mode, a 4K with 4096 carriers (Table 24.1).

The 4K mode is a compromise solution between the two other modes designed to double the SFN coverage area compared with the 2K mode. The SFN coverage area or size, known as a *cell*, is determined by the number of carriers of the OFDM mode used. By doubling the number of carriers, say from 2K to 4K, the cell size also doubles, but in doing so, it also makes the network more susceptible to the effects of the Doppler frequency shifts. Conversely, a decrease in the number of carriers from say 8K to 4K, improves its effectiveness to deal with the Doppler effect. In short, the 4K mode allows for a doubling of the transmitter distance in single-frequency networks (SFNs) compared to the 2K mode and, when compared to the 8K mode, it is less susceptible to the effects of Doppler shifts. It offers a new degree of network planning flexibility: 8K mode for small, medium and large SFNs with high-speed reception, 4K mode for small and medium SFNs with very high-speed reception and 2K mode for small SFNs with extremely high-speed reception.

In-depth interleaving

The in-depth interleaving may be viewed as a spin off of the fact that DVB-H terminals would incorporate 8K symbol interleaving capability as a standard. The type of interleaving requires a large memory size in the terminal in order to process the data transmitted in one complete 8K OFDM symbol. However, if a 4K mode is used, the size of the 8K mode memory is able to process two 4K frames or alternatively four 2K OFDM symbols. The new scheme results in an increased interleaving depth for the 2K and 4K modes and in improved performance. If the full amount of the available memory is used, the resulting method is called in-depth interleaving. In-depth interleaving provides an extra level of protection against short noise impulses caused by, e.g. ignition interference and interference from various electrical appliances.

Compatibility

DVB-H is fully compatible with DVB-T. It can be used in 6, 7 and 8 MHz channel environments. However, a 5 MHz option is also specified for use in non-broadcast environments. A key initial requirement, and an amazing feature of DVB-H, is that it can co-exist with DVB-T in the same multiplex. Thus, an operator can choose to have 2 DVB-T services and one DVB-H service in the same overall DVB-T multiplex.

Introduction to DAB

DAB was developed in the early 1990s by the European consortium Eureka 147, mainly to replace the widely used analogue frequency modulation (FM) broadcasting system. The VHF band is a scarce resource in many parts of the world, so there was a need for a spectrally more efficient modulation method than FM. In DAB, this is achieved by multiplexing several programmes into a so-called ensemble with a bandwidth of 1.536 MHz, where the number of programmes per ensemble is flexible and depends on individual programme bandwidth requirements. Further, conventional analogue techniques do not provide satisfactory performance in a mobile environment, because they are highly affected by multi-path propagation and thus fading. In DAB, orthogonal frequency division multiplex (OFDM) has been chosen to overcome the effects of multi-path propagation, enabling the system to operate in SFNs. DAB is designed to operate in any frequency band in the VHF and UHF range for terrestrial, satellite, hybrid (satellite with complementary terrestrial) and cable delivery.

DAB is a spectrum-efficient rugged system, not dissimilar to the digital terrestrial TV broadcasting in that the stereophonic audio is digitised, compressed and modulated using OFDM. Audio programme information is incorporated as a digital bitstream, with the system supporting a wide range of options for other data, either associated with or independent from the sound programmes. Compression is performed using MPEG-1 layers I, II or III described in Chapter 6. The encoder can operate in stereo or mono mode and the output bit-rate is selectable between 384 kb/s, for a stereo signal, down to 32 kb/s for a mono signal, with a corresponding reduction in the quality of the re-constructed audio signal. A value of 256 kbps has been judged to provide a high-quality stereo broadcast signal. However, a

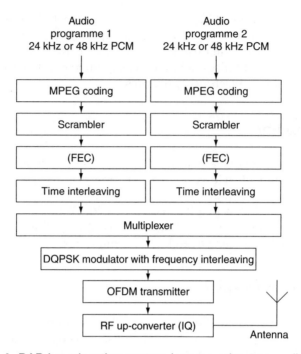

Figure 24.4 *DAB broadcasting system incorporating two audio programmes*

small reduction, to 224 kbps is often found acceptable, and in some cases it may be possible to accept a further reduction to 192 kb/s, especially if redundancy in the stereo signal is exploited by a process of *'joint-stereo'* encoding (i.e. some sounds appearing at the centre of the stereo image need not be sent twice). At 192 kb/s, it is relatively easy to hear imperfections in critical audio material.

The DAB broadcasting system (Figure 24.4) is not dissimilar than the DVB described in Chapter 8. The audio signal is MPEG layer-2 encoded and then scrambled. FEC is applied to the scrambled bitstream by employing punctured convolutional codes with code-rates ranging from 0.25 to 0.89. The bitstream is sent through a time interleaver before it is multiplexed with the other programmes to form an ensemble. The ensemble bitstream is then used to modulate the OFDM carriers using differential QPSK. To achieve orthogonality, the sub-carriers are spaced in frequency by the inverse of the symbol duration, theoretically resulting in zero inter-carrier interference. The relatively high ensemble bandwidth of 1.536 MHz gives good frequency diversity, since frequencies are not affected in the same way by fading. Adjacent bits within the MPEG bit-stream are made statistically independent with respect to bit errors by employing frequency and time interleaving, leading to good performance of the convolutional decoder (Viterbi) at the receiving end.

Table 24.2 *Properties of DAB transmission modes*

Parameter	Mode 1	Mode 2	Mode 3
Number of carriers	1536	384	192
Carrier frequency separation	1 kHz	4 kHz	8 kHz
Maximum radio frequency	375 MHz	1.5 GHz	3 GHz
Transmission frame duration	96 ms	24 ms	24 ms
Number of symbols/frame	76	76	153
Total symbol duration	1.246 ms	312 μs	156 μs
Guard interval duration	246 μs	62 μs	31 μs
'Active' symbol duration	1 ms	250 μs	125 μs
Null symbol duration	1.296 ms	324 μs	168 μs

Transmission modes

There are three different transmission modes, applicable to different ranges of radio frequency with the number of carriers and several other system parameters depending on the mode. DAB has three main modes with 1536, 768, 384 carriers and guard intervals between 246, 62 and 31 μs respectively (Table 24.2). There is a fourth mode between Mode 1 and Mode 2 with a symbol duration of 623 μs used in Canada. In each mode, the carriers occupy a total bandwidth of 1.536 MHz, they use DQPSK and time- as well as frequency interleaving. The maximum radio frequency that be used in each mode is that at which the system can overcome the Doppler effect while moving at speeds of up to 100 km/h. For mode 1, the maximum RF frequency is 375 MHz making it suitable for VHF transmission.

The total symbol duration consists of the principal symbol period and a guard interval. The latter prevents the echo of the previous symbol from interfering with the current symbol. By doing so, *inter-symbol interference (ISI)* is reduced to almost zero as long as the echoes from the various transmitters and propagation paths do not substantially exceed the guard interval. The maximum permissible difference in the length of the propagation path between two SFN transmitters D in meters can be calculated from the guard interval T_g and the propagation speed c:

$$D = T_g \times c \quad \text{where } c = 3 \times 10^8 \text{ m/s}$$

Mode 1 is intended for terrestrial transmission, particularly using SFNs. Its comparatively long symbol duration (1.246 ms) and guard period (246 μs) makes it most appropriate for a large network of terrestrial VHF (Band III) transmitters. Mode 2 is intended principally for small to medium coverage area (e.g. local radio) using UHF L-Band. The guard interval is sufficiently long to ensure immunity from multi-path propagation, but is not really suitable for SFN applications. Mode 3 is intended for cable and satellite transmission where there are no long echoes using the UHF L-Band.

For Mode 1, the available bit rate may be calculated as follows:

$$1536 \text{ carriers} \times 2 \text{ bits/carrier} = 3072 \text{ bits per symbol}$$

With a symbol duration of 1 ms, the number of symbols/s = 1000, resulting in a bit rate of

$$3072 \times 1000 = 3,072,000 \text{ bits/s} = 3.072 \text{ Mbps}.$$

However, not all the bit rate is available because of redundancy for error correction, control, synchronisation and guard period resulting in a useful bit rate of about 2.3 Mbps. This can provide, for example, five stereo programme services each at 224 kbps.

DAB frames

Although DAB is essentially dedicated to the transmission of 'audio' service, it may also deliver other services under the banner of 'general data' service, which may be data for the display of extended text (e.g. the contents of the 'Radio Times'). The partitioning of data into frames representing 24 ms periods of the application is retained but, generally, these are referred to as *'logical frames'*. It is helpful to consider each logical frame as a burst of data, because when the data for numerous services are multiplexed together they must be compressed in time, so each logical frame is transmitted in less than 24 ms and other data are transmitted between these bursts.

DAB-2

In November 2006, WorldDMB, the organisation in charge of the DAB standards, announced that the DAB system was in the process of being upgraded, and it will adopt the AAC+ audio codec to improve the efficiency of the system and stronger error-correction coding to improve the robustness of transmissions. This means there are two different versions of the DAB system: the older existing one, which was developed in the late 1980s, and an upgraded version, which has been dubbed *'DAB+'* or *'DAB version 2'*. Existing DAB receivers are incompatible with the new DAB standard, but receivers that will support the new DAB standard will become available in spring 2007.

DAB TV

As can be seen from the above, DAB easily lends itself to portable and handheld receivers as it was designed with mobile reception and SFNs in mind. It was not surprising therefore that it became a favourite in the delivery of digital multimedia broadcasting. Unlike DVB which had to be modified to incorporate the requirements for portable and handheld reception, DAB from its inception was designed for mobile reception with one antenna. With DAB, data is sent in bursts that are part of a frame which lasts 24 ms followed by a null frame using time interleaving to overcome the problem of fading. Another advantage of DAB is the use of *unequal error-protection (UEP)* technique in which bits are protected according to their importance in the decoding process. This is very important for mobile and portable reception where hostile reception conditions cannot be avoided.

The DAB system is capable of carrying IP packets (datagrams) using IP/UDP connectionless protocol. As these packets travel unidirectionally from a service provider to many users simultaneously, it is not necessary to establish a connection between the transmitter and the user prior to the transmission of data.

A DAB-TV system, also known as *DAB-IP* is illustrated in Figure 24.5. Live TV is encoded and encapsulated into an IP frame to be multiplexed with the normal digital radio broadcasts. The resulting transport stream is then fed into the DAB broadcast network and transmitted as a Mode 1 OFDM VHF signal. The IP interface provides an independent platform which supports a wide range of services and applications. TV encoding employs *Enhanced Packet Mode* (a WorldDMD Forum standard) which enables video and other services such as an electronic program guide (EPG) that are more sensitive to errors than the native audio services, to be carried.

At the receiving end, the handset decoder extracts the required TV channel from the multiplex, decodes it and feeds it into the small-screen

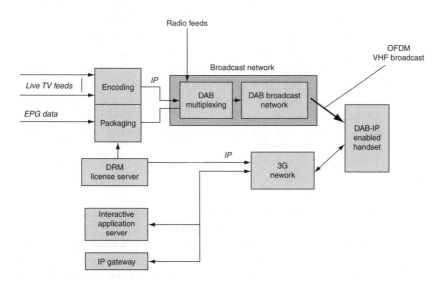

Figure 24.5 *DAB TV system architecture*

display. The handset also receives and decodes the normal interactive services through the third generation (3G) platform.

As can be seen, DAB-TV shares multiplex capacity with DAB digital audio services. This enables operators to use spare capacity on DAB networks to start offering mobile TV without waiting for a new spectrum to become available. This is the main reason for DAB-TV gaining ground over its rival DVB-H system.

On-line convergence

On-line convergence involves sending video broadcast services on traditional twisted-pair telephone lines. There is nothing new about video streaming, sending video clips down the line to be downloaded on a PC. However, sending live TV broadcasts down the line, usually known as IPTV, is of a qualitatively different scale. These services are often called *Broadband TV, ADSL TV, DSL TV* or *IPTV*. First let us look at the telephone system and at the technique known as ADSL.

In the UK alone there are over 30 million twisted-pair phone lines in operation between British Telecom exchanges and individual subscribers' premises. Originally the telephone lines were designed to carry simple command (dialling) pulses and frequency-restricted (300–3500 Hz) baseband voice signals. The remaining bandwidth of a copper wire was left unused. With the introduction of electronic exchanges, touch-tone dialling and routing functions became possible. The next step was to use a full digital system where the copper wires from each subscriber terminated in an interface or *'line card'* containing ADCs and DACs. This made a wide band of frequencies available to be divided into 4 kHz 'telephone channels'. The sampling rate of 8 kHz and 8-bit quantisation retained the traditional analogue bandwidth resulting in a 64 kbps per telephone channel. First, time-division multiplexing was used in what is known as *integrated services digital network* (*ISDN*) and by combining a number of 64 kbps channels, high bit rates were reached. One of the problems of ISDN is that the bit rate is limited and a new cabling from the subscriber to the exchange was needed. *Asymmetrical digital subscriber line* (*ADSL*) solved that problem by using frequency-division techniques and dynamic control of the bit rate.

ADSL

ADSL works on frequency-division multiplexing using 4 kHz wide channels (Figure 24.6). Twenty five channels are dedicated for the *upstream* or *back channel* (subscriber to provider) and 249 for the upstream (provider to subscriber). The different bandwidth allocation for the two streams is the 'A' for asymmetrical part of ADSL. As can be seen from Figure 24.6, the traditional analogue telephone is retained with the upstream occupying the frequency band above it and the downstream occupying the highest

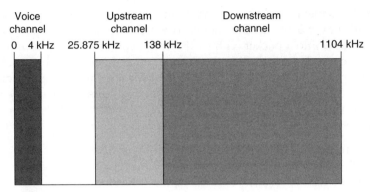

Figure 24.6 *ADSL bandwidth allocation*

frequency bands. To accommodate the varying quality of the telephone lines, ADSL constantly monitors the conditions of each channel and if a given channel has adequate level and low noise, the full bit rate is used. However, where a channel suffered from attenuation and noise, the bit rate is reduced. By independently coding the channels in terms of bit rate, the optimum data throughput for each telephone cable is obtained. Each ADSL channel is modulated using *discrete multitone technique* (*DMT*) in which combinations of discrete sub-carriers are used in a similar way to the OFDM scheme described in Chapter 8.

In the *downstream*, 249 sub-carriers are used, with each sub-carrier QAM-modulated with 0–15 bits. With a symbol rate of 4000, the maximum theoretical downstream bit rate = $15 \times 249 \times 4000 = 14.94$ Mbps. However, in practice, the maximum achievable downstream bit rate is 8.128 Mbps which itself is affected by the length of the telephone line from the subscriber to the exchange. Higher bit rates are obtained with other techniques such as *ADSL2*.

At the subscriber's home is a *band splitter* and a modem, and at the telephone exchange a modem and a *subscriber access multiplexer* (*SAM*), which sends the signal into a cable network on its way to the service provider. Unlike the ordinary telephone system with its line-grabbing and dial-up processes, ADSL is continuously alive and capable of two-way communication; data routing is directed by addressing information within the datastream, and each subscriber's terminal has an individual address.

IPTV

Internet protocol television (*IPTV*), also known as broadband television (BTV) involves accessing multimedia content via a broadband connection and viewing it on a normal TV. IPTV is not the same as Internet TV which accesses TV via a PC. IPTV is sometimes called ADSL TV or DSL TV.

The IP as mentioned above is a packet delivery system operating at the Network Layer 3 of the OSI model in which the data load (payload) is encapsulated into a packet with an IP header containing various information and control bits including the destination and source addresses. Since IP networks are bi-directional, IPTV can deliver not only live television but also interactive and on-demand TV. Telecom operators who have been traditionally interested in providing communication services between clients find their role is being expanded to provide what is known as *Triple Play*:

- communication services (including Voice over IP, VoIP);
- a high-speed internet connection;
- IP-based television and video-on-demand services

At the receiving end, playback requires only an Internet connection and an Internet-enabled device such as a personal computer, iPod, set-top-box connected to a TV receiver or even a 3G cell/mobile phone to watch the IPTV broadcasts. Apple's *iPhone* uses mobile phone *Global System for Mobile Communications* (*GSM*) quad-band (900-MHz, 1800 MHz for Europe and 850 and 1900-MHz for the Americas) to provide images and television shows and films, internet, email and text messages as well as mobile phone facility.

Bit rates

One of the main issues concerning broadband services delivered over the telephone network is their reach or coverage area. Generally, the higher the bitrate of the signal, the lower is the reach of the service. As the bandwidth required for minimum TV quality is relatively high when compared with broadband internet (2–4 Mbps in the case of MPEG-2), the TV reach is more limited than the traditional reach of broadband internet connections. The operators have several options to cope with this problem:

- They can reduce the bit rate and with it picture quality;
- they can use more advanced encoding schemes such as MPEG-4 AVC/H.264 or VC-1 (this will also help to pave the way to HDTV transmission), or
- they can upgrade their networks by introducing more efficient transmission technologies such as ADSL2+ or VDSL (very high bit rate DSL).

Upgrading the infrastructure requires significant capital investment and takes time to implement which leaves the second option as the favourite since reducing picture quality is not acceptable for the majority of customers.

The bandwidth problem becomes even more severe if more than one TV stream to the home is required. Such a need may arise if there are a number of TV sets in a house (e.g. one in the living room and another in a child's bedroom), each requesting a different TV programme at the same time. More than one TV stream is also required if there is a local personal video recorder (PVR) with one stream being recorded locally while another is being watched.

One of the main applications of IPTV is *'network-based private video recorder' (NPVR)*. Network Personal Video Recording is the ultimate time-shifted viewing where real-time broadcast television is captured in the network on a server allowing the end user to access the recorded programs on the schedule of their choice, rather than being tied to the broadcast schedule.

The drawback of IPTV is the inclination of IP connection towards packet loss and delays in cases where the IP link is not fast enough. While this may be overcome by the inclusion of a video buffer at the receiving end in which case, lost packets may be re-transmitted, to ensure sufficient picture and sound quality, IPTV requires a reliable network with a robust *Quality-of-Service (QoS)* mechanism. The required QoS can only be met by providers that are able to control all elements of the transmission path from the source to the user's premises. This is the reason why open Internet is not able to offer IPTV services, as it cannot guarantee QoS. In addition, streaming over open Internet would require some technical measures that address piracy, spoofing and network congestion. A comparison of the properties of IPTV and Internet video streaming is outlined in Table 24.3.

Closed IPTV network

A block diagram of the essential elements of a closed IPTV network is shown in Figure 24.7. It consists of four parts:

- Video head-end
- Packet core, transport and network edge
- Access network
- Home network

Content arrives from a satellite or antenna in digital or analogue format, with standard or high definition (or music), encrypted or unencrypted. Once the signal is *'down-linked'* or *'down-converted,'* it may need to be altered. Most digital signals use MPEG-2 encoding and transcoding to MPEG-4 or Microsoft's VC1 is now available to reduce the required bandwidth by up to 50%. Encoding methods intrinsically produce variable bit rates (VBR), in which fast-motion requires more bandwidth. With bandwidth at a premium, the operator usually limits the bandwidth that a channel can consume and converts the signal into a constant bit rate (CBR) packet stream. Cross-conversion may be used to change the resolution of the displayed picture. For example, a signal received in 1080i (resolution of 1920×1080) format may be converted to 720p (resolution of 1280×720) or into a mobile-friendly format before distribution. For analogue signals, an encoder digitises, compresses and packetises the signal.

Table 24.3 *Comparisons between IPTV and internet video streaming*

	IPTV	Internet video streaming
Footprint	Local (limited operator coverage)	Potentially supranational or worldwide
Users	Known customers with known IP addresses and known locations	Any users (generally unknown)
Video Quality	Standard or high-definition television QoS	Best effort quality, QoS not guaranteed
Connection bandwidth	Between 1 and 4 Mbit/s	Generally below 1 Mbit/s
Video format	MPEG-2 MPEG-4 Part 2 MPEG-4 Part 10 (AVC) Microsoft VC1	Windows media RealNetworks QuickTime Flash, and others
Receiver device	Set-top box with a television display	PC
Resolution	Full TV display	QCIF/CIF
Reliability	Stable	Subject to contention
Security	Users are authenticated and protected	Unsafe
Copyright	Media is protected	Often unprotected
Other services	EPG, PVR (local or network)	
Customer relationship	Yes; onsite support	Generally no
Complementarity with cable, terrestrial and satellite broadcasting	Potentially common STB, complementary coverage, common metadata	Pre-view and low-quality on-demand services

At the stage, high value content is encrypted so that it may be viewed by paying customers only. Encryption also protects against piracy.

The video and audio content from the various sources described above is fed into the multi-service interface where it is IP encapsulated before going into the internet routing and delivery network. The network must have capacity and QoS reliability to transport massive amounts of simultaneous video traffic from the video headend to central exchange from where it is fed to the subscriber via the *Access Infrastructure Unit*. The Access Unit is usually the bottleneck in terms of capacity to deliver a reasonable video service. The benchmark for IPTV service delivery over Broadband is

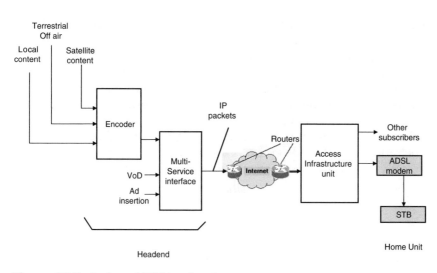

Figure 24.7 *A closed IPTV network*

20 Mbps per home based on providing two SDTV, one HDTV, voice, high-speed data and gaming. In addition, the access network must evolve to include features such a TV channel change and multicasting.

As was mentioned earlier, DSL is distance-sensitive. Whilst a TV channel can now be delivered over 2 Mbps most people consider that 20 Mbps down-stream is a requirement for offering an array of Video services to the home.

Fibre to the home with *passive optical networks* (*PON*) technology is used to deliver service using end-to-end fibre. Equally effective is Optical Ethernet used for point-to-point connection to customers.

Video-on-demand

The basic principle of video on demand (*VoD*) is very simple. Digitised video information stored on magnetic disks is retrieved by a video server and delivered to the home or office, where it is reconstructed using either a PC or a STP and displayed on a television set. One approach to video distribution is to use a number of channels to broadcast the same video piece such as a film with starting times staggered by say 5–10 min. Known as near video on demand (*NVOD*), this method will allow the viewer to choose the viewing time within specified limits. True VoD allows the viewer full choice of viewing time, together with VCR-type controls such as fast forward, rewind, replay and so on. Modern VoD is based on IP uni-cast stream, which means instead of broadcasting a signal to everyone, each consumer now has a personalised channel.

One of the greatest challenges to VoD is the amount of bandwidth required. Consider a network providing 150 standard definition multi-cast television channels (each requiring 1.5 Mbps) to 3000 consumers. Without VoD, the operator 'requires 225 Mbps (150 channels × 1.5 Mbps) on the network backbone to deliver television service to all consumers. The number of channels determine the required bandwidth. If every consumer is watching VoD programming, the operator requires 4.5 Gbps (3000 consumers × 1.5 Mbps) on the network backbone. Now, the number of consumers determines the required bandwidth.

Because of the bandwidth impact of VoD, understanding the expected penetration rate of VoD service is critical when designing the network. Typically the service mix is 90% Broadcast or multi-cast TV and 10% VoD.

25 Interconnectivity and ports

As the count of electronic 'black boxes' in the home increases, so do the number and complexity of the connections between them. The evolution of technology has brought with it progressively better and lower-loss signal linkages; over the years, we have moved from RF to composite video and then to S-video coupling for the picture and now use RGB links for some equipment. The advent of digital transmission and recording equipment, and the convergence of TV and computer technology and systems have spawned the domestic use of Firewire/IEE1394 data coupling. And increasingly we exchange control commands and data between boxes for the purposes of editing, automated recording, function switching and the like. This chapter examines the pros and cons of the various coupling and connection systems used in the domestic environment.

RF connections

For full versatility in recording and viewing of broadcast transmissions, the UHF aerial feed must be looped through the VCR and terrestrial DTV box on its way to the TV set. Some equipment (particularly satellite boxes) can generate interference in the UHF band, in which case the use of double-screened cable like the CT 100 type used for satellite downleads can be helpful when used for inter-equipment links working at UHF. Even though VCRs and terrestrial DTV receivers must have a UHF input, it is far better that their modulators are switched off, and that their signal outputs go via AV (usually SCART) cables to avoid carrier clashes and consequent interference in the UHF band; these boxes are often designed to act as 'switching centres', automatically routing the AV signals as required.

Where a VCR and/or DTV box is required to feed several screens simultaneously, perhaps at widely different points, RF distribution can be used; if more than two TV sets are involved, a multi-output UHF amplifier will be necessary, and by this means many receivers can be fed, at great distances if required. This system does not permit stereo sound to be conveyed, and may involve difficulty in setting carrier frequencies to avoid interference, but has the advantage of simplicity and convenience. Figure 25.1 shows a local UHF network.

Very often a household distribution system involves the main TV in the lounge and a second set in bedroom or kitchen, from where it is required to 'drive' the equipment in the living room. It can be achieved by the use of a remote control extender which has an infra-red receiving eye near the

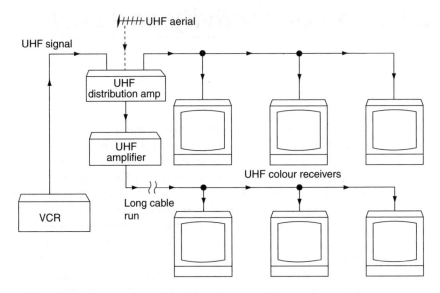

Figure 25.1 *RF signal distribution system for sound and vision*

second TV to relay command codes—either via the RF cable or by a wireless link—back to a mains-powered repeater in sight of the equipment to be controlled. Sky digital receivers have a built-in facility to power and operate a '*Remote Eye*' which is simple to connect using just a coaxial cable, but whose operation is limited to the sat-box itself.

Baseband distribution

Better picture and sound quality in a 'local network' comes from using a baseband distribution system as shown in Figure 25.2. This also gives the opportunity to distribute stereo and Dolby surround sound where it is

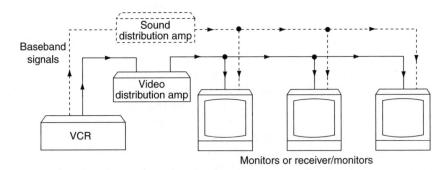

Figure 25.2 *Baseband video distribution system for better picture quality than that offered by an RF network*

present, while avoiding the risk of interference and patterning. Some applications outside entertainment TV may not require sound, dispensing with the dotted components in the diagram.

SCART connections

The 21-pin SCART coupling system is a versatile one. Its plug connection is shown in Figure 25.3 together with pin assignments. The SCART connector is a 21-pin non-reversible device. There are 20 pins available for

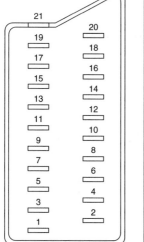

Pin No	Signal	Signal level
1	Audio output B (right)	Standard level: 0.5V rms Output impedence: Less than 1kohm*
2	Audio input B (right)	Standard level: 0.5V rms Output impedence: More than 10kohm*
3	Audio output A (left)	Standard level: 0.5V rms Output impedence: Less than 1kohm*
4	Ground (audio)	
5	Ground (blue)	
6	Audio input A (left)	Standard level: 0.5V rms Output impedence: More than 10kohm*
7	Blue input	0.7 +/- 3dB, 75 ohms positive
8	Function select (AV control)	High state (9.5–12V): AV mode Low state (0-2V): TV mode Input impedence: More than 10K ohms Input capacitance: Less than 2nF
9	Ground (green)	
10	Open	
11	Green	Green signal: 0.7 +/– 3dB, 75 ohms, positive
12	Open	
13	Ground (red)	
14	Ground (blanking)	
15	Red input	0.7 +/– 3dB, 75 ohms, positive
15	(S signal chroma input)	0.3 +/– 3dB, 75 ohms, positive
16	(Blanking input Ys signal)	High state (1–3V) Low state (0–0.4V) Input impedence: 75 ohms
17	Ground (video output)	
18	Ground (video input)	
19	Video output	1V +/– 3dB, 75 ohms, positive sync 0.3V (–3 + 10dB)
20	Video input	1V +/– 3dB, 75 ohms, positive sync 0.3V (–3 + 10dB)
20	Video input Y (S signal)	1V +/– 3dB, 75 ohms, positive sync 0.3V (–3 + 10dB)
21	Common ground (plug, shield)	

Figure 25.3 *SCART connector: layout, pinning and signal levels. Pins 10 and 12 are used by manufacturers for various purposes, mainly system control*

connections with pin 21 connected to the skirt and hence the chassis provided the overall screening for cable communication.

The SCART connector allows direct RGB connection to a TV receiver as well as CVBS together with independent stereo sound channel connections. It can also provide S- (separate Y/C) video. For a standard colour bar test signal, the expected waveforms at each of the video output (CVBS, red, green, blue and Y-C) are illustrated in Figure 25.4.

Pins 10 and 12 are used for intercommunications between devices connected to the SCART socket. Pin 8 carries out some control functions: when it goes high (9–12 V) it switches the TV to AV input, or a standing-by digital TV receiver to loop-through mode. At a level of about 6 V, it signals the presence of a wide-screen picture from VCR or set-top box (STB) to a suitably equipped TV which switches scan generators accordingly. In

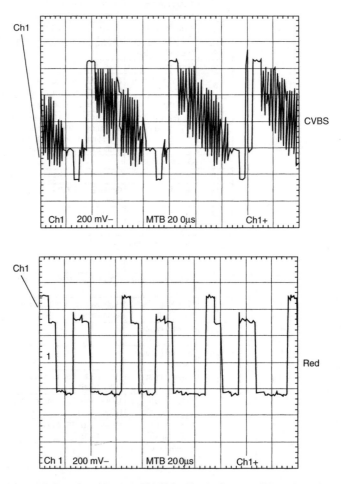

Figure 25.4 *Video waveforms: CVBS, Red, Green, Blue, Luminance Y and Chrominance C. Continued on pages 533 and 534*

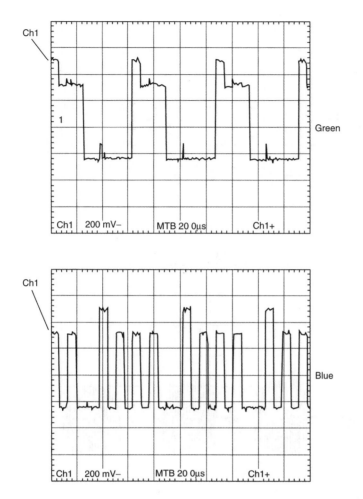

Figure 25.4 *Continued*

S-video mode, the luminance and sync signals are passed out on pin 19 and received on pin 20, while pin 15 conveys the C signal at a burst level of 300 mV. Switching between composite video and S-video is carried out at pin 16. TVs, DVD players and STBs can be programmed for the various signal modes, and it is important that the boxes at both ends of the cable are in agreement as to the mode in use; and that the best possible mode is selected, depending on the signal and source in use and the capabilities of the equipment involved. Thus RGB is appropriate to DVD players and large-screen TVs, while composite video is used by low-band (VHS, Video-8) machines.

Inexpensive SCART coupling leads have neither screened signal conductors nor separate ground paths for them, giving rise to cross-talk. This is manifested as *floating bars*, outlines and/or *colours on vision* and

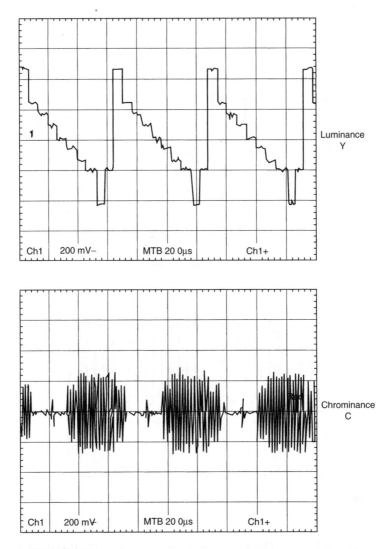

Figure 25.4 *Continued*

buzz/cross-talk on sound. The picture problems can often be solved by cutting off the connection to pin 19 in the SCART plug at the TV end, so long as no video-out signal is required from the TV tuner. The appropriation of SCART pins 10 and 12 by manufacturers for various purposes can lead to trouble, so it is best to either cut them or use leads in which they are not connected – unless the connection specifically requires their use. Indeed a fully wired SCART lead, heavy, thick and relatively expensive, is best avoided unless the application calls for it. Low-level applications involving just composite video and mono or stereo sound are best served by a lead with screened conductors at pins 1/2/3/6/19/20 (basically three crossed-over signal feeds) plus a link between pin 8. The control function of SCART pin 8, pulling the TV into AV mode, can be a nuisance where for example it is required to watch a TV programme while recording another from satellite. Often there is a TV/SAT key on the sat-box's remote control to overcome this, but if necessary the lead can be cut from SCART pin 8, when signal input selection is under the sole control of the TV set's remote handset.

Video coupling

There are three main modes of conveying an analogue video signal: composite video, S-video and RGB. We shall look at their characteristics in turn.

Composite video (CVBS) is the most common form of picture signal, representing the PAL (or other video signal) as broadcast by analogue terrestrial transmitters. While it can convey signals between any two pieces of video equipment it is mainly applicable to low-band VCRs and signals which come from analogue transmitters, satellite or terrestrial. The picture from a composite signal is limited in bandwidth/definition, and can display cross-colour and other spurious effects.

Similar to a composite video, the S-video coupling has separate leads and connections for luminance/sync and chroma components. It has the advantage of avoiding cross-talk effects like cross-colour, and it affords a little more picture definition by virtue of the absence of Y/C separating filters. It is applicable to high-band (S-VHS, Hi-8) camcorders and VCRs, also digital TV receivers and DVD players where an S-output port is provided. S-video is available in a specially designed plug/socket connector as shown in Figure 25.5. This cannot convey audio as well, so S-connection requires a separate audio link, typically in two (L, R) phono leads.

RGB offers the best possible video coupling mode for domestic applications, but it is only applicable where the signal is generated on the spot, as it were: DVD players, DTV receivers. Here, there are separate paths for each primary colour: red, green and blue, with virtually no bandwidth restriction inherent in the linking system or mode. In home-video applications, it is usually carried in SCART links as described above with sync at 300 mV peak on pin 20 (in) and pin 19 (out). VCRs do not deal in RGB signals, though some models have a loop-through SCART facility for them.

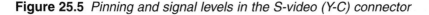

Pin No.	Signal	Level
1	Ground Y	
2	Ground C	
3	Y (S signal) input	1V ± 3dB 75 ohm, positive Sync. 0.3V − 3 + 10dB
4	C (S signal) input	0.3V ± 3dB 75 ohm, positive Sync.

Figure 25.5 *Pinning and signal levels in the S-video (Y-C) connector*

Audio coupling

Sound signals are coupled between boxes at a standard level of 500 mV rms, most commonly in the SCART connector, but also in phono links, using sockets for both input and output with a colour code of red for right channel and white for left channel or mono: a yellow phono socket generally carries video signals.

Dolby digital and other audio decoders and processors commonly form part of a home-cinema ensemble. Although they are concerned, for operational purposes, solely with the sound signal, it is necessary to include them in the video path on its way to the TV or monitor so that their captions and on-screen displays can be shown on screen. This enables the user to set up the system, check its mode and status, and get an indication of signal source. These processors/amplifiers are well equipped with input and output sockets for this purpose, and can form a convenient remote-controlled 'switching centre' for the various components of a complete AV system.

Speaker wiring

Apart from ensuring that each speaker is hooked to the correct socket—corresponding to its position in the room—the most important aspect of loudspeaker wiring is correct phasing, in which the + terminal of each speaker is connected to the + terminal on the box. Incorrect speaker phasing gives rise to strange and unnatural effects in the reproduced stereo or surround sound field. Loudspeaker power and impedance matching is also important: too 'small' a loudspeaker may get damaged at high volume levels, while too low a loudspeaker impedance – for instance by connecting two $4\,\Omega$ units in parallel—risks damage, at high volume level, to the output section of the amplifier in the box.

Cordless loudspeakers and headphones save the need to run audio cables in the viewing/listening room. They depend on infra-red or RF transmitters at the box, whose emissions are picked up by mains and battery-operated receiver/demodulator/amplifier systems, respectively. The carrier frequency for IR types is 2.3/2.8 MHz, and for RF systems 863–864 MHz, with the latter generally giving the best results.

Digital ports

There are several digital connections in common use in domestic electronic equipment. The digital video interface (DVI) and the high-definition multimedia interface (HDMI) were described in detail in Chapter 16. Some audio processor and video equipment such as digital TV receiver and DVD players are equipped to exchange multi-channel sound signals in digital bitstream format using optical fibre or RF/coaxial media. Here the complete audio bitstream is modulated onto an optical or RF carrier for lossless passage to a suitably equipped surround sound decoder.

Video graphic array

The *video graphic array* (*VGA*) interface was introduced in 1987 and since then it has become the standard connection between a PC and a display unit. The layout of the 15 pins and their functions is displayed in Figure 25.6. The original VGA connector was designed for a resolution of 640×350 with 8-bit colour coding giving 256 different colours. Pins 4, 11, 12 and 15 formed a unidirectional or a bidirectional communication medium. One of their functions is to identify the type of display and its capabilities. Pin 15 may also be used to turn the display on and off. The VGA connector was retained in its 15-pin format by later improved adaptors such as the super VGA (SVGA) and extended VGA (XVGA) designs. These added new features to the systems and also added support for higher resolutions such as SVGA at 800×600, XGA at 1024×768, super XGA (SXGA) at 1280×1024 and the ultra XGA (UXGA) at 1600×1200.

Universal serial bus

The *universal serial bus* (*USB*; pinout Figure 25.7) was first introduced in 1995 as a general-purpose low-cost connection between a PC and other peripheral devices such as a printer. Today, it is used in various devices including camcorders and projection systems. The original USB 1.0 had two levels of performance: 1.5 and 12 Mbps. It has four wires: 5 V supply with its ground and a bidirectional differential pair for data transmission (Table 25.1). The data transmission format is specified such that the serial bitstream is self-locking in that the timing information is derived from the data itself. A single USB host controller can support up to 127 peripheral

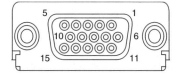

1	Red video	9	+5 VDC
2	Green video	10	Sync. return
3	Blue video	11	Unused
4	Unused	12	Data
5	Return	13	Horizontal sync
6	Red return	14	Vertical sync
7	Green return	15	Data clk.
8	Blue return		

Figure 25.6 *The VGA port and its pin function*

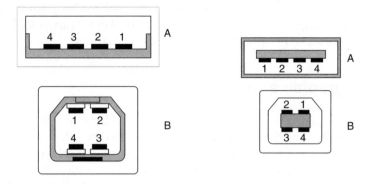

Figure 25.7 *USB connectors (male and female) pinout.*

Table 25.1   *USB Pin configuration*

Pin	Name	Description	Cable colour
1	Vcc	5 V d.c.	Red
2	D−	Data−	White
3	D+	Data+	Green
4	Gnd	Ground	Black

devices. Typically, though, a USB will have some half-dozen devices on its interface at any one time. The serial link performance was greatly improved with USB 2.0 which can support a bit rate up to 480 Mbps more than adequate to support compressed HDTV.

USB was designed from the start to permit what is known as *'isochronous'* data, such as video and audio data in which the timing of each data packet within the overall stream must be maintained for proper recovery at the receiving end. A block diagram of a USB driver IC is shown in Figure 25.8.

RS-232 connector

RS-232 (RS stands for recommended standard) is a common serial communication standard for PC and peripheral devices. It has been introduced in consumer electronics mainly for the purposes of connecting the device to a PC for the purposes of testing and upgrading the program of imbedded flash chips. The connector may be 25-way or more usually a 9-way D-type male. The pin format and pin functions are shown in Figure 25.9. Note that an RS-232 cable must have its pins 2 and 3 switched.

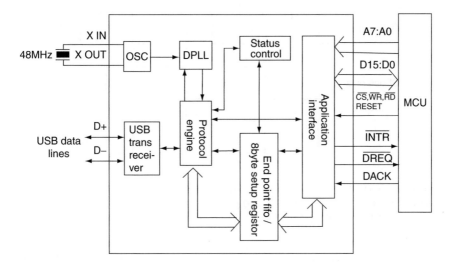

Figure 25.8 *USB driver IC*

CD   1   ● ● 6 DSR   RS-232 pinout
RD   2   ● ● 7 RTS
TD   3   ● ● 8 CTS
DTR   4   ● ● 9 RI
GND   5   ● ●

RS-232 pin functions

Pin no.	Function	
1	Carrier detect	CD
2	Receive data	RD or RX
3	Transmit data	TD or TX
4	Data terminal ready	DTR
5	Ground	GND
6	Data set ready	DSR
7	Request to send	RTS
8	Clear to send	CTS
9	Ring indicator	RI

Figure 25.9 *RS-232 pin assignment and functions*

Firewire

Other names for the Firewire link system are *IEE1394* and *i-Link*: they all
amount to the same thing in practice, but different equipment manufactures prefer different names! Firewire is a very fast (up to 400 Mbps) serial
data link originated by Apple Computers as a LAN (local area network)

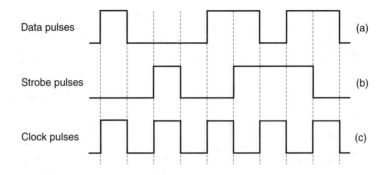

Figure 25.10 *Pulse trains in a Firewire link. The data pulses (a) are conveyed in one wire, and the strobe pulses (b), in another. Clocking pulses (c) are derived from (a) and (b)*

protocol for computers, and now adopted as a standard interface for domestic digital exchange between boxes: it can carry entertainment, communication and computing data.

The Firewire data is carried by a cable having two individually screened twisted-pair cables, one for NRZ (non-return-to-zero) data and one for strobe pulses as shown in Figure 25.10. Exclusive-OR gating of the two pulse trains provides the system clocking pulses. The pulse amplitude of each train is 220 mV, centred on a bias voltage of 1.86 V. The data stream has two components: a relatively slow asynchronous unidirectional pulse train for control purposes; and, time-interleaved with it, a very fast isochronous one which carries 'payload' data in the form of variable-length packets, with header, ident, address, data and error-check components. A negotiation and arbitration process is used for access control in a Firewire network.

The Firewire link system is bidirectional, with a capability of 63 devices on a single bus, and a vast number of them on a bridge-bus system. Firewire ports are provided on some computers, on DV-interface computer cards, digital TVs and STBs as well as DV and D-VHS cassette decks and camcorders.

Appendix A1
Teletext

Teletext is the oldest form of digital broadcasting technology, having been used in the UK for over 30 years, during which the decoder (an 'optional extra' in a TV set) has steadily shrunk in size and cost, though not in the complexity of the internal workings of its ICs or 'chips'. Teletext uses 'spare' lines in the broadcast field blanking interval to carry a digital pulse stream which when decoded can type out data on the TV screen in the forms of text and simple graphics like maps and diagrams. TV lines 7–18 in even fields and 320–331 in odd fields are used to carry the data, which can be seen as lines of twinkling dots above the 'live' picture if its height is turned down.

Each text line carries a colour burst to maintain the colour decoder's sub-carrier oscillator synchronisation, along with the usual line sync pulse and blanking interval. The line time given to text pulses is the same for that of picture signal: 52 µs. To prevent inter-carrier buzz with FM sound systems, the amplitude of the text pulses is limited to 66% of the peak white signal. In this two-state AM data transmission system, binary zero is signalled by the text signal hitting black level, and binary one by the signal rising to the 66% level, which represents 0.462 V with a standard 1 V peak-to-peak video signal. The pulses are not square: they are filtered before transmission and further rounded by the receiver's IF and demodulator circuits, and thus reduced to virtually sinusoidal shape by the time they reach the text decoder. The pulses are broadcast in non-return-to-zero (NRZ) form, so that the signal remains high to indicate a string of binary ones and low when a string of binary zeros are being conveyed.

Text data format

The text data rate is 6.9375 Mbps, corresponding to 444 times the line frequency, though the pulses are not necessarily locked to line scan rate. A data stream representing 01010101, etc., throughout the line will thus correspond to a quasi-sine wave at 3.46875 MHz. At this rate, there is room for 360 data bits on each TV line used for text conveyance: these 360 bits produce a single row of graphics or characters in the teletext screen display. The 360 bits per line are grouped into 45 eight-bit bytes, the first five of which synchronise and prime the decoder, and provide an 'address' to indicate where the row is to be placed in the text display. The remaining 40 bytes define the characters in the row, which can thus have a maximum of 40 letters, numbers or graphic blocks. Blanks or gaps, e.g. between words, have their own code so that data is always present.

The text display has 23 rows, permitting a maximum of $23 \times 40 = 920$ characters to be displayed on a single page to form simple maps, diagrams and alphanumeric text. Upper- and lower-case letters can be produced, as well as figures, symbols and 'building blocks' for graphics, in any of eight colours formed by the three primaries, plus black and white.

The teletext code table is shown in Figure A1.1. Starting at the top left-hand corner, the first four bits define the row number and the next three the column number. The codes are thus used to determine the character so that capital T, for example, would be called up by byte 0010101 (bits 1–7),

$b_7 b_6 b_5$ →	000	001	010		011		100	101	110		111	
$b_4\,b_3\,b_2\,b_1$ ↓ / Col → / Row	0	1	2	2a	3	3a	4	5	6	6a	7	7a
0 0 0 0 — 0	NUL[1]	DLE[1]			0		@	P			P	
0 0 0 1 — 1	Alpha^n red	Graphics red	!		1		A	Q	a		q	
0 0 1 0 — 2	Alpha^n green	Graphics green	"		2		B	R	b		r	
0 0 1 1 — 3	Alpha^n yellow	Graphics yellow	£		3		C	S	c		s	
0 1 0 0 — 4	Alpha^n blue	Graphics blue	$		4		D	T	d		t	
0 1 0 1 — 5	Alpha^n magenta	Graphics magenta	%		5		E	U	e		u	
0 1 1 0 — 6	Alpha^n cyan[2]	Graphics cyan	&		6		F	V	f		v	
0 1 1 1 — 7	Alpha^n[2] white	Graphics white	'		7		G	W	g		w	
1 0 0 0 — 8	Flash	Conceal display[2]	(		8		H	X	h		x	
1 0 0 1 — 9	Steady[2]	Contiguous graphics[2]	)		9		I	Y	i		y	
1 0 1 0 — 10	End box[2]	Separated graphics	*		:		J	Z	j		z	
1 0 1 1 — 11	Start box	ESC[1]	+		;		K	←	K		$1/4$	
1 1 0 0 — 12	Normal[2] height	Black[2] background	,		<		L	$1/2$	l		‖	
1 1 0 1 — 13	Double height	New background	-		=		M	→	m		$3/4$	
1 1 1 0 — 14	SO[1]	Hold graphics[2]	.		>		N	↑	n		+	
1 1 1 1 — 15	SI[1]	Release[2] graphics	/		?		O	#	o			

Figure A1.1 *Code table for broadcast teletext. This is held in a ROM look-up table in the character generator IC of the text decoder*

while byte 1111101 produces the symbol #. Note that to change from a capital to a lower-case letter only bit no. 6 changes, and that the codes for numbers are made by their binary equivalents followed by the code 011. The 32 codes in columns 0 and 1 are called control codes, and they determine whether the codes in the following columns represent alphanumeric characters or graphics and assign some 'attribute', such as a colour, to the characters to be displayed. These control codes all have bits 6 and 7 at zero: they produce no display themselves, occupying a single blank space in the displayed row—usually in the background colour of the preceding character.

Error protection

On its way from the transmitter to the decoder the text data is vulnerable to corruption by noise and interference, so a protection system is used, in the form of an odd-parity bit at the end of each of the 7-bit codes shown in Figure A1.1 so that each becomes an 8-bit byte. The parity bit added may be a binary one or a binary zero, such that there is always an odd number of 1 bit in the byte. This enables the decoder to check whether any bytes with an even number of ones have been received and reject them as being corrupt. This simple system has two limitations: it fails if an even number of bits in the byte are incorrect, and it cannot provide data correction. Even so, it is adequate for use with text symbols, where a minor error often goes unnoticed, and is corrected anyway when the page is cyclically updated.

Some of the text data requires greater protection than is afforded by this simple parity check. If the data conveying the page number is corrupt, the wrong page could appear in response to a viewer's request, while if the row address data is wrong the row will be printed out in the wrong place on the screen. To avoid this, the bytes that carry this information and the real-time clock data are heavily protected by means of a Hamming Code in which every alternate bit is a parity bit: bits 2, 4, 6 and 8 convey data while bits 1, 3, 5 and 7 provide protection, as shown in Table A.1. This permits single-bit errors to be detected and corrected by inversion. The penalty incurred is that the volume of information conveyed by the byte is reduced.

Row coding

As we have seen, the text page is organised in 23 rows. Of these, the most important is row 0 (top row) because it contains the magazine and page numbers, the service name and the date and time – it is called the header row. To capture a page for display, the viewer keys in the magazine and page numbers: when this keyed-in data corresponds with the transmitted code, the required page is selected and displayed on the screen.

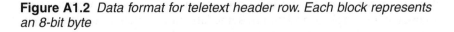

Clock run-in 10101010	Clock run-in 10101010	Framing code 11100100	Magazine and row-address group	Page number (units)	Page number (tens)	Page 'minutes' (units)	Page 'minutes' (tens)	Page 'hours' (units)	Page 'hours' (tens)	Control group A	Control group B	Character bytes

Figure A1.2 *Data format for teletext header row. Each block represents an 8-bit byte*

The data format for the header row is shown in Figure A1.2. First comes a clock run-in sequence consisting of 2 bytes carrying a series of ones and zeros, 101010, etc. It synchronises the decoder's bit-sampling clock oscillator. Next comes a framing-code byte whose purpose is to identify the start-point of the data bytes in the following pulse train. Its pattern is 11100100, chosen to give a reliable reset signal to the rest of the decoder and to be amenable to correction of single-bit errors by a simple check and correction circuit. The next 2 bytes, which are Hamming-protected, carry the magazine and row-address information. They are followed by page number and time-code bytes to enable the viewer to select a particular page at a specified time. Following these, the next two bytes provide information governing the whole page which is to follow: the control codes shown in columns 0 and 1 of Figure A1.1 for sub-title, update, news flash and similar functions, all with Hamming protection. Eight data bytes in the header row have thus been used, leaving 32 for the display of characters. Except for the page number, the header row text is the same for all the pages of any one transmission. It shows the name of the service (Ceefax, Skytext, etc.) and the day, date and real time in hours, minutes and seconds.

The following rows have the same clock run-in and framing codes as the header row. These are followed by individual magazine and row-address bytes, which are the only ones to have Hamming protection in the 'ordinary' rows. Bytes 6–45 contain 8-bit codes for the text or graphics in the line. Each byte is transmitted with the least significant bit (LSB) first and the parity bit last.

Transmission sequence and access time

We have seen that each text line transmitted during the field blanking period can produce one row of text on the screen. 24 TV lines are required to produce the 24 rows of a single page of text data, then, and at the normal transmission rate of 12 text lines per TV field it takes two fields to transmit a single teletext page. At a field rate of 50 per second, 25 pages a second, or 1500 pages a minute can be transmitted. If 750 pages are being transmitted, the worst-case access time will be 30 s and the average wait 15 s, assuming that the pages are transmitted repeatedly in sequence.

To maximise the user-friendliness of the system, high-priority pages such as indexes are transmitted more often than others. Also contributing to speed of access is a row-adoptive feature, in which the totally blank rows which occur in many text pages are omitted from the datastream. In addition to this, the text rows of several pages are interleaved, timewise, in the transmission to give a perceived reduction of the access time: this accounts for each row carrying its own magazine number. Sub-titles, assigned to page 888, have the highest priority: they are slotted into the first available TV line after the cueing point. These techniques are limited in their effectiveness in that most of them shorten the waiting time for some pages while lengthening that for others. For the user, the most effective reduction in page access time comes from the use of the Fastext system.

Text data memory and Fastext

There is no need to store the parity bits, so the random access memory (RAM) capacity required for a single page of text is 7 (bits) × 40 (characters) × 24 (rows), calling for 6720 bits, the minimum required in the decoder. Most decoders have a two-page storage capacity, in practice 2k × 7 bits, so that they can automatically capture the page following the one being displayed. If, for instance, page 183 is called up, page 184 is also captured and written into memory for immediate display. The action of the viewer stepping forward to 184 prompts the decoder to seek, acquire and store the data for the next page (185) in the now unused half of the memory, overwriting stored page 183 with the page 185 data.

Fastext decoders incorporate larger memories which can store the requested page and the next seven, automatically loaded in sequence by the decoder; as each page is discarded, the next page in the sequence is written into the memory. This is based on the premise that viewers usually step through pages sequentially and give instantaneous new-page readout when they do. The broadcaster's text editor can enhance the Fastext function by anticipating the user's requirements and sending additional instructions to the decoder, based on the current page's contents, for extra page acquisition. It works thus: the bottom row of the text page has four colour-coded 'prompts' which may typically be news, sports, travel and weather. The remote control handset has four keys with corresponding colours. Pressing any one of them selects the magazine and page numbers simultaneously, loading the required pages into the memory for instant selection. If, for instance, travel is selected the on-screen coloured prompts change to air, sea, road and 'continue', the latter providing an escape from the travel menu.

A 'packet' system, in which additional data rows are transmitted but not necessarily displayed on screen, is used for Fastext data acquisition. The packets contain additional control data rather than characters and are sent in advance of the teletext pages (which now become packets 0–23) so that the control data arrives ahead of the page to which it relates. Packet 24 is displayed on the screen as a row because it contains the Fastext prompts, while packet 27 provides the page-linking data. There are eight packets in the text specification, used for various purposes including linked character sets (primarily for other languages) and general-purpose data transmission. Packet 30 is assigned to programme delivery control (PDC) for videocassette recorders; the data it contains is captured and stored in a suitably equipped VCR, and used to initiate a recording at any required time on any channel which radiates PDC data. Its advantage is that it compensates for any unexpected alterations to the broadcast timing schedules.

Also contained in the teletext data stream packet is a widescreen switching (WSS) flag for application to the scan generators. It automatically selects the best scanning amplitudes for the programme being broadcast. This automatic function can be cancelled by the viewer if required.

Decoder operation

Figure A1.3 shows a text decoder in the simplest possible block diagram form. The first block gates the data pulses out of the video signal and decodes them. The next block selects the data requested by the viewer, according to instructions sent via the control logic block, and passes it to the page-memory RAM, whose write-in control signal also comes from the control block. Read-out from the RAM is again governed by the control block: it takes place at a slower rate, a complete field scan period being required for a page in the memory to be read out. It was written in during short $52\,\mu s$ bursts at varying intervals.

The data from the RAM is fed to a character-generator read only memory (ROM) which uses it to produce the character graphics and symbols shown in columns 2–7a of Figure A1.1, and to give them the colours and attributes assigned by the control data stored with the character codes. The outputs from the ROM consist of RGB signals in serial data form, plus a blanking pulse train.

Data from the viewer's infra-red control handset is fed into the control logic section of Figure A1.3 in serial form, most often along the two-wire

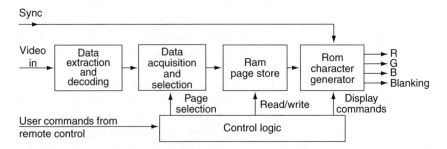

Figure A1.3 *The basic elements of a text decoding system*

control bus described in the previous chapter. After serial-to-parallel conversion, it is stacked in registers ready to act on the transmitted text data and the character-generator ROM.

Memory extension

We have seen that there is little scope for increasing the number of teletext lines transmitted during the blanking interval, and that the Fastext system is only capable of reducing access time when pages are called up in a logical sequence. For fast access to a large and random selection of text pages, it is necessary to provide a large data memory capable of storing scores or hundreds of sets of page data so that page selection, generally taking less than a fifth of a second, becomes a process of memory search and read rather than wait for a coincidence between requested and transmitted page numbers.

These large data memories are called background memory systems, and consist of an interface/data router and up to 4 Mb of DRAM (dynamic RAM), capable of storing a maximum of 512 pages. The same decoding and display techniques as outlined in Figure A1.3 is used, and the data is systematically stacked in the memories as it is received. When a page request is made by the viewer, the DRAM is rapidly scanned in the full-field mode: time is saved by restricting the search to the memory area assigned to the relevant magazine and page header information. To ensure that the most up-to-date page is presented the memory is scanned in reverse order, that is from the most recently stored page to the oldest one. When the required page is located, the memory is searched forward to read the data out to the one-page display memory within the text decoder IC.

Multilingual teletext

The advent of satellite transmissions which cover many countries simultaneously, and the acceptance of the UK-designed teletext system as a world standard (used in over 35 countries) has led to the development of character generators capable of holding up to 192 characters, the extras being selected by the control bits in the header row shown in Figure A1.2 and by packet 26 which specifies the supplementary character that overrides the basic one at any required point in a row and column. Similar means are used to sub-title programmes in many languages simultaneously; the desired translation being selected by the viewer for insertion in the picture. For text pages where two languages may be required on the screen at the same times, e.g. educational purposes, there are ICs with dual-ROM character generators capable of producing two alphabets. Switching between them during a single row of the display is achieved by using a control character to shift the readout as required, in similar manner to the selection of alphanumeric or graphic characters.

With suitable software in the TV's control system, the teletext ROM can be used to generate on-screen menus and captions for such purposes as programme/channel number readout, analogue control setting displays and viewer-programmed status titles.

In current TV receiver, the functions of the text decoder are incorporated in the system-control microprocessor. On-screen captions and graphics can also be generated here, and this concept simplifies making teletext-equipped sets.

Appendix A2
I²C serial control bus

A serial bus is invariably used by microprocessors and microcontrollers to control other units in the system. While manufacturers may use their own proprietary serial control bus systems, the most popular standard control bus is the 2-line inter IC, IIC or I²C bus.

The I²C bus has two bidirectional lines, a *serial clock, SCL* and *serial data, SDA*. Any unit connected to the bus may send and receive data. Data is transmitted in 8-bit words or bytes as shown in Figure A2.1. The first byte contains the 7-bit address of the device for which the information is intended, while the eighth bit is a read/write bit to signify whether the data is required from, or being sent to, the device. A number of data bytes follow. The total number in a message depends on the nature of information being transferred. Each data byte is terminated by an acknowledge, *ACK* bit. Like all other bits, the ACK bit has a related clock pulse on the clock line as shown.

The first byte of any data transfer is preceded by a start condition and is terminated by a stop condition. To ensure that two devices do not use the bus simultaneously, an arbitration logic system is used. The clock, which operates only when data is transferred, has a variable speed. Data may then be sent at a slow or a fast rate of up to 100 kbps.

A number of peripheral chips are available in the market including tuners, channels decoders, EEPROMs, ADCs and a variety of digital signal processors (DSPs) for operation with the I²C control bus.

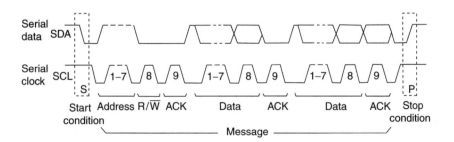

Figure A2.1 *The I²C serial bus*

Appendix A3
Error control techniques

In all types of communication systems, errors may be minimised, but they cannot be avoided completely. Hence the need for error detection and/or correction techniques. If an error is detected at the receiving end, it can be corrected in two different ways: the recipient can request a repeat of the transmission or the recipient may attempt to correct the errors without any further information from the transmitter, known as forward error correction (FEC). Retransmission is commonplace in communications system where it is possible. However, if distances are large, such as communications with a space probe, or real time signals are involved such as the case in audio and video streaming or broadcasting, retransmission is not an option.

Information messages invariably contain what is known as redundancy. In ordinary English text, the U following the Q is quite unnecessary and 'At this moment in time' can be easily reduced to 'At this moment' or even 'Now'. Redundant letters or words play a very important role in communication. They allow the recipient to make sense of distorted information. This is how we can make sense of badly spelled seaside postcards, a corrupted fax message or a text message.

As far as digital communication is concerned, redundancy is unnecessary data that occupies precious bandwidth space. For this reason, compression techniques are used to ensure that only necessary data is transmitted. In video and audio processing, raw data is compressed by 100 times or more to reduce the bandwidth requirements. To provide for error correction capability, controlled redundancy, i.e. extra bits are added to enable messages corrupted in transmission to be corrected at the receiving end.

Parity

The most basic error control technique is parity, which provides a rudimentary error detection technique. Parity involves a single parity bit at the end of a digital word to indicate whether the number of 1s is even or odd (Figure A3.1). There are two types of parity checking: even parity (Figure A3.1a) is when the complete coded data, including the parity bit contains an even number of 1s; odd parity (Figure A3.1b) is when the complete coded data contains odd number of 1s. At the receiving end, the number

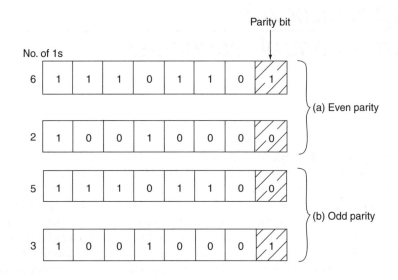

Figure A3.1 *Odd and even parity*

of 1s is counted and checked against the parity bit; a difference indicates an error. This simple parity check can only detect an error occurring in a single bit. An error affecting two bits will go undetected. Furthermore, there is no provision for determining which bit is actually faulty.

Block error coding

In the simplest form of block coding, serial data are first rearranged into blocks of rows and columns (Figure A3.2a). Parity bits for instance even parity are then added to each row and column as shown. A single bit error will cause two parity bits, to indicate an error. In Figure A3.2b, bits R and C indicate an error in the relevant row and column which points to an erroneous bit X.

Two main factors determine the type of error correction scheme used: the bit error rate (BER) and the type of errors expected. The latter refers to whether errors are expected to occur in single bits or in bursts. BER is the ratio of the number of bits that are likely to be corrupted in a specified bit stream. For instance, a BER of 10^{-10} means that 1 bit in 10^{10} bits may be corrupted. If the errors occur in bursts, then interleaving must be used.

Interleaving

There are two types of interleaving that may be used in digital A/V applications: bit interleaving and block or symbol interleaving.

Bit interleaving is the process of rearranging the sequence of bits before transmission takes place. The principles of bit interleaving is illustrated in

Row parity
bits

0	0	1	0	1	0	1	**1**
1	1	1	1	0		0	**0**
0	1	0	0	1	0	1	**1**
0	0	1	1	0	1	1	**0**
0	1	1	0	1	0	0	**1**
1	0	0	1	1	1	1	**1**
0	0	0	0	0	1	0	**1**
1	**1**	**1**	**1**	**0**	**0**	**1**	**1**

Column parity bits →

(a)

0	0	1	0	1	0	1	**1**
1	1	1	1	0		0	**0**
0	x	0	0	1	0	1	**R**
0	0	1	1	0	1	1	**0**
0	1	1	0	1	0	0	**1**
1	0	0	1	1	1	1	**1**
0	0	0	0	0	1	0	**1**
1	**C**	**1**	**1**	**0**	**0**	**1**	**1**

(b)

Figure A3.2 *Block error coding*

Figure A3.3 in which the order of an 8-bit word (b0–b7) is rearranged by interleaving to a new order b2, b5, b7, b1, b4, b6, b0 and b3. If after trans-mission three adjacent bits (b1, b4, b6) went faulty, then de-interleaving at the receiving end would restore the original order of bits and separate the

Original bit sequence	0	1	2	3	4	5	6	7
After interleaving	2	5	7	1	4	6	0	3
Wrong bits due to errors				X	X	X		
Damaged bit sequence	2	5	7	X	X	X	0	3
Restored original bit sequence after deinterleaving	0	X	2	3	X	5	X	7

Figure A3.3 *Effect of interleaving on burst errors*

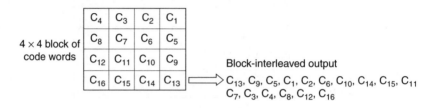

4 × 4 block of code words
C4 C3 C2 C1
C8 C7 C6 C5
C12 C11 C10 C9
C16 C15 C14 C13
Block-interleaved output
C13, C9, C5, C1, C2, C6, C10, C14, C15, C11
C7, C3, C4, C8, C12, C16

Figure A3.4 *Block interleaving*

error burst into single bit errors as shown which may then be corrected by such coding techniques as Hamming or Reed–Solomon.

In block interleaving (illustrated in Figure A3.4), code words produced by the encoder are written in a memory buffer, row by row: C1, C2, C3 and so on. When the rows have been filled, the code words are extracted column by column: C13, C9, C5, C1, C14 and so on.

Appendix A4
Processing devices

Modern consumer equipment employs one or more processing devices to carry out such functions as programming, synchronisation and control. Four types of processor chips may be used:

- General purpose microprocessors for general system programming and control
- Dedicated microprocessors such as video decoders or RF processor
- Microcontrollers
- System-on-a-chip

Three main technologies are employed in the fabrication of integrated circuits: TTL, CMOS and NMOS. The latter two types are normally used because of their high component density.

General purpose microprocessor system

The architecture of a microprocessor-based system is shown in Figure A4.1. It consists of the following component parts:

- Central processing unit (*CPU*)
- Memory chips: various types of RAM and ROM
- Address decoder chip
- Processing devices
- Bus structure

The central processor unit

The CPU is usually a single VLSI (very large-scale integration) or ULSI (ultra large-scale integration) chip containing all the necessary circuitry to interpret and execute program instructions such as data manipulation, logic and arithmetic operations and timing and control. The capacity or size of a microprocessor chip is determined by the number of the data bits it can handle. A 16-bit chip has a 16-bit data width; a 32-bit processor has a 32-bit data width and so on. Eight-bit and 16-bit processors are generally

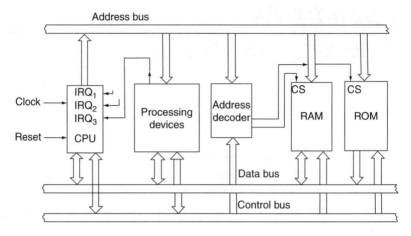

Figure A4.1 *The microprocessor system*

employed as dedicated controllers in industrial applications and domestic appliances such as robots, washing machines and TV receivers.

Microprocessors differ in the speed by which they execute instructions. CPU speeds are indicated by the frequency of the system clock in MHz or even GHz. While the bit width or size determines the quantity of information that may be transferred in any one cycle, the speed determines the number of transaction that may be executed per second.

Memory chips

Microprocessor system requires a certain amount of data storage capacity where programs such as start up routines and other processing software reside. DRAM (dynamic RAM) and ROM are two types of memory chips that are normally used to provide the necessary memory storage space. Other types such as *SDRAM, DDR, SRAM, PROM, EPROM, EEPROM* or *FLASH* may also be used.

Address decoder chip

The address decoder receives a group of address lines and depending on their combinations, enables one of its outputs, normally by taking it to logic low. If this line is connected to the *chip select* (*CS*) pin of a memory chip, then that memory chip will be enabled, i.e. selected. With two address lines, four (2^2) outputs are available. With three address lines, eight (2^3) outputs are available and so on. Figure A4.2 shows a typical 2–4 address decoder with its truth table.

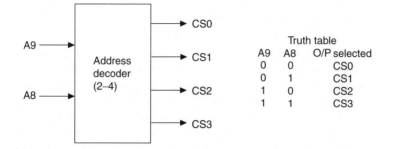

Figure A4.2 *The address decoder pinout and truth table*

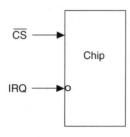

Figure A4.3 *Symbols for active low signals*

The bus structure

The hardware elements described above are interconnected with each other by a bus structure consisting of three types of buses: address, data and control as shown in Figure A4.1. The address and data buses provide a parallel highway along which multi-bit addresses and data travel from one unit to another. The control bus incorporates the lines that carry the various control signals to and from the CPU.

The data bus is used to transfer data between the CPU and other elements in the system. The address bus is used to carry the address of memory locations from which data may be retrieved or read, from memory devices, or to store or write data into memory locations. It is also used to address other elements in the system such as the input/output ports. The control bus carries the control signals of the CPU such as the clock, RESET, Read (RD) and Write (WR).

Control signals

The number and type of control signals depends on the microprocessor used and the design of the system. Control signals are normally active low, i.e. active when at logic 0. Active low signals are signified by a bar or by a small circle at the chip pinout as illustrated in Figure A4.3. The main control signals of a CPU are CLOCK READ/WRITE (*RD/WR*), RESET (*RST*) and *INTERRUPT* (INT or IRQ).

A clock pulse is an essential requirement for the operation of the processor. The clock synchronises the movement of the data around the various elements of the system and determines the speed of operation without which the system comes to a halt. A crystal-controlled oscillator is used to provide accurate and stable timing clock pulses. Stable clock frequency is essential and a small drift of about 10% may cause the processor to malfunction. The *READ* and *WRITE* control signals determine the direction of data transfer to or from the microprocessor chip. In a READ operation when the CPU is receiving data from memory, the READ line is active allowing data to be transferred to the CPU. In a WRITE operation when the CPU is sending data to memory, the WRITE line is active enabling data transfer from the CPU to memory.

When a peripheral device such as a channel decoder or a transport demultiplexer needs attention, a hardware *interrupt request signal* (IRQ) is sent to the CPU. When such a signal is received, the main programme is interrupted temporarily to allow the CPU to deal with the request. After servicing the device, the CPU returns to the original programme at the point where it left it.

Reset (RES or RST)

This is a type of interrupt, which overrides all other interrupts. The RESET input pin is normally held high. If it is taken low, it immediately stops the CPU program, and the processor is reset. To restart the microprocessor operations, the RESET pin must be taken high again at which point the start-up routine is executed and the CPU and other devices are initialised.

Microprocessor types

There are two basic types of microprocessors: complex instruction set code (CISC) and the faster reduced instruction set code (*RISC*). RISC processors are capable of performing fast mathematical operations using fewer or 'reduced' number of instructions. Examples of *CISC* processors are all the Intel 80XXX and the Pentium series. Examples of RISC processors are the power PC and OAK.

Dedicated processors

Dedicated processors are those processors designed for one particular task such as servo control and video/audio encoding/decoding. They are programmed and controlled by the resident system processor via the address/data bus and control signals or by a serial bus or both. Processing chips have their own individual chip clock which determines the processing speed. They normally require their own dedicated memory store which is accessed by a dedicated address/data bus. The chip clock is distinct from the system clock, which provides the necessary system synchronisation.

Recent advances in microprocessor technology have resulted in the development of very fast *digital signal processor* (*DSP*) chips. The introduction of faster and denser silicon chips with the increasing sophistication of processing technology has resulted in many applications previously handled by board level systems, to be taken over by DSP chips. DSPs can handle intensive amounts of data; carry out very fast data manipulation, multiplication, conversion from digital to analogue and vice versa and complex processing algorithms. They are available in 16- and 24-bit architecture and are increasingly being used at consumer level devices such as CD and DVD players as well as digital TV receivers.

Microcontrollers

Microcontrollers also known as *central control units* (*CCUs*) are dedicated single-chip computers. They contain the elements of the microprocessor itself as well as RAM, ROM or other memory devices and a number of input/output ports. A variety of microcontrollers are available from various manufacturers (Intel 8048/49 and 8051 series, Motorola 6805 and 146805, Texas TMS1000 and Ziloc Z80 series) for use as dedicated computer systems for such applications as car engines, washing machines, VCRs and TV receivers. The difference between one type of microcontroller and another lies in the type and size of memory, instruction set, operating speed, number of available input and output lines and data width, e.g. four, eight or sixteen bits. In the majority of cases, microcontrollers have their program stored permanently into an internal ROM at the manufacturing stage, a process known as mask programming. Some chips have an external EPROM available for user programming.

System-on-a-chip

System-on-a-chip (*SoC*) is one of the latest advances in chip technology, which is rapidly replacing Application Specific Integrated Circuit (ASIC) technology. It is widely used in digital television receivers, DVD player/writers and flat screen panels. SoC combines the core of a microprocessor with embedded memory space (DRAM, SRAM or FLASH), I/O ports, serial *UART* and external bus interface (Figure A4.4). A special

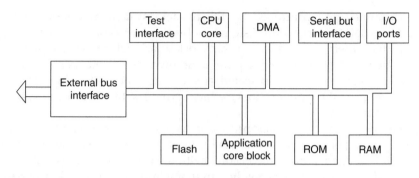

Figure A4.4 *System-on-a-chip (SoC)*

interface for testing purposes is usually provided designed for use at the testing and development stage. Apart from reducing the number of chips in a chip set, SoC reduces the power requirements of the system. The core is based on a powerful RISC processor chip such as the ARM or OAK families.

Appendix A5
The OSI seven-layer reference model

For a decade or two, networks were built using different hardware and software specifications. As a result they were incompatible and it became difficult for networks to communicate with each other. There was thus a need for a standard network communication model which the *International Organisation* for *Standardisation (ISO)* met by approving a network model in 1984 to help manufacturers create interoperable network implementation. That model was called the *Open System Interconnection (OSI)* model. It describes how information such as a spreadsheet or a video clip or data makes its way from an application by user A, through a network medium such as a pair of wires, a radio or a satellite link to another user B located on a remote network.

The process is divided into seven stages called layers (Figure A5.1) with each layer specifying a particular function. The upper three layers, layers 7, 6 and 5 (application, presentation and session) are concerned with service to the application used. The lower four layers, layers 1–4 (transport, network, data link and physical) are concerned with the flow of data from one user, through the network to another user.

The information generated at the application layer is divided into blocks of data, known as *data units (DUs)*. Each DU consists of a header and a payload. The header contains ID, sync and control information. The next layer, *presentation*, receives the DUs and carries out the necessary data formatting to ensure that the data is in a form that can be understood and read by subsequent layers. It then adds its own header to form a new DU, a process known as *encapsulation*, and passes it to the next layer. The *session* layer carries out the necessary programming to define the start and end of the conversation across the network, known as a session and introduces the necessary elements of control. It then adds its own header and passes the newly constructed block of data on to the next layer. Having defined a session at layer 5, the following four layers ensure that the remote host is identified, bi-directional communication takes place and the data is delivered in a form that can be deciphered by the receiving host. Layer 4, the *transport* layer, establishes direct end-to-end communication between the source and the remote host. It generally provides at least two services, a 'reliable' connection-orientated service (e.g. *TCP*, transmission control

Layers

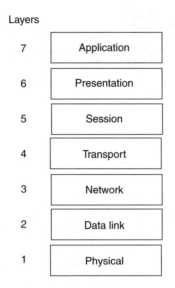

Figure A5.1 *The OSI 7-layer reference model*

protocol) and an 'unreliable' connectionless service (e.g. *UDP*, user data protocol). The first ensures that a connection is made and an acknowledgment is received before data is exchanged. With the unreliable service a network connection is not necessary prior to sending a message and no acknowledgment is required. With this service, there is no guarantee that messages sent out are actually received at the intended destination. Once again, the transport layer adds its own header to the DU it receives from the session layer and produces its own DU known as *segment* for the reliable service and *datagrams* for the unreliable service. The DUs move to the next layer, layer 3, the *network* layer which encapsulates them into packets by adding its own header. These packets are then passed down to the *data link* (layer 2) which encapsulates them into frames. The network layer functions at the network address level. It achieves this by including the source and destination network addresses in its header. The following layer, the data link layer, is responsible for the transmission of data from one host to another, hence the name data link. It receives the packets from the Network Layer and encapsulates them into frames. *A frame* usually has three parts: a header containing control/information, a data payload which is the packet from the previous layer and a trailer which usually contains a *checksum* for error correction. Finally, physical layer 1. This is responsible for converting the frames from the layer above into a sequence of bits which may be transmitted along a communication media such as a cable. The manner in which the bits, the type of modulation and frequencies used depends on the type of communication network.

At the receiving remote end, the process is reversed and DUs are *de-encapsulated* as they move from one layer to the next.

Appendix A6
HDMI–DVI compatibility

All HDMI sources shall be compatible with DVI 1.0 compliant sink devices (i.e. monitors or displays) through the use of a passive cable converter. Likewise, all HDMI sinks shall be compatible with DVI 1.0 compliant sources (i.e. systems or hosts.) through the use of a similar cable converter.

Type A-to-DVI-D cable wire assignment

Type A pin	Signal Name	Wire	DVI-D pin
1	TMDS Data2+	A	2
2	TMDS Data2 Shield	B	3
3	TMDS Data2–	A	1
4	TMDS Data1+	A	10
5	TMDS Data1 Shield	B	11
6	TMDS Data1–	A	9
7	TMDS Data0+	A	18
8	TMDS Data0 Shield	B	19
9	TMDS Data0–	A	17
10	TMDS Clock+	A	23
11	TMDS Clock Shield	B	22
12	TMDS Clock–	A	24
15	SCL	C	6
16	DDC Data	C	7
17	DDC/CEC Ground	D	15
18	+5V Power	5V	14
19	Hot Plug Detect	C	16
13	CEC	N.C.	
14	Reserved (in cable but N.C on device)	N.C.	
	TMDS Data 4–	N.C.	4
	TMDS Data 4+	N.C.	5
	TMDS Data 3–	N.C.	12
	TMDS Data 3+	N.C.	13
	TMDS Data 5–	N.C.	20
	TMDS Data 5+	N.C.	21
	No Connect	N.C.	8

Type B to DVI-D cable wire assignment

Type B pin	Signal Assignment	Wire	DVI-D pin
1	TMDS Data2+	A	2
2	TMDS Data2 Shield	B	3
3	TMDS Data2−	A	1
4	TMDS Data1+	A	10
5	TMDS Data1 Shield	B	11
6	TMDS Data1−	A	9
7	TMDS Data0+	A	18
8	TMDS Data0 Shield	B	19
9	TMDS Data0−	A	17
10	TMDS Clock+	A	23
11	TMDS Clock Shield	B	22
12	TMDS Clock−	A	24
13	TMDS Data5+	A	21
14	TMDS Data5 Shield	B	19
15	TMDS Data5−	A	20
16	TMDS Data4+	A	5
17	TMDS Data4 Shield	B	3
18	TMDS Data4−	A	4
19	TMDS Data3+	A	13
20	TMDS Data3 Shield	B	11
21	TMDS Data3−	A	12
25	SCL	C	6
26	DDC Data	C	7
27	DDC/CEC Ground	D	15
28	+5V Power	5V	14
29	Hot Plug Detect	C	16
22	CEC	N.C.	
23	Reserved	N.C.	
24	Reserved	N.C.	
	No Connect	N.C.	8

Appendix A7
The decibel (dB)

This is a unit of measurement that gives a comparison or the ratio of two powers such as the gain of an amplifier or the S/N ratio of a television cable.

Thus if power P1 is to be compared with power P2, then in terms of decibels we have

$$10 \ \log_{10} \frac{P1}{P2} dB$$

For example, if P1 is twice the value of P2, then

$$\frac{P1}{P2} = 10 \ \log \frac{P1}{P2} = 10 \ \log 2 = 3 \ dB$$

Conversely, a ratio of 1:2 is $10 \ \log 0.5 = -3 \ dB$.

Since it is more usual to measure the voltage rather than power, the decibel can be re-defined as

$$20 \ 10 \ \log_{10} \frac{V1}{V2} dB$$

The above is *true only* if V1 and V2 are measured across the same impedance.

It follows that a 2:1 voltage ratio is 6 dB and a 10:1 voltage ratio is 20 dB and so on. Conversely, a 1:2 voltage ratio = −6 dB.

The following is a list of some dB values and their corresponding power and voltage ratios:

dB	Power ratio	Voltage ratio
0	1	1
3	2	1.4
6	1.4	2
10	10	3.16
12	15.8	4
15	31.6	5.6
20	100	10
30	1000	31.6
40	10,000	100
50	100,000	316

In many applications, dB is used to indicate the power or voltage level at a test point. To do this, a reference power or voltage level has to be assumed and this is normally 1 mW, 1 mV or 1 µV. The units are dBmW (or dBm), dBmV and dBµV (or dBµ) respectively. For example, 3 dBmW means the power level is 3 dB above the reference 1 mW. With 3 dB representing a power ratio of 2:1, power level of 3 dBmW = 2 × 1 mW = 2 mW. Equally, a value of 10 dBµV = 3.16 µV and so on.

Appendix A8
Amplitude and frequency modulation

Amplitude modulation

Amplitude modulation (AM) is amongst the earliest forms of modulation to be used, and is probably the easiest to understand. In AM, the amplitude of a sine wave carrier is made to change in accordance with a modulating signal. For instance, if the modulating signal was the sine wave shown in Figure A8.1a, then the modulated carrier will change its amplitude as shown in Figure A8.1c. When the modulated carrier is analysed, it is found to contain three sine wave components: the carrier f_c and two *side frequencies*: f_1 and f_2 which are the difference and sum of the carrier and the modulating frequency respectively (Figure A8.2).

The information to be transmitted is duplicated in the two *sidebands* with the carrier itself containing no information at all. On top of that, the carrier has the highest amplitude and hence energy compared with the sidebands. Energy is thus wasted since the carrier contains no information. The carrier may be *suppressed* thus saving in power requirements without compromising the integrity of the transmitted information. As would be expected, a modulation waveform devoid of its carrier wave differs significantly from that with the carrier wave intact since the former is composed of only the upper and lower sidebands as illustrated in Figure A8.3.

It will be seen that the envelope has effectively 'collapsed', so that the top and bottom parts intertwine. Further, the high-frequency signal inside the collapsed envelope has also changed character. The frequency, however, is just the same as the original carrier suffering a phase reversal at the cross over point when the envelope pass through the datum line.

Normally the signal is not one single sine wave, but a range of sine wave signals with a particular bandwidth such as a video signal with a bandwidth of 0–5.5 MHz. In this case, two identical upper and lower sidebands together with the carrier are produced extending 5.5 MHz on each side resulting a total bandwidth of $2 \times 5.5 = 11$ MHz (Figure A8.4).

At the transmitting end, a stable oscillator is used to generate the carrier wave. For TV broadcasts, the crystal oscillator runs at a sub-multiple of the station frequency, and its output is multiplied up to the required

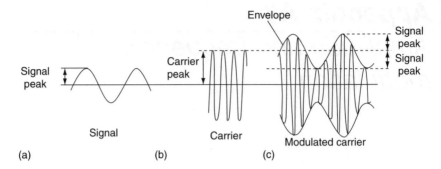

Figure A8.1 *Amplitude modulation*

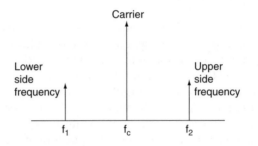

Figure A8.2 *Side frequencies*

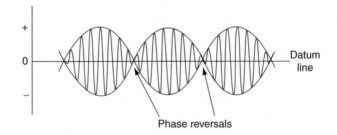

Figure A8.3 *Suppressed carrier AM signal*

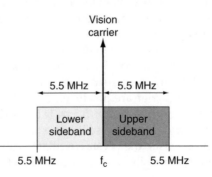

Figure A8.4 *Sidebands*

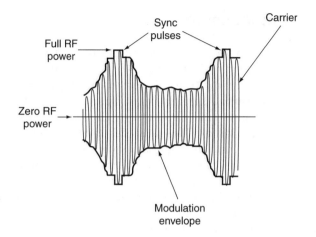

Figure A8.5 *Modulated UHF carrier*

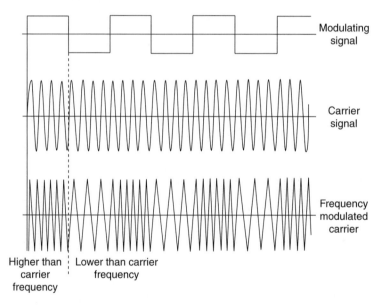

Figure A8.6 *Frequency modulation*

frequency (corresponding to the specified channel number). The modulating signal, CVBS is made to vary the amplitude of the carrier wave, by feeding it to a modulator which controls the gain of an RF amplifier which is handling the carrier wave. Thus the video signal is impressed onto the carrier as illustrated in Figure A8.5 which shows the waveform fed to the UHF broadcasting aerial. The polarity of the composite video signal is arranged to give what is known as negative modulation whereby the tips of the sync pulses give rise to maximum carrier power, and whites giving a lower power of about one fifth of the maximum.

Frequency modulation

Frequency modulation (FM) is the second most popular mode for broadcast use, and is used for high-fidelity (hi-fi) sound transmissions in VHF Band II. Most of these radio transmissions are multiplex-stereo-encoded. So far as we are concerned here, the most significant use of FM is for the TV sound transmissions which accompany the UHF picture broadcasts, and for satellite television broadcasts. In this type of modulation the carrier amplitude is kept constant and the frequency of the wave is varied at a rate dependent on the frequency of the modulating signal and by an amount (*deviation*) dependent on the strength of the modulating signal (see Figure A8.6). Thus the 'louder' the signal the greater the deviation, while the higher its frequency, the greater the rate at which the carrier frequency is varied either side of its nominal frequency.

Maximum deviation of the TV sound carrier is about $\pm 50\,\text{kHz}$ corresponding to maximum modulation depth of this particular system, commonly described as 100% modulation.

Index